REMOTE SENSING OF PLANET EARTH

Edited by **Yann Chemin**

Remote Sensing of Planet Earth
Edited by Yann Chemin

Published by InTech

Janeza Trdine 9, 51000 Rijeka, Croatia

Publishing Process Manager Maja Kisic

Technical Editor Teodora Smiljanic

Cover Designer InTech Design Team

Additional hard copies can be obtained from orders@intechopen.com

Remote Sensing of Planet Earth, Edited by Yann Chemin
p. cm.
ISBN 978-953-307-919-6

For EU product safety concerns:
IN TECH d.o.o., Prolaz Marije Krucifikse Kozulić 3, 51000 Rijeka, Croatia,
info@intechopen.com or visit our website at intechopen.com.

Meet the editor

Dr Oreste Capelli was born in 1956 in Bologna (Italy). After graduating in Medicine and Surgery at Modena University (1981), he became a specialist in Pulmonary (1985) and Internal Medicine (1992). From 1985 to 1990 he worked as a Pneumologist at Modena University Hospital, teaching clinical methodology in the graduate school of pulmonary medicine. From 1991 to 2003 Dr Capelli worked as a General Practitioner in a rural setting. In this period he collaborated with Local Health Authority to implement local guidelines and clinical audit. In 2004, after he completed a Master in Evidence Based Medicine and Research Methodology, Dr Capelli began working at the NHS Centre for the Evaluation of Effectiveness of Health Care (CeVEAS) in Modena, in charge of EBM courses for GPs. Unrtil 2011 he planned several Clinical Audit and observational researches in Primary Care. Since January 2012 Dr Capelli is the Head of the Unit for Integrated Management of Chronic Diseases, in the Primary Care District Service at the Regional Health Authority of Emilia Romagna (Bologna, Italy).

Contents

Preface

When seen from space, "Planet Earth" is a mix of clouds, with the majority (two thirds) of the total surface under ocean water with the remaining third of land forming what we call continents, with various degrees of increasing albedo from open water bodies, vegetation, bare soil, rocks, deserts, and snow/ice packs.

In a very short time (relative to Earth's age), the modern human civilization has conquered its neighboring space with probes, satellites, and vehicles carrying humans for exploration. From the range of observing platforms (airborne or space-borne) circumventing our inner atmosphere to its boundary, in low Earth orbit up to geostationary orbit, a large number of Earth observation sensors and satellites are monitoring the state of our home planet.

Monitoring of water and land objects enters a revolutionary age with the rise of ubiquitous remote sensing and public access. Earth monitoring satellites permit detailed, descriptive, quantitative, holistic, standardized, global evaluation of the state of the Earth skin in a manner that our actual Earthen civilization has never been able to before.

The water monitoring topics covered in this book include the remote sensing of open water bodies, wetlands and small lakes, snow depth and underwater seagrass, along with a variety of remote sensing techniques, platforms, and sensors.

The Earth monitoring topics include geomorphology, land cover in arid climate, and disaster assessment after a tsunami. Finally, advanced topics of remote sensing cover atmosphere analysis with GNSS signals, earthquake visual monitoring, and fundamental analyses of laser reflectometry in the atmosphere medium.

Remotely yours,

Dr. Yann Chemin
International Water Management Institute
Sri Lanka

Part 1

Water Monitoring

On the Use of Airborne Imaging Spectroscopy Data for the Automatic Detection and Delineation of Surface Water Bodies

Mathias Bochow[1,2] et al.*
[1]*Helmholtz Centre Potsdam – GFZ German Research Centre for Geosciences*
[2]*Alfred Wegener Institute for Polar and Marine Research in the Helmholtz Association*
Germany

1. Introduction

There is economical and ecological relevance for remote sensing applications of inland and coastal waters: The European Union Water Framework Directive (European Parliament and the Council of the European Union, 2000) for inland and coastal waters requires the EU member states to take actions in order to reach a good ecological status in inland and coastal waters by 2015. This involves characterization of the specific trophic state and the implementation of monitoring systems to verify the ecological status. Financial resources at the national and local level are insufficient to assess the water quality using conventional methods of regularly field and laboratory work only. While remote sensing cannot replace the assessment of all aquatic parameters in the field, it powerfully complements existing sampling programs and offers the base to extrapolate the sampled parameter information in time and in space.

The delineation of surface water bodies is a prerequisite for any further remote sensing based analysis and even can by itself provide up-to-date information for water resource management, monitoring and modelling (Manavalan *et al.*, 1993). It is further important in the monitoring of seasonally changing water reservoirs (e.g., Alesheikh *et al.*, 2007) and of short-term events like floods (Overton, 2005). Usually the detection and delineation of surface water bodies in optical remote sensing data is described as being an easy task. Since water absorbs most of the irradiation in the near-infrared (NIR) part of the electromagnetic spectrum water bodies appear very dark in NIR spectral bands and can be mapped by simply applying a maximum threshold on one of these bands (Swain & Davis, 1978: section 5-4). Many studies took advantage of this spectral behaviour of water and applied methods like single band density slicing (e.g., Work & Gilmer, 1976), spectral indices (McFeeters, 1996, Xu, 2006) or multispectral supervised classification (e.g., Frazier & Page, 2000, Lira, 2006). However, all of

* Birgit Heim[2], Theres Küster[1], Christian Rogaß[1], Inka Bartsch[2], Karl Segl[1], Sandra Reigber[3,4] and Hermann Kaufmann[1]
[1]*Helmholtz Centre Potsdam – GFZ German Research Centre for Geosciences, Germany*
[2]*Alfred Wegener Institute for Polar and Marine Research in the Helmholtz Association, Germany*
[3]*RapidEye AG, Germany*
[4]*Technical University of Berlin, Germany*

these methods have the drawback that they are not fully automated since the analyst has to select a scene-specific threshold (Ji *et al.*, 2009) or training pixels. Moreover there are certain situations where these methods lead to misclassification. For instance, water constituents in turbid water as well as water bottom reflectance and sun glint can raise the reflectance spectrum of surface water even in the NIR spectral range up to a reflectance level which is typical for dark surfaces on land such as dark rocks (e.g., basalt, lava), bituminous roofing materials and in particular shadow regions. Consequently, Carleer & Wolff (2006) amongst others found the land cover classes water and shadow to be highly confused in image classifications. This problem especially occurs in environments where both, a high amount of shadow and water regions can exist, such as urban landscapes, mountainous landscapes or cliffy coasts as well as generally in images with water bodies and cloud shadows.

In this investigation we focus on the development of a new surface water body detection algorithm that can be automatically applied without user knowledge and supplementary data on any hyperspectral image of the visible and near-infrared (VNIR) spectral range. The analysis is strictly focused on the VNIR part of the electromagnetic spectrum due to the growing number of VNIR imaging spectrometers. The developed approach consists of two main steps, the selection of potential water pixels (section 4.1) and the removal of false positives from this mask (sections 4.2 and 4.3). In this context the separation between water bodies and shadowed surfaces is the most challenging task which is implemented by consecutive spectral and spatial processing steps (sections 4.3.1 and 4.3.2) resulting in very high detection accuracies.

2. Optical fundamentals of water remote sensing

For the spectral identification of water pixels and the separation from other dark surfaces and shadows it is necessary to understand the influencing factors contributing to the surface reflectance of water bodies and especially to the optical complexity and variability of coastal and inland waters. The spectral reflectance of water (its apparent water colour) is a function of the optically visible water constituents (suspended and dissolved) and the depth of the water body (Effler & Auer, 1987, Bukata *et al.*, 1991, Bukata *et al.*, 1995). The concentration and composition of (i) phytoplankton, (ii) suspended particulate matter (SPM) and (iii) dissolved organic matter loading dominate the optical properties of natural waters. Shallow coastal and inland waters may also contain the spectral signal contribution from the bottom reflectance that significantly differs with the various materials (mainly sands (different colours), muds (different colours), macrophytes (different abundances, groups and compositions), reefs (different structures, different colours).

Smith & Baker (1983) and Pope & Fry (1997) provide absorption spectra of pure water derived from laboratory investigations. The Ocean Optic Protocols (Müller & Fargion, 2002) propose the absorption spectra of Sogandares & Fry (1997) for wavelengths between 340 nm and 380 nm, Pope & Fry (1997) for wavelengths between 380 nm and 700 nm, and Smith & Baker (1983) for wavelengths between 700 nm and 800 nm. Buiteveld et al. (1994) investigated the temperature dependant water absorption properties. Morel (1974) provides spectral values of the pure water volume scattering coefficient at specific temperatures and salinity, and the directional phase function. Gege (2005) used the data from the afore listed publications to construct the WASI absorption spectrum of pure water. This absorption spectrum formed the basis of the knowledge-based algorithm for water identification presented in Section 4.3.1.

Specular reflection of direct sunlight at the water surface into the sensor should be avoided by choosing a different viewing geometry. Specular reflection of the diffuse incoming sky radiation at the water surface can not be avoided and accounts up to 2 to 4 % of the overall surface reflectance that is measured by a sensor. Thus, most of the incoming radiation penetrates the water. Wavelengths larger than 800 nm are entirely absorbed by a large water column of pure water, so reflectance and transmission are no more significant in those longer wavelengths. As solar and sky radiation transmits into the water, the scattering by suspended particles and the absorption by suspended and dissolved water constituents are the water colouring processes. The wavelength peak of the spectral reflectance from transparent waters lies in the blue wavelength range and in this case energy may be reflected from the bottom up of up to 20 meters deep. If waters are less transparent due to higher concentrations of phytoplankton and sediments, and if the back-reflected signal from the bottom in shallow water bodies reach back to the air/water interface, there is significant reflectance from the water body also at the longer wavelength ranges (green to red) and there is a rise of the water-leaving reflectance even in the NIR wavelength region. In the case of phytoplankton blooming, high sediment loads or shallow waters with a bright bottom reflectance the water leaving signal significantly rises in the NIR and the overall reflectance may reach near 10 to 15 %. Therefore, there is no mono-type of the shape and the magnitude of the spectral water-leaving reflectance (Fig. 1). Inland and coastal waters may exhibit bright, turbid waters due to phytoplankton and sediments or bottom reflectance of their shallow areas, and in these cases simple thresholding techniques are no solution for the extraction and delineation of water bodies.

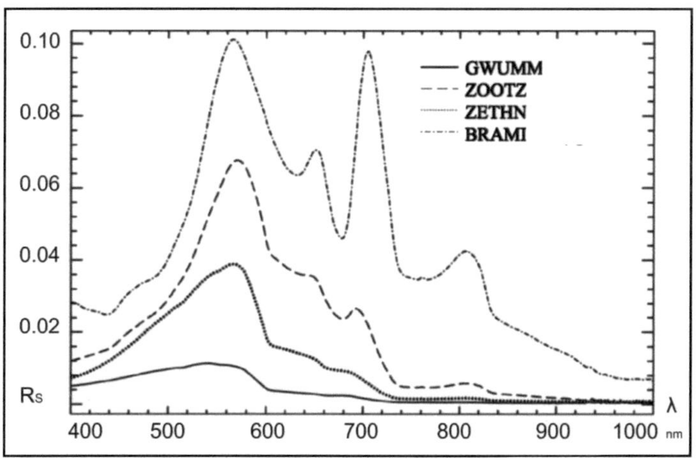

Fig. 1. Surface reflectance spectra, R_S (scale 0-1), of different inland waters (Rheinsberg Lake District, Germany) representing different water colours (Reigber, in prep). GWUMM, Grosser Wummsee, highly transparent, oligotrophic (nature reserve, densely forested); ZOOTZ, Zootzensee, mesotroph (rural, forested); ZETHN, Zethner See, turbid, mesotroph-eutrophic (rural); BRAMI, Braminsee, highly turbid, polytrophic (fish farming, rural)

3. Overview of existing methods for water body mapping

In the majority of algorithms for water body mapping a spectral band in the NIR spectral region plays an important role due to the high absorption of water and resulting high contrast in NIR bands to many other surface types. However, Manavalan et al. (1993) found that optimal cut-of gray values for individual spectral bands have to be carefully adjusted and are varying between different images. Band ratios or spectral indices are often used to mitigate spectral differences between images and also to enhance the contrast between surface types. Consequently, indices like the NDWI (McFeeters, 1996) (Equation 1) and MNDWI (Xu, 2006) (Equation 2) have been developed. Basically, the authors suggest a default threshold value of zero for these indices, i.e. gray values greater than zero represent water pixels. However, the comparative study of Ji et al. (2009) showed that an image and landscape specific adjustment of threshold values can improve results. Therefore, these methods are not fully suitable for automation. Further, NDWI shows high false positives in build-up areas (Xu, 2006). Xu developed the MNDWI to enhance the separation between water and built-up areas using Landsat ETM+ images. However, in high spatial resolution images there is no single spectral profile for the class "built-up areas" (Roessner *et al.*, 2011) and many man-made materials have positive NDWI and/or MNDWI values (Fig. 2 and Tab. 1). This is also true for shadow over non-vegetated areas. Fig. 3 shows that indices like the NDWI are not suitable for water body mapping in urban areas using high spatial resolution images since no threshold value can be found for which both, false positives and false negatives are low.

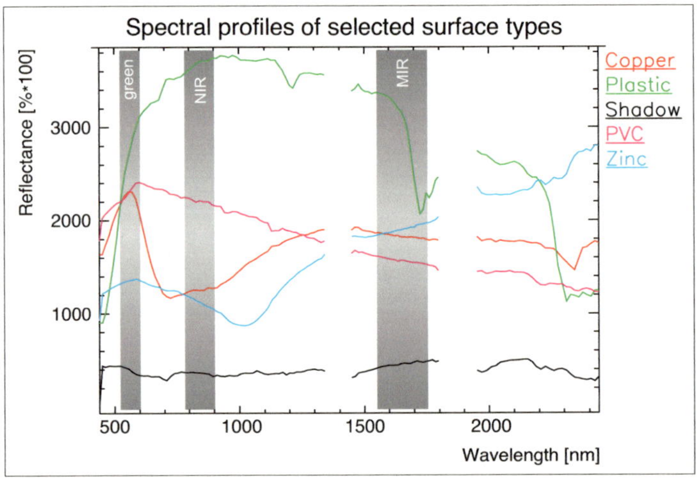

Fig. 2. Reflectance spectra of man-made materials with positive NDWI and/or MNDWI values. The gray bars indicate Landsat TM bands which are typically taken for calculating the NDWI and MNDWI. The spectra were collected from the test site Potsdam

Surface type	NDWI	MNDWI
Copper	0.28	0.10
Plastic	-0.13	0.01
Shadow	0.03	-0.10
PVC	0.03	0.20
Zinc	0.09	-0.17

Table 1. Corresponding index values of the spectra in Fig. 2

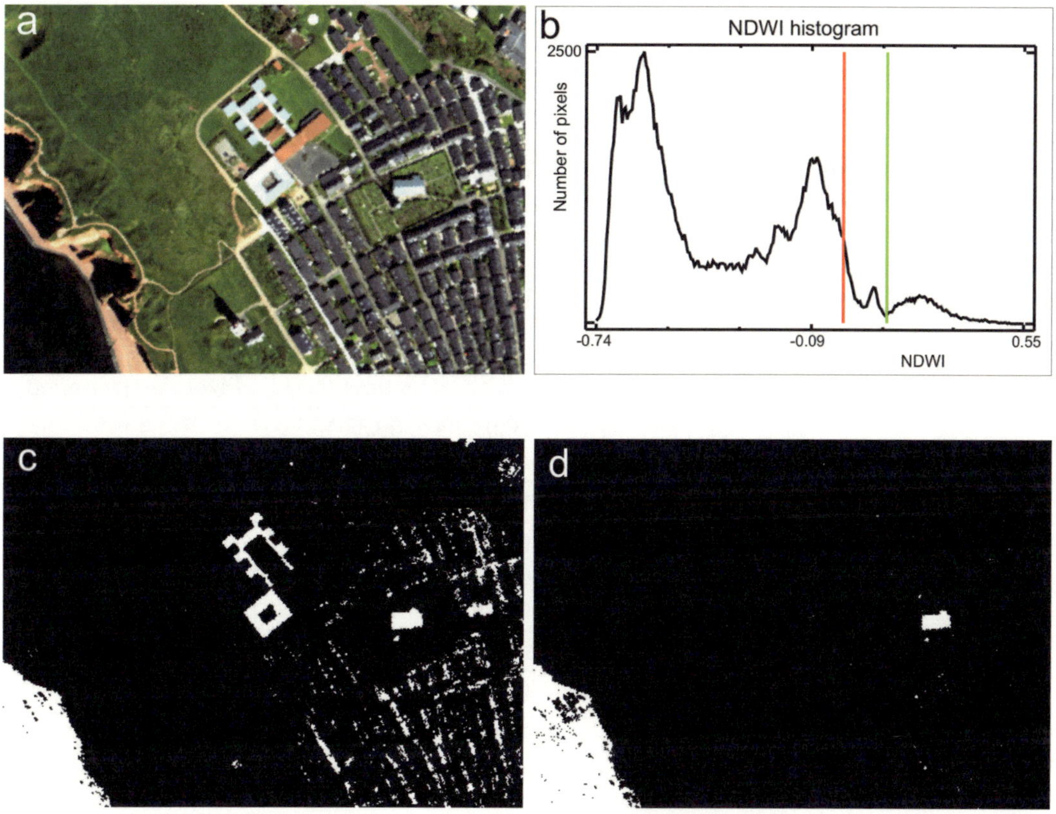

Fig. 3. True colour composite of an AISA image of Helgoland, Germany, with (b) histogram of the NDWI, (c) Water mask by threshold 0 (red line in histogram) on the NDWI; (d) Water mask by threshold 0.13 (green line in histogram) on the NDWI. In image c the water body (bottom left side) is almost totally included in the water mask but many urban features are so, too. In image d some parts of the water body are already lost but still some urban features are present

$$NDWI = \frac{(green - NIR)}{(green + NIR)} \tag{1}$$

where *green* is a green band and *NIR* is a NIR band

$$MNDWI = \frac{(green - MIR)}{(green + MIR)} \tag{2}$$

where *green* is a green band and *MIR* is a middle infrared band

In addition to the spectral-based approaches object-oriented methods have been developed for water body mapping (e.g. Xiao & Tien, 2010). However, since these methods use size and shape features they have to be adjusted individually for each application and can not be used for mapping ponds, rivers and coastal waters with the same configuration at the same time.

4. Material and methods

In this investigation a knowledge-based algorithm for the automated mapping of water bodies was developed based on a spectral database from five airborne hyperspectral datasets from the two German cities Berlin (two datasets) and Potsdam, and the German island Helgoland (two datasets) (Tab. 2). Five independent datasets were used for validation (Tab. 2). The selected scenes comprise urban, rural and coastal landscapes as well as different sensors to prove the wide applicability of the developed approach. The AISA Eagle sensor is an airborne VNIR pushbroom scanner (400 - 970 nm) with 12 bit radiometric resolution and variable spatial and spectral binning options, the latter resulting in mean spectral sampling intervals between 1.25 nm and 9.2 nm (Spectral Imaging Ltd., 2011) and

Test site	Sensor	Acquisition date, time (UTC)	Pixel size (rounded)
Berlin (*urban*)	HyMap	20.06.2005, 09:38 * 20.06.2005, 10:12 *	4 m 4 m
Potsdam (*urban*)	HyMap	07.07.2004, 10:29 *	4 m
Helgoland (*coastal*)	AISA Eagle	09.05.2008, 08:32 * 09.05.2008, 09:26 ° 09.05.2008, 09:41 *	1 m 1 m 1 m
Rheinsberg (rural)	HyMap	20.06.1999, 10:46 °	10 m
Dresden (*urban*)	HyMap	07.07.2004, 09:39 °	4 m
Mönchsgut (*coastal*)	HyMap	03.09.1998, 13:47 °	6 m
Döberitzer Heide (*rural*)	AISA Eagle	19.08.2009, 11:42 °	2 m

* Datasets analyzed during algorithm development
° Independent datasets for validation

Table 2. Dataset-specific characteristics

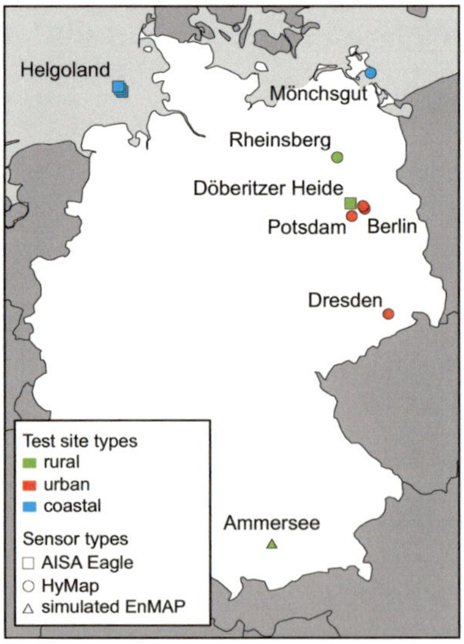

Fig. 4. Location of the test sites within Germany

488 to 60 spectral bands, respectively. The mean spectral sampling interval of the analyzed datasets is 2.3 nm for "Döberitzer Heide" and 4.6 nm for "Helgoland". The HyMap sensor is an airborne VNIR-SWIR whiskbroom scanner with 16 bit radiometric resolution consisting of four detector modules with mean spectral sampling intervals of 15 nm (VIS and NIR), 13 nm (SWIR1) and 17 nm (SWIR2) (Cocks *et al.*, 1998). The 128 spectral bands cover the spectral region from 440 nm to 2500 nm.

Water detection is a trivial task as long as there are no other dark surfaces present in the image. Unfortunately, the most prominent spectral characteristic of water pixels - water pixels are very dark - also applies to a couple of other surfaces such as dark rocks (e.g., lava, basalt) or bituminous roofing materials and especially to pixels covered by shadow. To account for this, we developed a two-step approach that firstly masks low albedo pixels as potential water pixels (section 4.1) and secondly applies a process of elimination to consecutively remove false positives (sections 4.2 and 4.3).

4.1 Masking potential water pixels

Masking of potential water pixels is done by thresholding a spectral mean image of all NIR bands between 860 nm and 900 nm of a sensor. As pointed out before water absorbs most of the incident energy in the NIR spectral region exhibiting a high brightness contrast to the majority of other surfaces. However, since every scene is different a scene-specific threshold has to be found. This is done automatically based on the histogram of the NIR spectral mean image (Fig. 5). After finding the histogram peak of low albedo surfaces (first local

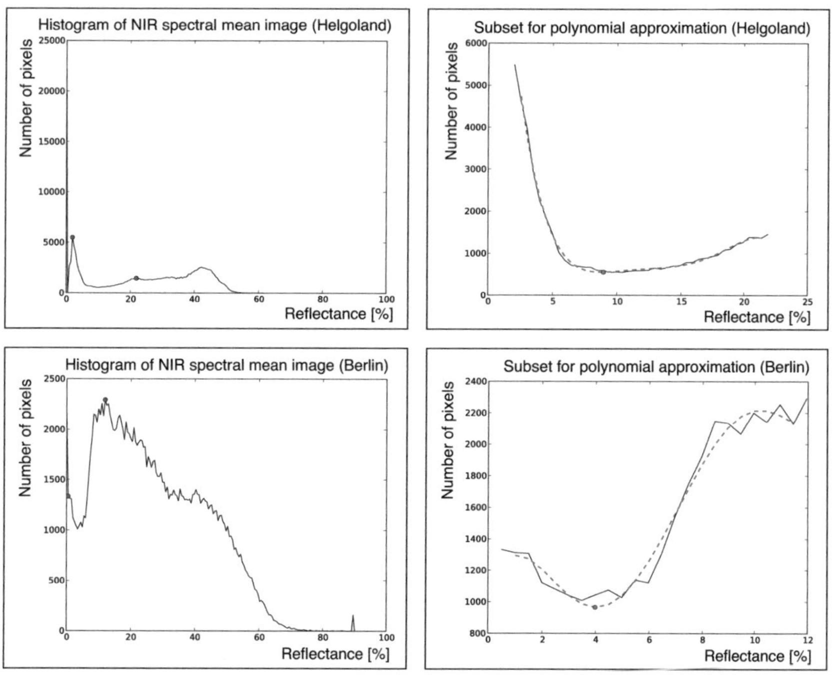

Fig. 5. Histograms (left: full, right: subset) of the NIR spectral mean images of two test sites (top: Helgoland, bottom: Berlin)

maximum) and a point near to the second local maximum (red dots in Fig. 5) the histogram between these two points is approximated by a polynomial of degree 5 (magenta dashed lines in Fig. 5). Then, the x value at the local minimum of the polynomial plus a safety margin of 2 is taken as the maximum reflectance threshold to be applied on the NIR spectral mean image. This results in a low albedo mask shown exemplarily for the test site Potsdam in Fig. 6. From this mask the water pixels have to be identified and other low albedo surfaces (mostly shadow) have to be removed.

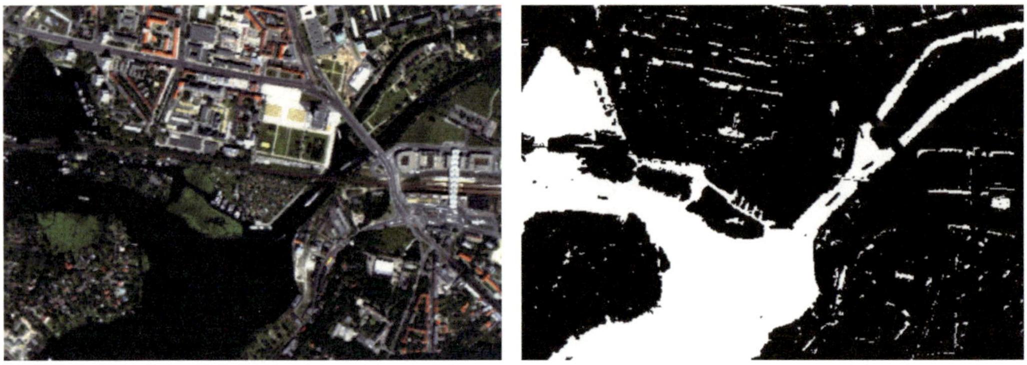

Fig. 6. Low albedo mask (right-hand) for the test site Potsdam

4.2 Differentiation between macrophytes in water and vegetation under shadow on land

Reflectance spectra of macrophytes (big emergent, submergent, or floating water plants) are characterized by spectral features of vegetation, such as the chlorophyll absorption features in the blue and red wavelength regions and the red edge in the NIR wavelength region. The light absorbing properties of water result in reflectance spectra exhibiting a comparably low albedo to those of shadowed vegetation on land (Fig. 7). Therefore, shadowed vegetation cannot be removed from the low albedo mask by simply thresholding an NDVI image.

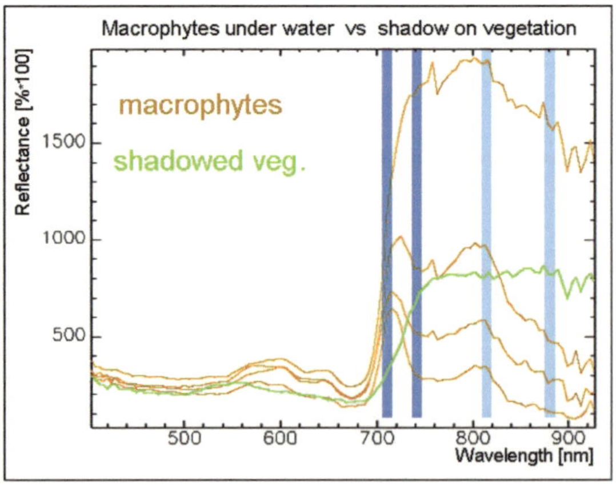

Fig. 7. Reflectance spectra of macrophytes in comparison with a reflectance spectrum of shadowed vegetation on land. The blue bars mark the wavelength of the two ratios used for distinguishing both surface types

However, a diagnostic spectral difference between both surfaces can be found in the NIR spectral region where the increasing water absorption causes the reflectance spectra of macrophytes to decrease between 710 - 740 nm as well as 815 - 880 nm. Therefore, pixels of shadowed vegetation can be removed from the low albedo mask using the condition:

$$VI^* > 1.0 \text{ AND } (R_{740} - R_{710} / 740 - 710 < -0.001 \text{ OR } R_{880} - R_{815} / 880 - 815 < -0.01) \quad (3)$$

where

VI^* = modified vegetation index = $\max(R_{710}, R_{720}) / R_{680}$
R_{740} = reflectance at wavelength 740 nm
Reflectance values must be scaled between 0 - 100

4.3 Removal of shadow pixels

Water and shadow reflectance spectra are on average both very dark. The reflectance level of both decreases with wavelength due to a decreasing proportion of diffuse irradiation (case of shadow) and due to the increasing light absorption (case of water). Additionally, both show a high spectral variability due to different types of shadowed surfaces (case of shadow) and due to varying water constituents and bottom reflection (case of water). However, despite this variation all water reflectance spectra have one thing in common: the pure water itself. Therefore, spectral features of pure water, especially absorption features, can be seen in every reflectance spectrum of water. However, the presence of these spectral features depends on the spectral superimposition of the water constituents and bottom coverage. Section 4.3.1 describes how these aspects can be considered in the development of a knowledge-based classifier for spectrally distinguishing water and shadow. Section 4.3.2 then continues with a spatial analysis.

4.3.1 Spectral analysis for water-shadow-separation based on spectral slopes

Fig. 8 shows the absorption spectrum of pure water (logarithmic scale) in comparison with selected surface reflectance spectra of different water bodies of the analyzed datasets. It can be seen that the increasing absorption within specific wavelength intervals (1st, 2nd, 4th and 5th light red bar) results in decreasing reflectance for most of the reflectance spectra. The 3rd light red bar represents a short wavelength interval of stagnating absorption where some water reflectance spectra temporarily rise due to increasing reflectance of water constituents or water bottom before decreasing again. However, these effects are not present within all wavelength intervals of all water reflectance spectra because they can be superimposed by the reflectance of the water constituents and water bottom. In order to find the slope combinations that occur for typical water bodies we analyzed 112.041 surface reflectance spectra from five datasets (two from Helgoland, two from Berlin, one from Potsdam). The selected datasets contain several types of water bodies (rivers, lakes, ponds, North Sea; transparent to productive and turbid waters). A first-degree polynomial was fitted to the spectra within each of the five wavelength intervals using the least squares method. If the algebraic sign of the slope within a wavelength interval met the expectation it was coded to 1 otherwise to 0. This resulted in a five-digit binary vector for each analyzed water reflectance spectrum representing the co-occurrence of slopes within the respective diagnostic wavelength intervals that met the expectation. The 25 possible binary vectors were numbered from 0 to 31 whereas the 0 vector (none of the 5 slopes met the expectation) was excluded from further analysis. The numbered combinations are shown in Fig. 9 in comparison with the numbered combinations of 33.721 analyzed shadow spectra. It can be seen that many combinations are occupied either by water or by shadow spectra and thus provide a clear separation between water and shadow. These combinations are implemented in the developed approach so that applied to an image many pixels of the low albedo mask can either be identified as water or rejected as shadow. The other combinations marked by the orange arrows are ambiguous. Pixels that fall into these combinations need a consecutive spatial processing described in Section 4.3.2.

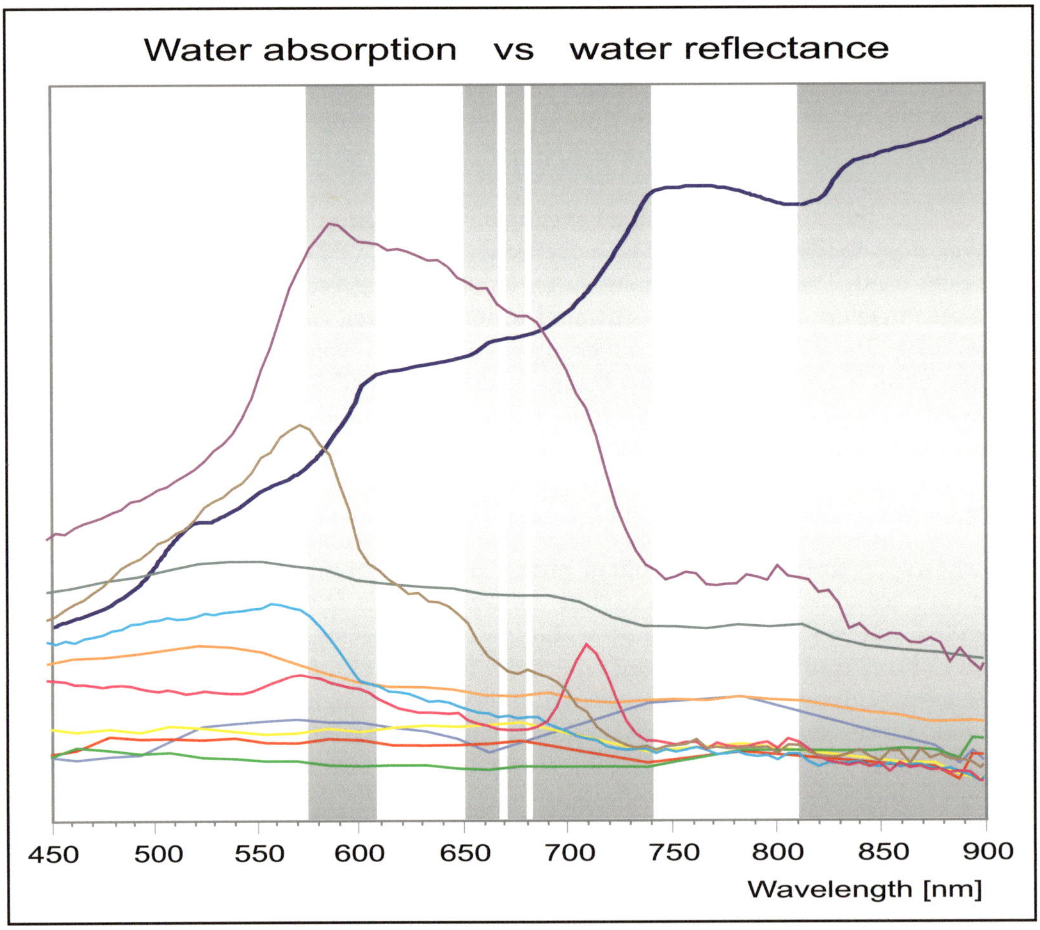

Fig. 8. Absorption of pure water (thick blue line, logarithmic scale, source: WASI (Gege, 2005)) in comparison to water surface reflectance spectra from different water bodies of the analyzed datasets. The increasing absorption within specific wavelength intervals (light red bars) results in decreasing reflectance for most of the reflectance spectra but is partly superimposed by the reflectance of the water constituents and water bottom

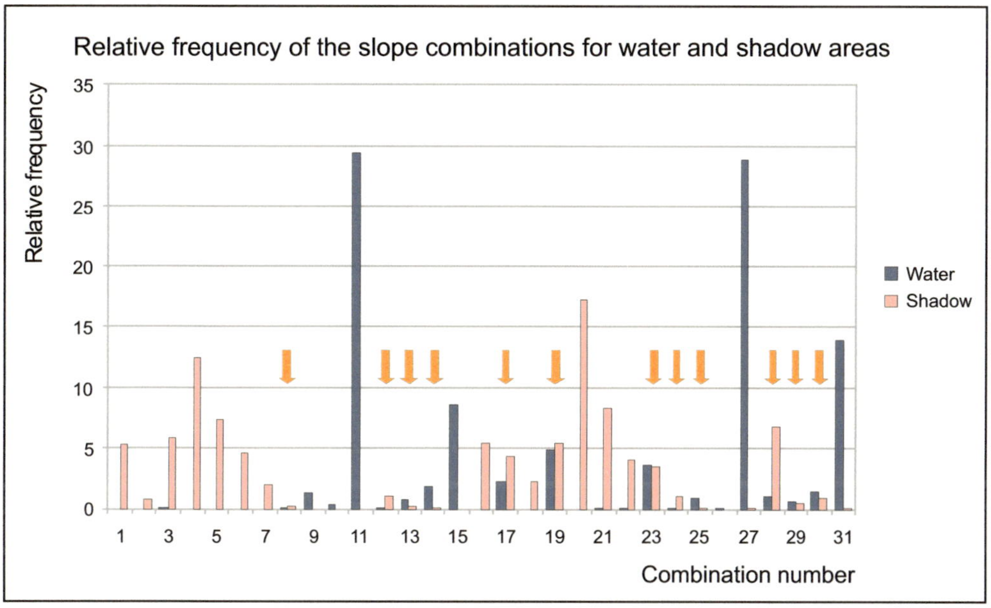

Fig. 9. Numbered slope combinations for water and shadow reflectance spectra. Due to the different amount of analyzed pixels of water and shadow (112.041 and 33.721) the *relative* frequency per land cover class is given. Combinations that are occupied by only one bar (or one very big and one very small bar) provide a clear separation between water and shadow. The combinations marked by the orange arrows are spectrally ambiguous

4.3.2 Spatial analysis for water-shadow-separation

Pixels of the low albedo mask that have not been identified as water or shadow based on the unambiguous spectral slope combinations are subjected to a consecutive spatial analysis. In this processing the idea is to decide according to the dominating spectral decision (see previous section) made within the neighbourhood of the ambiguous pixels (Fig. 10). The spectral decisions in the neighbourhood are counted using a 3x3 filter kernel resulting in a water score and a no-water score for each ambiguous pixel. If one of the two scores is more than three times higher than the other the ambiguous pixel is either identified as water or as no-water and is written into the respective image of confirmed water or no-water areas. If this is not the case the filter kernel iteratively grows up to a size of 33x33. Thereby, the identified water and no-water pixels are written into the respective image of identified water or no-water areas after each iteration so that they can be counted by the filter of the following iterations. When the filter kernel has reached a size of 33x33 and there are still ambiguous pixels left the decision threshold is reduced to two times higher than the other score and the filter kernel is reset to a size of 3x3. When the filter kernel reached a size of 33x33 for the second time it is again reset to a size of 3x3 and the decision is then simply related to the higher score. At this stage the filter starts growing again without a limit and until a decision was made for every ambiguous pixel. The graduation of the decision threshold has the advantage that pixels with an unambiguous neighbourhood are confirmed first and then accounted for in the following iterations. Finally, after all pixels have been identified either by spectral or spatial processing, the spectrally or spatially identified water pixels are combined into the final water mask. A last

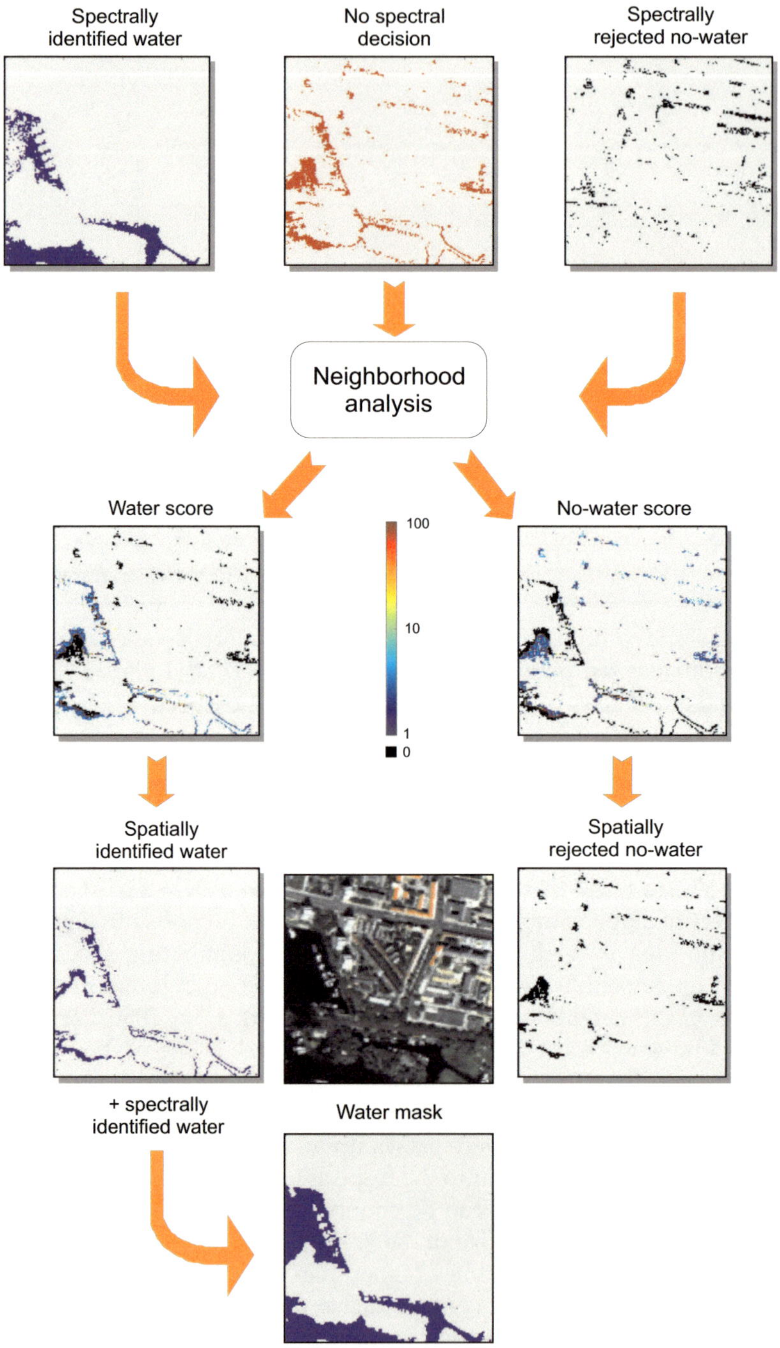

Fig. 10. Spatial processing illustrated by an exemplary subset of the Potsdam test site

aesthetic correction is done by filling up one pixel wholes within water areas which are considered as errors induced by noise. The filling of wholes can optionally be extended onto larger wholes (up to a certain size) which are likely to be boats (see Fig. 11).

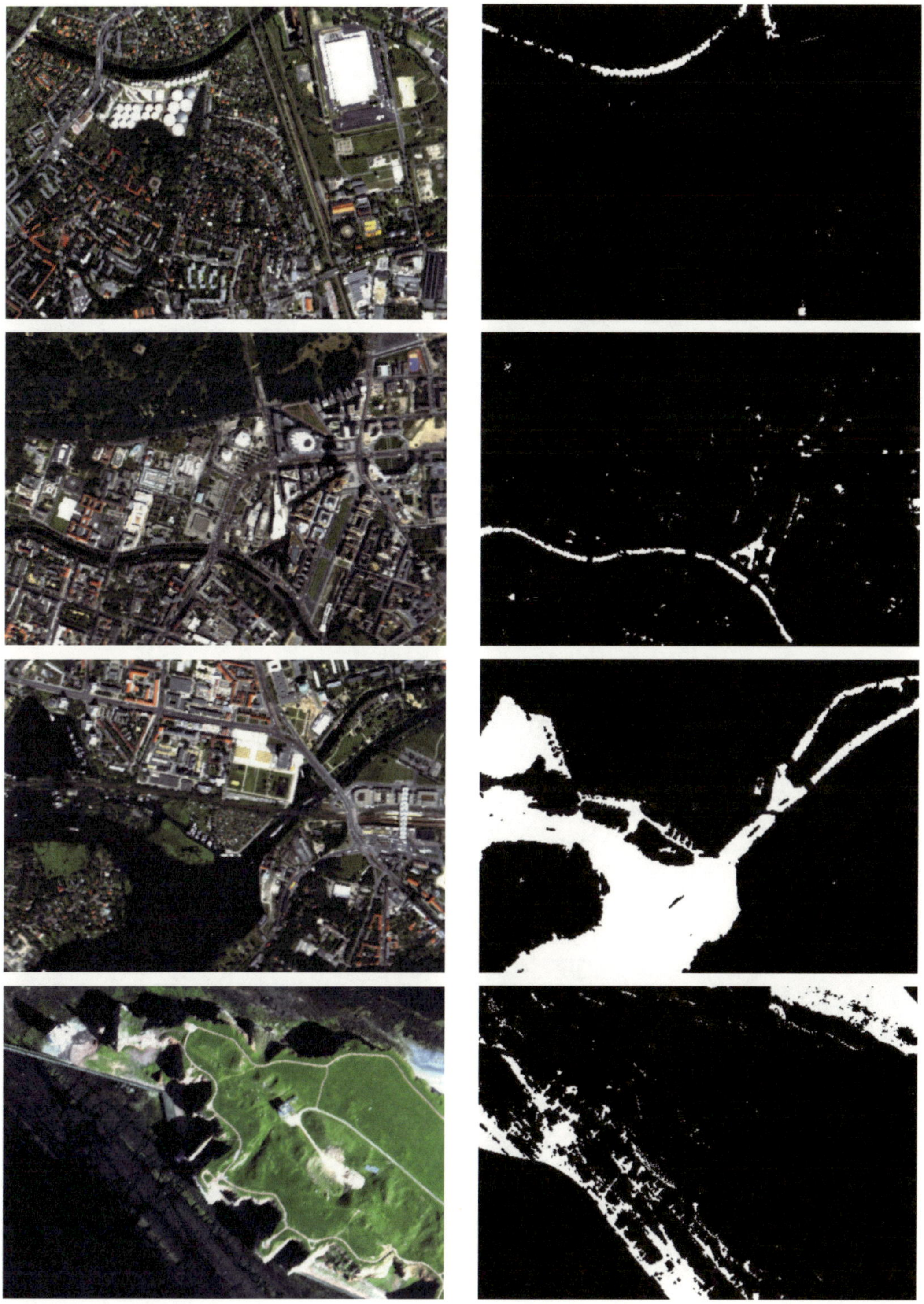

Fig. 11. (continued)

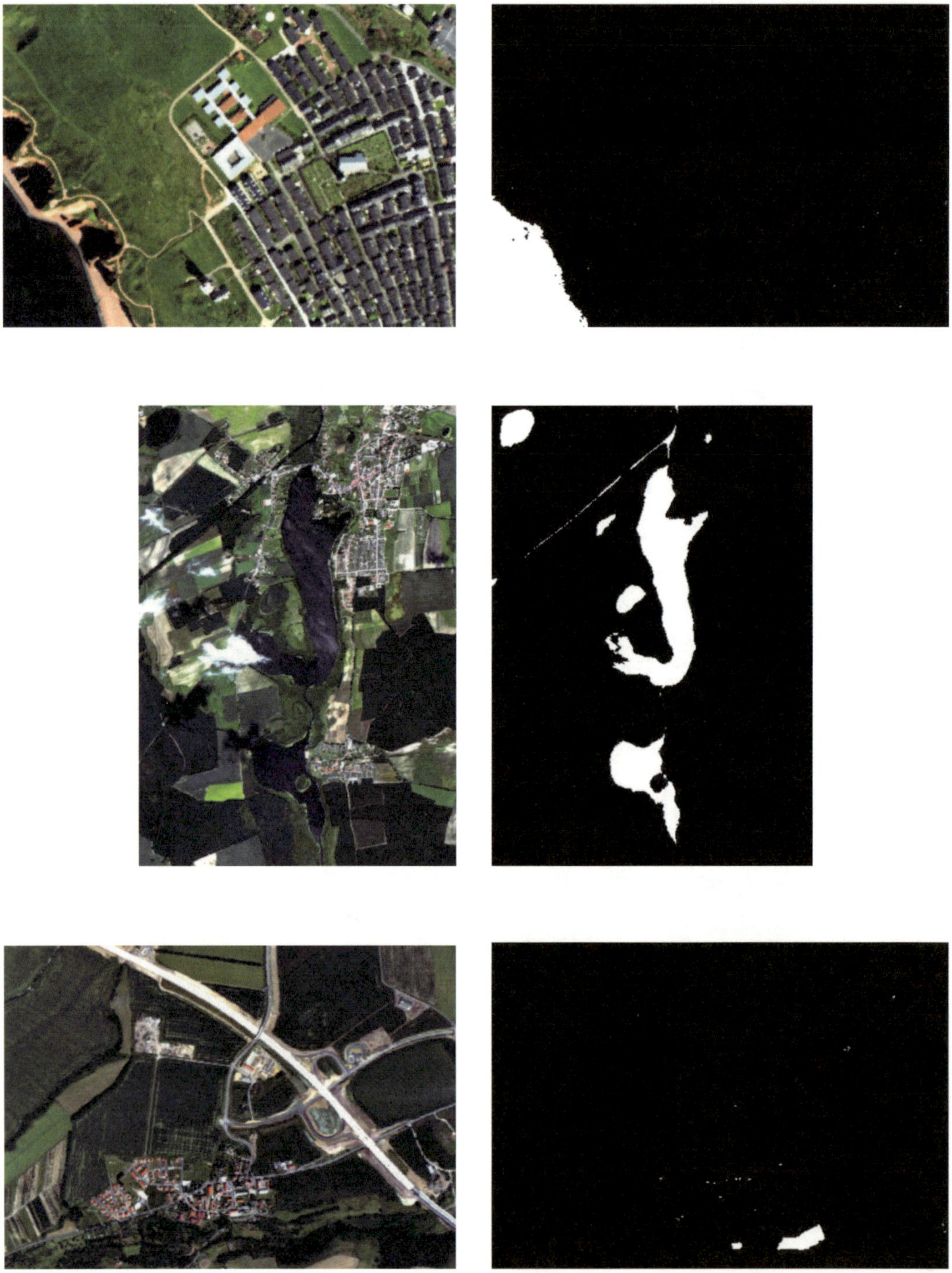

Fig. 11. (continued)

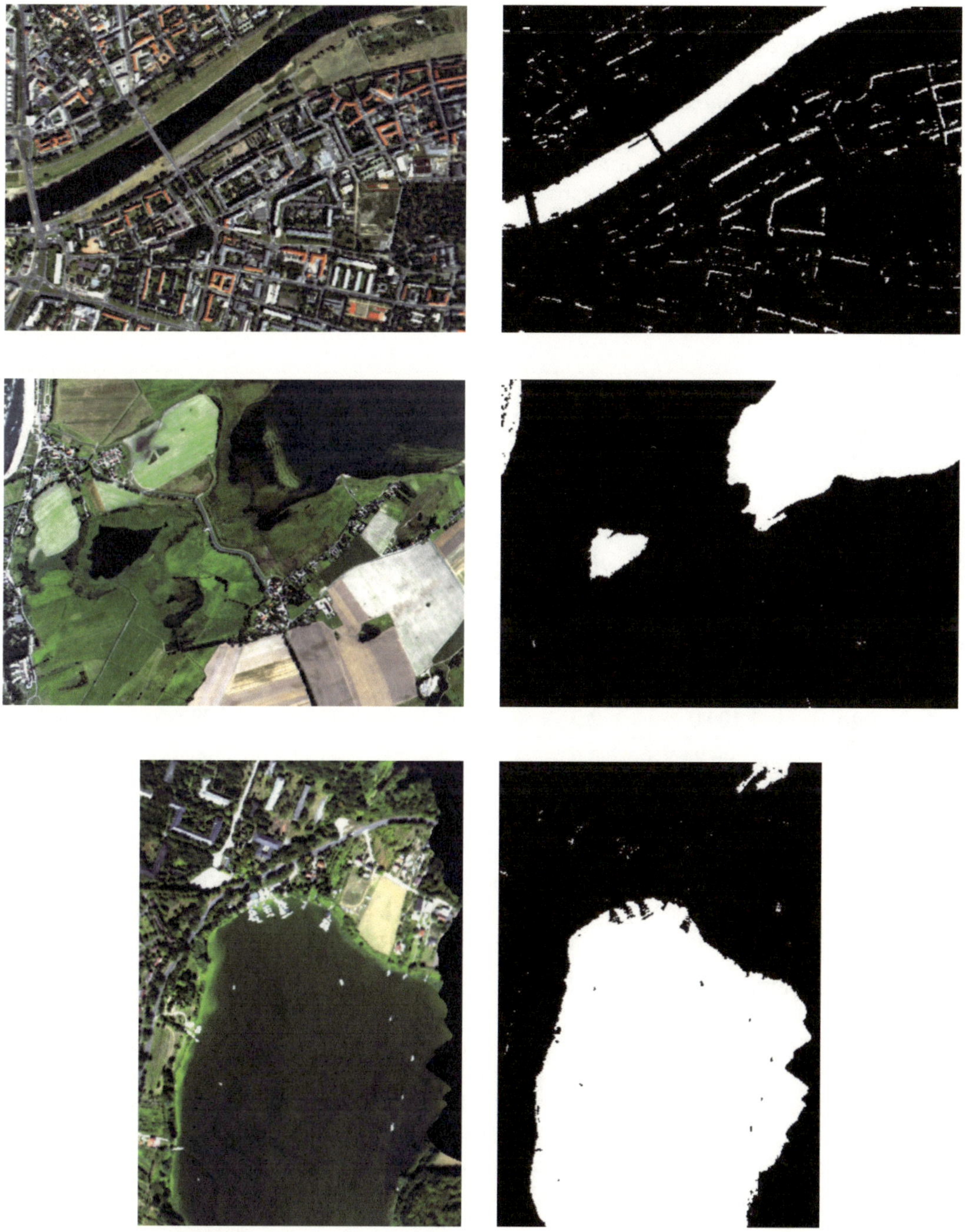

Fig. 11. Automatically detected water areas for the ten test sites Berlin_09:38, Berlin_10:12, Potsdam, Helgo_08:32, Helgo_09:26, Rheinsberg, Dresden_sub1, Dresden_sub2, Mönchsgut, Döberitzer (top to bottom; same order as in Tab. 3)

5. Results and discussion

In order to assess the accuracy of the developed approach water areas have been comprehensively digitized on-screen for selected validation sites of 175.000 pixels in size (350 by 500 or 500 by 350). The subsets have been chosen to contain as many challenging surface types as possible and to represent all different landscape types and sensors. Based on the digitized water reference areas no-water reference areas have been created by buffering the water reference areas with a two pixel buffer because of the mixed pixel problem and inverting the buffered areas. Using the reference areas of water and no-water several error metrics based on confusion matrices have been calculated. These are the probability of detection (POD), probability of false detection (POFD), false alarm ratio (FAR), overall accuracy (OA), average accuracy (AA) and kappa coefficient given in Tab. 3.

Test site	POD	POFD	FAR	OA	AA	Kappa
Berlin_09:38	79.5	0.1	9.6	99.8	89.7	0.845
Berlin_10:12	71.8	0.6	34.7	99.0	85.6	0.679
Potsdam	98.2	0.5	1.8	99.2	98.9	0.977
Helgo_08:32	99.8	2.3	25.3	97.8	98.8	0.843
Helgo_09:26	99.6	0.2	3.1	99.8	99.7	0.982
Rheinsberg	98.2	0.3	3.0	99.6	99.0	0.974
Dresden_sub1	98.7	0.0	7.9	100.0	99.3	0.953
Dresden_sub2	100.0	2.5	25.1	97.7	98.8	0.844
Mönchsgut	98.8	0.0	0.2	99.7	99.4	0.991
Döberitzer	100.0	1.8	1.9	99.1	99.1	0.981

Table 3. Results of the accuracy assessment. The first four test sites are subsets of datasets from which reflectance spectra have been analysed during the algorithm development. The last six test sites are subsets from independent validation datasets. The largest errors are highlighted in gray and discussed below.

The overall accuracy (a common error measure for classification results) amounts to 97% or above for all the test sites. However, to evaluate the detection accuracy of an underrepresented class the overall accuracy is not the best measure because it credits correct detections and correct not-detections equally and it is strongly influenced by the dominating class, i.e. the no-water class in this study. The overall measures average accuracy and especially kappa coefficient – although very high, too - reveal the remaining problems of the algorithm much better (highlighted in gray in Tab. 3). However, the most sensitive measures are the class-specific measures POD and FAR.

POFD, POD and FAR are typical measures for evaluating the accuracy of forecasting methods (Jolliffe & Stephenson, 2003) as well as two-class classification problems like detection tasks (one class of interest and one background class). The POFD of a class, also known as the false alarm rate, measures the fraction of false alarm pixels in relation to the background class, i.e. the number of false alarm pixels divided by the total number of ground truth pixels of the background class (= omission error of the no-water class). The achieved POFDs for the test sites are very low (usually below 1 %) showing that water can be well distinguished from no-water surfaces. This is a big step forward compared to the NDWI and MNDWI which applied to high spatial resolution data result in many false positives for urban surface materials (see Fig. 3).

The POD of a class, also known as hit rate, measures the fraction of the detected pixels of the class of interest that were correctly identified, i.e. the number of correctly identified pixels divided by the total number of ground truth pixels of the class (= producer accuracy of the water class). The achieved PODs for most of the test sites are very high (> 98 %) showing that the developed algorithm usually detects almost all water pixels. False negatives occur only for small water bodies (small ponds within the park at the top left in Berlin_09:38, parts of the river in Berlin_10:12, and narrow rivers in Rheinsberg). Possible explanations are the adjacency effect (light from neighbouring pixels that is scattered into the instantaneous field of view by the atmosphere) and diffuse illumination of the water surface by surrounding trees. These two effects might be the reason for the spectral shape of the water spectra of small water bodies with surrounding trees that looks much more like a reflectance spectrum of vegetation than one of water (Fig. 12) and do not show the typical decreasing slopes that enabled the spectral identification of water as shown in section 4.3.1.

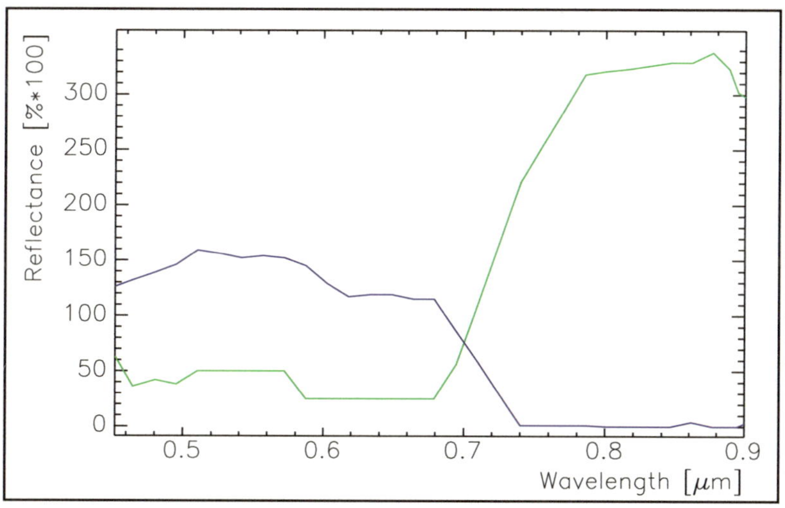

Fig. 12. A typical surface reflectance spectrum of water (blue) compared to a reflectance spectrum of a small water body with surrounding trees (green)

The false alarm ratio (FAR) gives the fraction of false alarm pixels in relation to the number of detected water pixels in the image, i.e. the number of false alarm pixels divided by the total number of classified water pixels (= commission error of water class). This error measure reveals clearly if to much water pixels have been falsely identified. This is the case for the test sites Berlin_10:12, Helgo_08:32, and Dresden_sub2 as well as in a weakened form for Berlin_09:38. In all of these test sites the confusion is related to shadow areas classified as water. For the test site Helgo_08:32 this can be explained by the intertidal zone which is wet even when the water is gone. Therefore, it is possible that there are some small water influenced areas under the shadow which is a problem that has not yet been regarded in the water-shadow separation (section 4.3.1) and is still an open issue for the future.

Another open issue is the detection of white water pixels which are usually to bright to be included in the low albedo mask (section 4.1). This can be seen in the top left side of the test site Mönchsgut.

Overall, it can be seen from Tab. 3 that the accuracies of the independent datasets is not less than the accuracies of the datasets analyzed during the algorithm development. Thus, the algorithm seems to be robust and generalizes well to unknown datasets.

6. Conclusion

A new algorithm for the detection and delineation of surface water bodies based on high spatial resolution airborne VNIR imaging spectroscopy data has been developed. In contrast to existing methods the proposed approach does not require *a priori* knowledge nor user input, manual thresholding or fine-tuning of input parameters and is able to automatically detect and delineate surface water bodies with a very high accuracy. Thus, the developed algorithm is suitable for implementation in automated processing chains. The algorithm was tested on different sensor data (AISA Eagle and HyMap), works for different types of landscapes (tested: urban, rural and coastal) and is not influenced by different atmospheric correction methods (tested: ATCOR-4 (Richter, 2011), MIP (Heege & Fischer, 2004), ACUM-R (unpublished in-house development by K. Segl), the method of L. Guanter et al. (Guanter *et al.*, 2009), and empirical line correction). Future issues will be to improve the detection of small and narrow water bodies, the detection of white water and of water under shadow. Furthermore, the proposed method will be tested on hyperspectral VNIR satellite data.

7. Acknowledgement

This work was made possible by several flight campaigns carried out by the Deutsches Zentrum für Luft- und Raumfahrt (DLR) Oberpfaffenhofen, Germany. We further thank the people of the Geomatics Lab of the Humboldt University of Berlin for providing the HyMap data of Berlin. We also acknowledge financial support for AISA flight campaigns at Helgoland of the BIS Bremerhaven and the WFB Bremen in the framework of the projects 'Innohyp' and 'CoastEye'.

8. References

Alesheikh A.A., A. Ghorbanali & N. Nouri (2007). Coastline change detection using remote sensing. *International Journal of Environmental Science and Technology*, Vol. 4, No. 1, pp. 61-66

Buiteveld H., J.H.M. Hakvoort & M. Donze (1994). The optical properties of pure water. In: *Ocean Optics XII Proc. Soc. Photoopt. Inst. Eng.*, Vol. 2258, 174-183 pp.

Bukata R.P., J. Jerome, K.Y. Kondratyev & D.V. Pozdnyakov (1991). Optical properties and remote sensing of inland and coastal waters. *J. of Great Lakes Res.*, Vol. 17, pp. 461-469

Bukata R.P., J. Jerome, K.Y. Kondratyev & D.V. Pozdnyakov (1995). *Optical properties and remote sensing of inland and coastal waters*. CRC Press, Boca Raton, FL

Carleer A.P. & E. Wolff (2006). Urban land cover multi-level region-based classification of VHR data by selecting relevant features. *International Journal of Remote Sensing*, Vol. 27, No. 6, pp. 1035-1051

Cocks T., R. Jensen, A. Stewart, I. Wilson & T. Shields (1998). The HyMap airborne hyperspectral sensor: the system, calibration and performance. In: *Proc. of the 1st EARSeL Workshop on Imaging Spectroscopy*, Zürich.

Effler S.W. & M.T. Auer (1987). Optical heterogeneity in Green Bay *Water Resources Bulletin of the Geological Institutions of Uppsala*, Vol. 23, pp. 937-941

European Parliament and the Council of the European Union (2000). *European Water Framework Directive, Directive 2000/60/EC*. Vol. Official Journal L 327, European Union (Hrsg.), 0001-0073 p.

Frazier P.S. & K.J. Page (2000). Water body detection and delineation with Landsat TM data. *Photogrammetric Engineering and Remote Sensing*, Vol. 66, No. 12, pp. 1461-1467

Gege P. (2005). The Water Colour Simulator WASI - User manual for version 3. DLR-Interner Bericht, No. DLR-IB 564-1/2005, DLR, 83 p.

Guanter L., R. Richter & H. Kaufmann (2009). On the application of the MODTRAN4 atmospheric radiative transfer code to optical remote sensing. *International Journal of Remote Sensing*, Vol. 30, No. 6, pp. 1407-1424

Heege T. & J. Fischer (2004). Mapping of water constituents in Lake Constance using multispectral airborne scanner data and a physically based processing scheme. *Canadian Journal of Remote Sensing*, Vol. 30, No. 1, pp. 77-86

Ji L., L. Zhang & B. Wylie (2009). Analysis of Dynamic Thresholds for the Normalized Difference Water Index. *Photogrammetric Engineering and Remote Sensing*, Vol. 75, No. 11, pp. 1307-1317

Jolliffe I.T. & D.B. Stephenson (2003). *Forecast verification : a practitioner's guide in atmospheric science*. J. Wiley, Chichester, West Sussex, England, Hoboken, NJ

Lira J. (2006). Segmentation and morphology of open water bodies from multispectral images. *International Journal of Remote Sensing*, Vol. 27, No. 18, pp. 4015-4038

Manavalan P., P. Sathyanath & G.L. Rajegowda (1993). Digital image-analysis techniques to estimate waterspread for capacity evaluations of reservoirs. *Photogrammetric Engineering and Remote Sensing*, Vol. 59, No. 9, pp. 1389-1395

McFeeters S.K. (1996). The use of the normalized difference water index (NDWI) in the delineation of open water features. *International Journal of Remote Sensing*, Vol. 17, No. 7, pp. 1425-1432

Morel A. (1974). Optical properties of pure water and pure seawater, In: *Optical aspects of oceanography*, Jerlov N.G. & E. Steeman Nielsen (eds.), pp. 1-24, Academic, London

Müller J.L. & G.S. Fargion (2002). Ocean Optic Protocols for Satellite Ocean Colour Sensor Validation. Edited by NASA, Sensor Intercomparison and Merger for Biological and Interdisciplinary Ocean Studies (SIMBIOS) Project Technical Memoranda, 308 p.

Overton I.C. (2005). Modelling floodplain inundation on a regulated river: Integrating GIS, remote sensing and hydrological models. *River Research and Applications*, Vol. 21, No. 9, pp. 991-1001

Pope R.M. & E.S. Fry (1997). Absorption spectrum (380-700 nm) of pure water .2. Integrating cavity measurements. *Applied Optics*, Vol. 36, No. 33, pp. 8710-8723

Reigber S. *Erfassung limnologischer Parameter aus Gewässern des Norddeutschen Tieflandes mit hyperspektralen Fernerkundungsdaten. "Investigation of limnological parameters of inland waters in the North German Plain (Germany) using hyperspectral remote sensing data"*. PhD thesis, Computer Vision and Remote Sensing, Berlin University of Technology, Berlin.

Richter R. (2011). Atmospheric / topographic correction for airborne imagery. In: *ATCOR-4 user guide,* Software user guide, DLR - German Aerospace Center, Wessling, 167 p.

Roessner S., K. Segl, M. Bochow, U. Heiden, W. Heldens & H. Kaufmann (2011). Potential of hyperspectral remote sensing for analyzing the urban environment, In: *Urban Remote Sensing: Monitoring, Synthesis and Modeling in the Urban Environment,* Yang X. (ed.), pp. 49-62, Wiley, Oxford

Smith C.S. & K.J.S. Baker (1983). The analysis of ocean optical data. In: *7th SPIE, Ocean Optics,* Vol. 478, 119-126 pp.

Sogandares F.M. & E.S. Fry (1997). Absorption spectrum (340-640 nm) of pure water .1. Photothermal measurements. *Applied Optics,* Vol. 36, No. 33, pp. 8699-8709

Spectral Imaging Ltd. (2011). aisaEAGLE hyperspectral sensor.

Swain P.H. & S.M. Davis (1978). *Remote sensing : The quantitative approach.* McGraw-Hill International Book Co., New York

Work E.A. & D.S. Gilmer (1976). Utilization of satellite data for inventorying prairie ponds and lakes. *Photogrammetric Engineering and Remote Sensing,* Vol. 42, No. 5, pp. 685-694

Xiao G. & D. Tien (2010). An object-based classification approach for surface water detection. *Int. J. Intell. Syst. Technol. Appl.,* Vol. 9, No. 3/4, pp. 218-227

Xu H.Q. (2006). Modification of normalised difference water index (NDWI) to enhance open water features in remotely sensed imagery. *International Journal of Remote Sensing,* Vol. 27, No. 14, pp. 3025-3033

2

Remote Sensing for Mapping and Monitoring Wetlands and Small Lakes in Southeast Brazil

Philippe Maillard[1], Marco Otávio Pivari[2] and Carlos Henrique Pires Luis[1]
[1]*Universidade Federal de Minas Gerais*
[2]*Instituto Inhotim*
Brazil

1. Introduction

Wetlands and small lakes are areas with great ecological value that are increasingly threatened through excessive pressure on water resources. In some cases, this pressure can lower the aquifer and result in a significant reduction of the area of small lakes or the drying out of wetlands. In other cases, logging, road building and other degradations of the surroundings of lakes can increase nutrients loads that reach the water and alter the state of these lakes towards eutrophication and reduction of the open water surface through colonization by aquatic plants. The first requirement to help protect these areas is a thorough mapping and monitoring of the changes that affects them: past, present and future. Many of these areas are poorly known and have not been mapped thoroughly and most have never been monitored. Remote sensing is the only effective means to perform both tasks by enabling rapid mapping of their situation both past and present. While images from the recent generations of Earth observing satellite and sensors come in a wide range of spatial resolution up to about half a meter, historical data at medium-scale resolution can provide a record of past situations and help determine an evolutionary trend.

This chapter is dedicated to the description of methods for the cartography of small lakes using high-resolution data for actual or near-actual mapping and medium-resolution historical data for determining the evolutionary path of these areas in the last three decades. In particular, the accent is given to two distinct approaches: 1) the use of region-based unsupervised segmentation and classification to delineate small lakes, and 2) multi-temporal image analysis of long sequences of images to assess changes of both small lakes and wetlands communities. Two case studies are described to illustrate these methods.

2. First case study: The Rio Doce lake system

It as been observed in the Brazilian Pantanal, that the process of aquatic plant succession starts with the emergence of free floating macrophytes followed by colonization of epiphytes. The latter can be subtituted by paludian plant of higher stature. Eventually, if this process is pursued without interruption, it can culminate by the emergence of floating island and the constitution of an organic soil (Pivari et al., 2008; Pott & Pott, 2003). Pantanal wetlands are subject to alternate flooding and drought that cause these floating islands to drift with the current and wind or to dry out causing the death of its vegetation (Junk & Silva, 1999). Conversely, in the "Rio Doce" lake system of the present study (Figure 1), the water level is almost constant throughout the year (average variation of less than 1 m), therefore the floating islands that form tend to perpetuate and grow and can eventually occupy the whole area of the lake.

Fig. 1. Location of the Rio Doce study area including the Rio Doce State Park (black thick line).

Although the process of floating island formation is a natural one, in certain cases it can be initiated or accelerated by human interference. Our hypothesis is that a significant degradation of the surroundings of the lakes can cause an increase in sediments and nutrients load that can alter the state of the lake from oligotrophic to eutrophic. This new chemical balance is known to be beneficial for the development of free floating macrophytes species. If the aquatic environment is lentic, isolated and perennial (without seasonal flooding pulses) the emergence of macrophyte tend to colonize an ever increasing area of the lake and will eventually lead to the formation of floating islands. These floating island can, in turn grow indefinitely until the whole lake is covered. There are a number of these completely covered lakes in the Rio Doce lake system. Although we speculated that it is the degree of human interference (logging, agriculture, fertilizers, road construction, etc.) that is the main factor responsible for causing some lakes to be colonized by floating islands and others not, a clear trend could not be verified. Some lakes appear to have seen their open water area increased despite the degradation of their surroundings.

The objective of this study is to verify if the history of recent human interferences can help explain the formation of large areas of floating islands within the Rio Doce lake system. To do so, we have used a 20 years temporal series of Landsat images to assess the behavior of these lakes in terms of their area of open water and determine if it can be associated with the degree of human interference. A high resolution Ikonos[1] mosaic of images and a RapidEye[2] mosaic were also used to complement our field data for the initial delineation of the lakes.

2.1 Material and method

2.1.1 Study area

The Rio Doce valley is located in the eastern part of the state of Minas Gerais and is an important physiographic feature of Southeastern Brazil. The relief is strongly ondulating at an average altitude of about 250 m a.s.l. and varies between 195 and 525 m a.s.l. with many depressions occupied by lakes (Gilhuis, 1986). Annual rainfall ranges between 1000 and 1250 mm and the climate by is hot and humid (Köppen: Aw) megathermic, with a distinct dry (April-September) and rainy season (October to March).

The Rio Doce lake district is the third largest lake system in the Brazilian territory (Tundisi et al., 1981). According to Esteves (1988), these water bodies originated in the Pleistocene through a blocking of the mouth of former tributaries of the Doce and Piracicaba rivers under the influence of an epirogenetic shift. This also explains the continuity and depth (up to about 30 m) of the lakes, meandering their ways.

The Rio Doce lake system is situated in the Atlantic Forest domain (*Mata Atlântica*), where the vegetation is classified as mesophilous semi-deciduous forest (Veloso et al., 1991). The dense native forest that naturally surrounds the lakes prevents the entry of large quantities of allochthonous material (sediments), allowing the limnological characteristics of these water bodies to sustain over time without large fluctuations in their physicochemical characteristics and in the chemical composition of their sediments (Meis(de) & Tundisi, 1986). Under these conditions, the lakes generally present an oligotrophic state and a low diversity of dominant macrophytes (Ikusuma & Gentil, 1985).

However, these lakes are in various states of health and those within the boundaries of Rio Doce State Park (RDSP) are generally well preserved. In 2009 some of the lakes located in this protected area have been recognized internationally as a Ramsar Site (site 1900 http://www.ramsar.org/), with an important wetland area for the conservation of biodiversity as well as economic, cultural, scientific and recreational resources (SMASP 1997). Most of the lakes located outside the RDSP boundaries have had their surrounding native vegetation devastated, a factor that changed their original oligotrophic status to eutrophic. Since the 1950s these areas have suffered from various human activities, beginning with the removal of vegetation for charcoal production to supply metallurgical plants. Today, these areas are used for extensive plantations of eucalyptus and are intertwined by an extensive network of paved and unpaved roads. Other sources of threat include residential and industrial pollution, hunting and predatory fishing, fragmentation of remaining habitat and introduction of exotic species.

[1] An American commercial satellite operated by Space Imaging Corporation and producing panchromatic and multispectral images with ground resolutions of one and four meters respectively.

[2] A German-owned constellation of five satellites producing five meter resolution multispectral imagery.

2.1.2 Satellite and cartographic data

The imagery data available for this project came in the form of an Ikonos mosaic of 2006, a RapidEye set of images of 2010, Landsat historical data and out-of-date cartographic data (last updated in the 70's). The Ikonos mosaic was already pan-sharpened [3] and was made available by the Forest Institute of Minas Gerais (*Instituto Estadual de Florestas - Minas Gerais*). The RapidEye images with a ground resolution of 5 m were also made available by the IEF and were used to complete the Ikonos mosaic to the North, South and East. The Landsat database was constituted of 17 Landsat-5 TM images covering the 1989-2010 period, two of which had to be excluded because of their poor quality (Table 1). The cartographic data consisted mainly of the hydrographic network which was added to map products.

Date	Quality	Date	Quality	Date	Quality
04/07/1985	good	05/07/1997 *	rejected	24/07/2004	good
04/05/1986	good	08/07/1998	good	14/05/2007	good
15/07/1989	good	28/08/1999	good	05/09/2008	good
27/08/1993	good	27/06/2000	good	07/08/2009	good
01/10/1994	good	27/04/2001	good	26/08/2010	good
18/07/1996	good	20/06/2003 *	rejected		

Legend: *images with too many clouds or haze

Table 1. List of Landsat-5 TM images (orbit/scene 217-73 and 74) used in this study along with a quality assessment.

Field work was conducted over a period of four years in which as many as 20 lakes were visited and over 200 species of aquatic plants were collected and identified (Pivari et al., 2008). Positional data was also acquired using a navigation GPS to register the images to a common cartographic projection (UTM 23 South).

Because no survey of the lakes was done, our approach was to use the Ikonos and RapidEye images as basis for the contouring of all the lakes while accounting for possible positioning inaccuracies by applying a buffer of 75 meters outside the interpreted vectors. These vector would subsequently be used to eliminate undesirable classified pixels and areas. At the same time, based on the knowledge acquired in the field, the wetland areas were divided into four different classes: 1) macrophytes with visible open water, 2) bogs, 3) peatland, and 4) floating islands. Figure 2 shows examples of these wetland classes.

Our main goal being to determine if the formation of floating islands can be related to the degradation of the surroundings, these wetland classes were considered as a whole and it was assumed that what was not classified as open water belonged to the wetland class, that is within the vicinities of the lakes. The main reason for not considering these different types of wetlands was that they were not spectrally separable from the tests we conducted. We also had insufficient validation data to do a full scale classification of aquatic communities.

2.1.3 Lake classification with MAGIC

To classify the open water areas of the lakes, a region-based unsupervised classification approach was adopted where two classes were sought: water and non-water. The MAGIC

[3] Pan-sharpening involves resampling the 4 m multispectral imagery to 1 m using the panchromatic channel.

Fig. 2. The four main types of wetland encountered in the study region: (a) floating island, (b) peatland, (c) macrophytes with open water and (d) bogs.

(©2010 Systems Design Engineering, University of Waterloo, Canada) program (Clausi et al., 2010) is the product of an ongoing research (actually in version 2) and was chosen to segment and classify the images for having yielded excellent results in several other studies (Barbosa & Maillard, 2010; Maillard et al., 2008). MAGIC is an acronym that stands for "MAp Guided Ice Classification" because it was originally developed as a tool for classification of sea ice types. With new applications being tested and implemented, the "I" in MAGIC might eventually stand for "image".

The classification of MAGIC is unique in its implementation and the principles it embodies. It is an hybrid segmentation-classification approach that uses two different paradigms: "watershed" and Markov Random Fields (MRF). The segmentation is started by applying a "watershed" algorithm that produces a preliminary segmentation and generates segments (areas) of 10-30 pixels depending on the noise level in the image. The "watershed" algorithm implemented in MAGIC was developed by Vincent & Soille (1991) and divides an image into segments with closed boundaries. The "watershed" algorithm first looks for local minima and then works by region growing until it finds a divide line with another "catchment" area. However, it tends to oversegment the image, a characteristic that MAGIC takes advantage of in order not to "miss" any object.

Conversely, the MRF model (Li, 1995) assumes that the conditional probability of a pixel given its neighbors is equal to the conditional probability of that pixel given the rest of the image. This makes it possible to consider every pixel within its neighborhood as an independent process (Tso & Mather, 2001) and to compute the conditional probability of a pixel belonging to a given class using the Bayes rule:

$$P(Y_i|x) = \frac{p(x|Y_i)P(Y_i)}{\sum_i [p(x|Y_i)P(Y_i)]} \tag{1}$$

where $p(x|Y_i)$ is the conditional distribution of vector x given class/segment Y_i and $P(Y_i)$ is the prior probability of the Y_i class. Suppose that the energy associated to the prior probability is E_r and that E_f represents the energy of the spatial context $p(x|Y_i)$, then the general energy formula is given by Geman et al. (1990):

$$E = E_r + \alpha E_f \tag{2}$$

where E_f is the energy form of feature vector f having k dimensions. Assuming a Gaussian distribution E_f can be modeled as:

$$E_f = \sum_{s,m=Y_s} \left\{ \sum_{k=1}^{K} \left[\frac{(f_s^k - \mu_m^k)^2}{2(\sigma_m^k)^2} + \log(\sqrt{2\pi}\sigma_m^k) \right] \right\} \tag{3}$$

where μ_m and σ_m are the mean and standard deviation of mth class in the kth feature vector. E_r represents the energy of the labels (classes) in the neighborhood of the pixel being analyzed based on a system of *clique* (generally pairs or triplets of contiguous pixels):

$$E_r = \sum_s \left[\beta \sum_{t \in N_s} \delta(y_s, y_t) \right] \tag{4}$$

where y_s and y_t are the respective class of pixels s and t (inside the *clique*), and $\delta(y_s, y_t) = -1$ if $y_s = y_t$ and $\delta(y_s, y_t) = 1$ if $y_s \neq y_t$. β is a constant. In the absence of training samples to determine the labels of the pixels of the *clique*, these are initially randomly determined and gradually stabilize by iteration.

In equation 2, α is a parameter that sets the proportions of the relative contribution of E_r and E_f within E. The adaptation of Deng & Clausi (2005) adopted in MAGIC makes α iteratively change the weighting between the spectral (global) and spatial (local) components; early iterations favor the spectral component and increased iterations gradually increase the weight on the spatial component.

MAGIC is unique in the sense that instead of working on pixels, it uses the actual segments produced by the "watershed" algorithm. These segments are arranged topologically, so that all contiguous segments can be determined through an adjacency graph or RAG (Region Adjacency Graph). MAGIC will then merge contiguous segments if the union produces a decrease in the total energy of the neighborhood defined above.

The advantage of the MRF model is its inherent ability to describe both the spatial context location (the local spatial interaction between neighboring segments) and the overall distribution in each segment (based on parameters of distribution of spectral values for example). This new approach was entitled "Iterative Region Growing Using Semantics" or

IRGS and is described in Yu & Clausi (2008). Because MAGIC associates the segments to a predefined set of classes, it is considered a region-based unsupervised classification system.

MAGIC incorporates a number of innovative features such as 1) importing vector polygons to guide or restrict the classification (hence the "map-guided"), 2) a number of other segmentation approaches both traditional (*e.g.* K-means, gaussian mixture) and MRF-based, 3) the ability to compute texture features (grey level co-occurrence matrix and gabor) and 4) a functional graphical user interface (GUI). Figure 3 illustrates the GUI of MAGIC with the classification results for the Landsat 2010 image and the pop-up window for the IRGS algorithm.

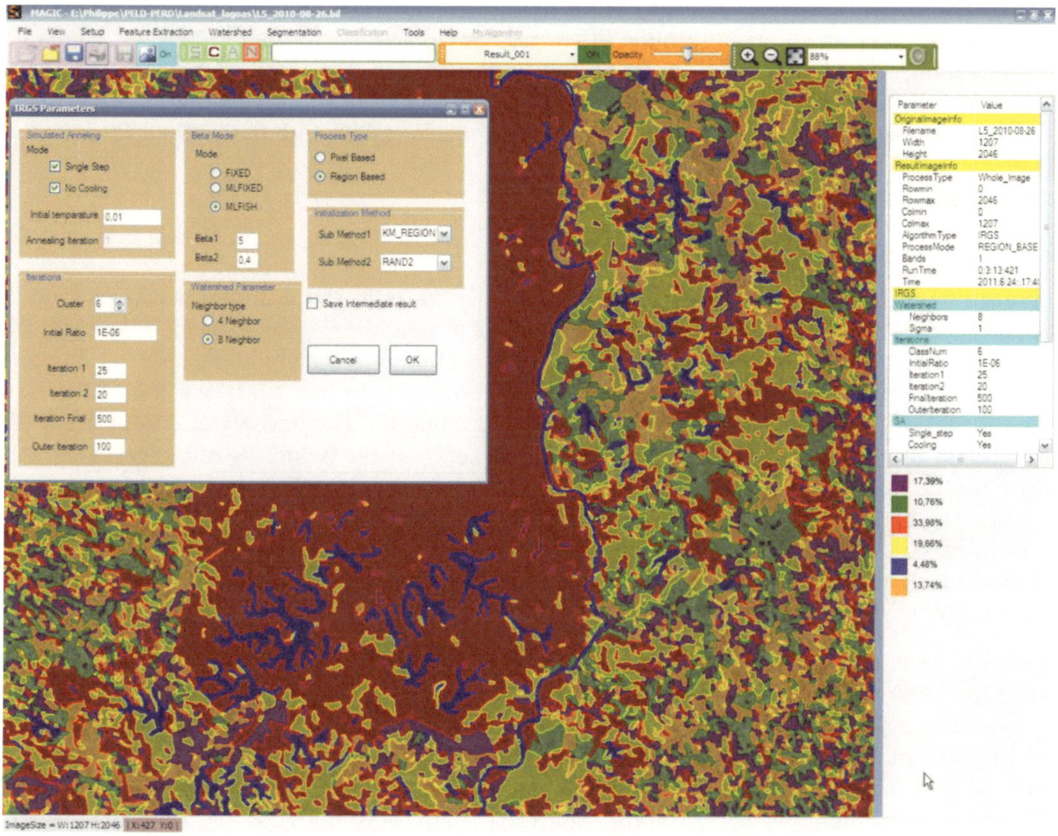

Fig. 3. The graphical user interface (GUI) of MAGIC also showing the pop-up window for the IRGS segmentation / classification.

The unsupervised classification was performed on all Landsat images using exclusively the mid-infrared band (band 5) generally considered the best option for separating land from water (Ji et al., 2009; Xu, 2006). This approach also included rivers in the classification results which were eliminated using the lake buffers. Other "misclassified" pixels (dark shadows, tiny reservoirs) were also eliminated by the process.

2.1.4 Area calculation and statistical modeling

The calculation of the open water area in each lake was a straight forward operation performed by simply counting the water pixels within each individual lake contour (plus buffer). Each lake was treated as an individual "area of interest" for witch a statistics calculation yielded the total number of non-zero pixels. The area of each lake and each year was organized into a worksheet and processed using MiniTab (Copyright ©2011 Minitab Inc.). Because many small lakes "disappeared" during the period analyzed, some of which could "reappear" some years after, their inclusion into the regression processed posed an analysis problem and so it was decided to retain only the lakes larger than 10 hectares. This was also partly due to the resolution of the Landsat images (0.9 ha) that did not allow a satisfactory precision for very small areas. Simple linear regression was performed between the area of all remaining lakes and the time represented by the year of the Landsat images. The slope parameter of the regressions (provided it was statistically significant) was used to determine the trend in the behavior of the open water areas of the lakes through the 1989 - 2009 period.

2.2 Results

2.2.1 Open water classification

Because MAGIC is unsupervised and the user only feeds in the number of classes (and a region weight parameter that controls the merging of neighboring segments), it is normally better to specify more classes than actually needed so that the clusters in the spectral domain are more restrictive and more consistent. In this case, after a few trials, we found that six classes worked best and could be adopted for all 15 images. The non-water classes are then eliminated by defining which class number represents water (which is not necessarily the same all the time since class numbers are attributed randomly). The next step consisted in eliminating lakes smaller than 10 ha, rivers and any pixel being wrongly attributed the same class as water like very dark shadows (very rarely). The vectorized lakes interpreted from the Ikonos and RapidEye images with a 75 m buffer was used as a mask to retain only the 147 lakes larger than 10 ha. Figure 4 illustrates this process.

Between lakes, peatbogs, and swamps, there were 765 interpreted "objects", more than half of which (399) did not have open water at any time, or did not pertain to the Rio Doce lake system leaving some 366 "objects" with open water. However, only 173 had open water in all 16 years analyzed. The graph in Figure 5 shows the number of lakes with open water for each year of the 16 Landsat images as well as the number of lakes considering the number of years without open water. From the subjective analysis of both curves, we estimate that there are usually between 240 and 260 lakes. We also found that this number appears to be slowly increasing with time, which might be the results of more restrictive land use and more protective measures from both the authorities and the forestry companies.

2.2.2 Regression: Open water area *vs* year

Despite the fact that an average of $\approx$ 250 lakes have open water, only 107 lakes were left after the elimination of the lakes that had more than four years without open water because of the negative effect it would have of the regression analysis. One hundred and seven (107) regressions were done using the area of the lake as dependent and the year as independent. Of these, only the regressions with a coefficient of determination above 0.5 were retained and only when a clear trend (growing or shrinking) could be identified ($| slope | > 0.003$). This left

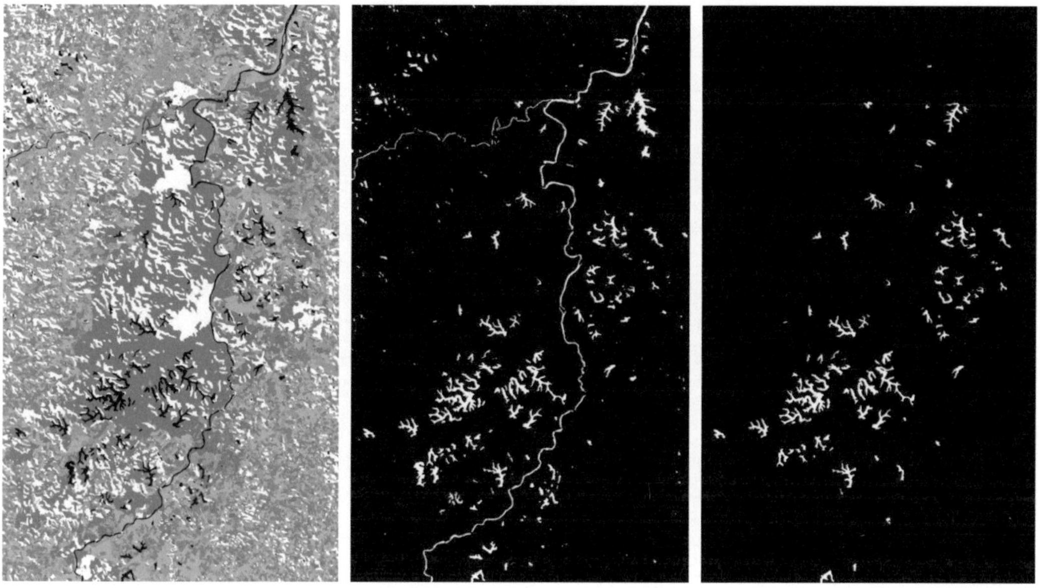

(a) (b) (c)

Fig. 4. Results of the open water classification for lakes larger than 10 ha and for the year 2009: a) the classification results of the unsupervised region-based classification using 6 classes, b) after elimination of the non-water classes, c) after eliminating small lakes and rivers.

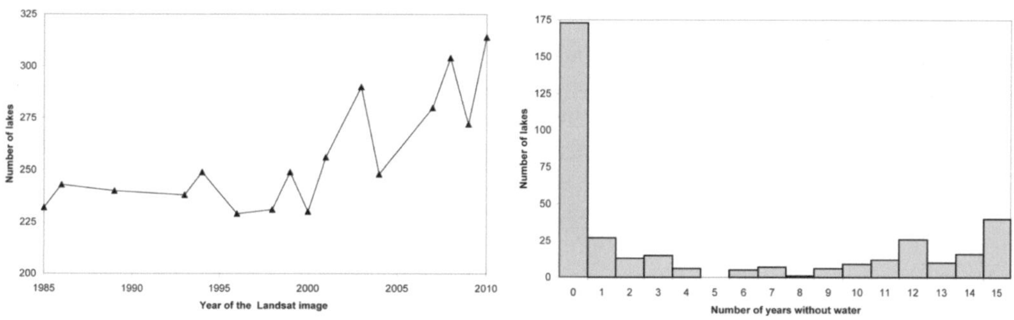

(a) (b)

Fig. 5. Graph showing in a) the number of lakes for each year studied and, in b) the number of lakes considering the number of years without open water (zero meaning that the lakes have open water in all years, 15 meaning that these lakes had no open water in 15 of the 16 years).

a total of 27 lakes for which the slope of the regression, the coeficient of determination and the area are listed in Table 2. The table also outlines which lakes are shrinking (left column) or growing (right column) and which are inside or outside the protected area of the State Park. These results are also illustrated graphically in Figure 6.

SHRINKING Inside State Park				GROWING Outside State Park			
Lake	slope	r^2	Area (ha)	Lake	slope	r^2	Area (ha)
141	-0.0039	65.00	31.72	116	0.0032	62.30	63.78
140	-0.0039	78.60	303.24	113	0.0059	57.40	20.26
110	-0.0170	65.90	86.79	111	0.0079	63.30	15.58
109	-0.0058	92.00	68.42	97	0.0120	50.50	224.68
108	-0.0373	68.90	10.27	94	0.0260	74.50	11.86
103	-0.0055	73.00	24.03	87	0.0183	76.20	21.40
45	-0.0093	73.50	24.64	84	0.0070	53.30	20.26
Outside State Park				79	0.0032	70.50	161.77
120	-0.0071	73.10	48.58	28	0.0033	61.30	20.07
118	-0.0068	67.30	31.22	24	0.0049	52.30	64.70
115	-0.0099	61.60	23.93	4	0.0025	51.50	62.85
114	-0.0062	62.50	62.53	2	0.0119	65.60	28.89
85	-0.0347	71.00	15.57				
72	-0.0404	93.60	23.64				
69	-0.0363	60.20	15.75				
55	-0.0350	91.30	21.85				

Table 2. Slope and coefficients of determination for the lakes that have seen their open water area significantly changed during the period of study. The lakes are separated as being inside or outside the Rio Doce State Park. The areas correspond to the vectors interpreted form the 2010 RapidEye image.

2.3 Discussion and future research

The results generated were directly usable to establish an historical progression of the situation of the open water area of over one hundred lakes. The region-based unsupervised classification of the water / non-water classes proved to be fast and accurate when compared with the digitized contours extracted from the Ikonos and RapidEye mosaics. This new approach saved the time and effort that would have been needed to create training samples. Because the study was based on historical data, no validation could be made available. Still, based on the comparison with the visual interpretation, the extraction of the open water area of these lakes proved very accurate, especially when considering their spatial consistency (*i.e.* the contiguousness of the water pixels). Some questions remain open like the density of aquatic plants needed for a pixel to fall in or out of the open water class. It is clear that this parameter depends on the plant species, on the quality of the water and on the time of year (phenology). Because of the large number of similar lakes involved in the study, it stands out that this parameters does not affect the overall results.

However, the results obtained do not agree with our initial hypothesis stating that the degradation of the surroundings of the lakes tend to have a shrinking effect on the open water area of the lakes. Instead, the study shows that the problem is far more complex than we originally expected and that only a thorough and constant monitoring of some of the lakes in various situations could lead to a better understanding. One important observation is that even the lakes in the protected areas are not necessarily safe from the accelerated process of being transformed into bogs or being covered by floating islands. In particular, it has been

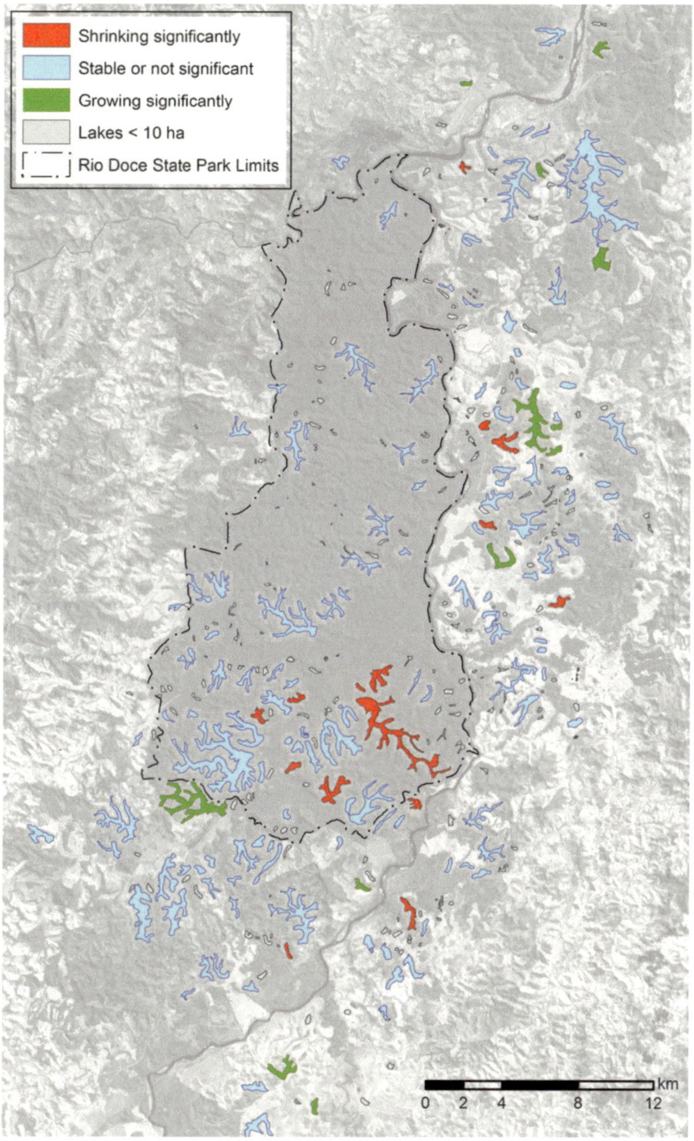

Fig. 6. Image map showing the dynamics of the lakes in the Rio Doce lake system. Only lake larger than 10 ha have been color-coded according to their dynamic state (shrinking or growing). The map overlays band 5 of the 2010 Landsat image (faded).

observed that *Nymphaeaceae caerulea* (a kind of water lily), an exotic plant from Africa, is propagating even in remote lakes, probably through the actions of aquatic birds.

Future research will be focused on acquiring more thorough validation data to classify aquatic communities like it was done in another study of the Pandeiros (Barbosa & Maillard, 2010). It is also planned to implement a program to monitor the dynamic of aquatic plant communities for some lakes that appear to be shrinking or growing both inside and outside the State Park and in different environments (eucalyptus, pasture, forest). Since the Government of Minas

Gerais has implemented a program to acquire RapidEye images of the whole State once or twice a year, this new perspective will facilitate a constant monitoring program with good ground resolution (5 m).

3. Second case study: The receding of six small lakes in the upper Peruaçu watershed

The Cerrado biome as a whole covers over one and a half million square kilometers in Brazil, about 9% of which can be considered semiarid mostly in Northeast Brazil. In Southeastern Brazil the only patch of semiarid Cerrado is at the northernmost tip of Minas Gerais in an ecological tension zone between Cerrado and Caatinga (thorn shrub). The Peruaçu river watershed, a $\approx 1500km^2$ area falls within this zone and has attracted much attention because of its natural beauty and its archeological and cultural heritage. Receiving less than 1000 mm of precipitations yearly, and having up to seven months without any rain, water resources in the region is critical to the survival of populations but has been suffering from excessive exploitation. In particular, the Veredas do Peruaçu State Park is apparently seeing the continuous lowering of its aquifer in the last few decades. This is mostly observable from the receding of a few small lakes (Figure 7) inside the park and one larger lake outside. Although the phenomenon is obvious to the local population, over exploitation of water resources appears to continue undisturbed.

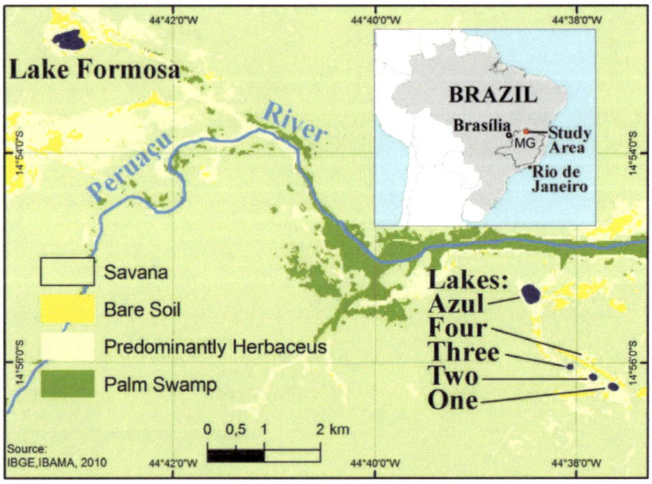

Fig. 7. Location of the six lakes in the Study area near the head waters of the Peruaçu River.

Even though human occupation can be considered sparse, because the Peruaçu watershed is small (1450 km^2) and the region is considered semiarid, we stipulate that the pressure of irrigation, eucalyptus plantations and wells is too great for its supporting capacity. It could be argued that the lowering is caused by local changes in the precipitation and water balance (WB), but since no records of the level of the aquifer or the lakes are available for the past, we had to develop a methodology entirely based on historical remote sensing and meteorological data to unambiguously demonstrate and quantify the phenomenon.

A multi-temporal remote sensing approach was used to create a time sequence of images to monitor the size of the Peruaçu lakes. Landsat images stood as the most logical choice for analyzing the dynamics of these lakes for being the longest record of systematical remote sensing data available for civil use. Passive optical infrared images are also considered the most effective type of data for delineating water bodies since they absorb almost totally the incoming radiation and produce a sharp contrast with the surrounding vegetation and soil (Bonn & Rochon, 1992; Jensen, 2005).

Considering the small size of the lakes under investigation, the resolution of Landsat TM images is somewhat marginally acceptable because of the mixed pixel problem. The fact that water bodies are smooth continuous surfaces let us postulate that mixed border pixels have a predictable behavior and could be sub-sampled using some interpolation technique.

The objective of this study is to infer the dynamics of the fluctuations of the water level of the aquifer through past monitoring of the successive shrinking and growing of the open water surface of six lakes found in the *Veredas do Peruaçu State Park* and surroundings. To achieve this, we created a methodology for extracting the open water surfaces of these lakes from an historical series of Landsat TM images using interpolation to overcome the mixed pixel problem. We also computed the water balance record for the same period to verify if the behavior of the aquifer can be attributed to modifications in the climate record.

3.1 Material and method

3.1.1 Study area

The study area (Figure 7) is located in Northern Minas Gerais - Brazil, a savannah region that can be marginally classified as semiarid with less than 1000 mm of rain per year. Five of the six lakes under study are inside the limits of the *Veredas do Peruaçu State Park*. The sixth and largest lake Formosa is outside the protected area but still in its immediate vicinities. The hydrographic network is part of the Peruaçu River Basin being a left tributary of the *São Francisco* River. Rainfall is unevenly distributed during the year and is mostly concentrated between November and March. The whole region is mostly flat with deep soils composed mostly of sand and less than 15% of clay that have a low capacity of water retention.

The lakes themselves are small with the largest having an average area of around ten hectares. Because there is no general agreement about the names of the four smaller lakes, they were given the genereic names One, Two, Three and Four while the two larger ones are called Formoza and Azul. Although there has been a few hypothesis to explain the genesis of these lakes and their relative alignment, no conclusive results were ever presented. Lake Four had open water until 2000 but has dried up and is now but an intermittently saturated herbaceous round field. Unofficial reports by the local population all outline the gradual decrease of the open water surface of most of these lakes but no actual study was ever undertaken.

Until the 1970s the region was occupied by small family groups descended from the Indian tribe *Xacriabá*. In the middle of that decade the Brazilian government offered subsidies and incentives to companies that were willing to invest in eucalyptus plantations for wood supply. This was also the beginning of a much denser occupation of the area by workers and farmers. The impacts of the plantations were reflected in the decrease of biodiversity, both in terms of fauna and flora, and also by an increased pressure on water resources. Eucalypt planting ceased in the early 1990's, then the companies abandoned their activities in the region due to low productivity and a practice that was not well adapted to the natural conditions. The region was recognized as having unique biological characteristics and the Brazilian authorities created a national park (*Cavernas do Peruaçu*) and a state park (*Veredas do Peruaçu*) to protect the natural beauties and the archeological heritage (rock paintings) of the Peruaçu watershed (Maillard et al., 2009). Although the area is now protected by law, the effect of the previous uses can still be observed and the area surrounding the parks still suffer from human pressure, especially on water.

3.1.2 Data and data pre-processing

A total of 51 images from Landsat-5 TM (World Reference System: orbit/scene 219/70) were chosen. Landsat-5 has been continuously collecting image data since 1984 which constitutes the beginning the period considered by this research and ends in 2009. Two images from Landsat-7 ETM+ were also acquired to complete the dataset for the year 2002 for which the Landsat-5 scenes were too cloudy. The dates of the images (Table 3) correspond ideally to the end of the wet season (first image) and the end of the dry season (second image) but had to be slightly shifted in cases where images were either of low quality (clouds) or unavailable. Three images also had to be excluded because they presented calibration problems.

Year	1st image	2nd image	Year	1st image	2nd image	Year	1st image	2nd
1984	13/jun	13/oct	1985	31/may	06/oct	1986	15/mar	09/oct
1987	02/mar	12/oct	1988	21/apr	30/oct	1989	Excluded	Excluded
1990	10/mar	20/oct	1991	30/apr	07/oct	1992	18/may	23/sep
1993	18/mar	12/oct	1994	22/apr	12/aug	1995	24/apr	02/oct
1996	26/mar	20/oct	1997	09/feb	07/oct	1998	20/jun	26/oct
1999	19/mar	11/sep	2000	24/apr	15/oct	2001	24/mar	01/oct
2002	20/apr*	13/oct*	2003	20/jul	08/oct	2004	01/apr	24/sep
2005	04/apr	13/oct	2006	20/jun	30/sep	2007	Excluded	03/oct
2008	24/feb	05/oct	2009	14/mar	06/sep	2010	4/may**	

Table 3. List of Landsat images (* indicates Landsat-7, the rest are Landsat-5; ** the 2010 image was only used to validate the lake contour extraction method).

The images were geometrically and radiometrically corrected and an atmospheric effect compensation was also applied. The geometric correction was done in an "image-to-image" approach using a one-meter Ikonos image as basis (which was geometrically adjusted using control points from a geodetic GPS survey). The atmospheric and radiometric correction were applied using an in-house program build for that purpose: *Corat_Landsat*. The program takes as input a worksheet containing 1) the name of the image file, 2) the digital number value for the dark object substraction (Chavez Jr., 1988) for bands 1, 2, 3, 4, 5 and 7, 3) the sun elevation angle and 4) the sun-earth distance in astronomical units. The output is a 16 bit reflectance image (reflectance values were redistributed between 0 and 10000).

The calculation of the water balance was based on the method proposed by Thornthwaite & Mather (1955) which consists in determining the hydraulic characteristics of a given region without direct measurements on the ground. The water balance is the simple budget between input and output of water within a watershed:

$$\Delta S = \left(\underbrace{P + G_{in}}_{Inflow} \right) - \left(\underbrace{Q + ET + G_{out}}_{Outflow} \right) \tag{5}$$

where P is the precipitation, G_{in} and G_{out} represents the ground water flow, Q is the runoff water and ET is the evapotranspiration.

The procedure simplifies the calculation by estimating all its components from only two input parameters: average daily temperature and precipitation:

$$AW_t = AW_{t-1} exp\left(-\frac{PET_t}{AWC}\right) \tag{6}$$

where AW_t is the available water at time t, AW_{t-1} is the available water at time $t-1$, PET_t is the potential evapotranspiration at time t and AWC is the soil's water holding capacity. The water balance can be summarized in three situations.

- $\Delta P < 0$; net precipitation (precipitation - potential evapotranspiration) is less than zero: the soil is drying.
- $\Delta P > 0$ but $\Delta P + AW_{t-1} \leq AWC$; net precipitation is more than zero but net precipitation plus the available water from time $t-1$ is less or equal than the soil's water holding capacity: soil is wetting.
- $\Delta P > 0$ but $\Delta P + AW_{t-1} > AWC$; net precipitation is more than zero and net precipitation plus the available water from time $t-1$ is more than the soil's water holding capacity: soil is wetting above capacity and water goes to runoff.

3.1.3 Interpolation and lake contours extraction

Because the lakes are all very small, the 30 m spatial resolution of Landsat TM images became restrictive in terms of contour definition of the lakes. To overcome this limitation, we decided to exploit the very stable behavior of water in the optical infrared region of the electromagnetic spectrum that simply absorbs almost all energy in that part of the spectrum (Ji et al., 2009). In fact, water reflection is almost zero beyond 760 nm (McCoy, 2005). Conversely, the surroundings of all these small lakes is composed of sand and vegetation in large proportion which both reflect much more than water even in the absorbtion bands caused by water content in the leaves as can be seen in Figure 8. In many cases, a simple threshold in an infrared image histogram can reliably separate water from the other land covers with a relatively good rate of success and investigators have developed simple techniques for doing so in a systematical manner (Bryant & Rainey, 2002; Jain et al., 2005). Histograms of near infrared images containing a fair amount of open water surfaces are usually bimodal with the first peak directly related to water. Yet, when one looks closer, the water-land limit is often blurred by a varying width occupied by aquatic plants that can fluctuate over various time scales (yearly or seasonally). Using a sequence of historical Landsat images for which we had no validation data, we needed to have a very strict definition of the water-land interface. We defined the lake "water-edge" as the point at which water overwhelmingly dominates the surface and estimated that point to correspond to 70-80%.

Scale (or spatial resolution) can have various effects on image classification accuracy. A finer resolution can usually decrease the proportion of pixels falling on the border of objects (hence less mixed pixels) which can result in less classification confusion. Conversely, a finer resolution will generally increase the spectral variation of objects that can, in turn increase classification confusion (Markham & Townshend, 1981). Fortunately, water (especially clear and deep) is a spectrally smooth surface for which a finer resolution will bring more benefit

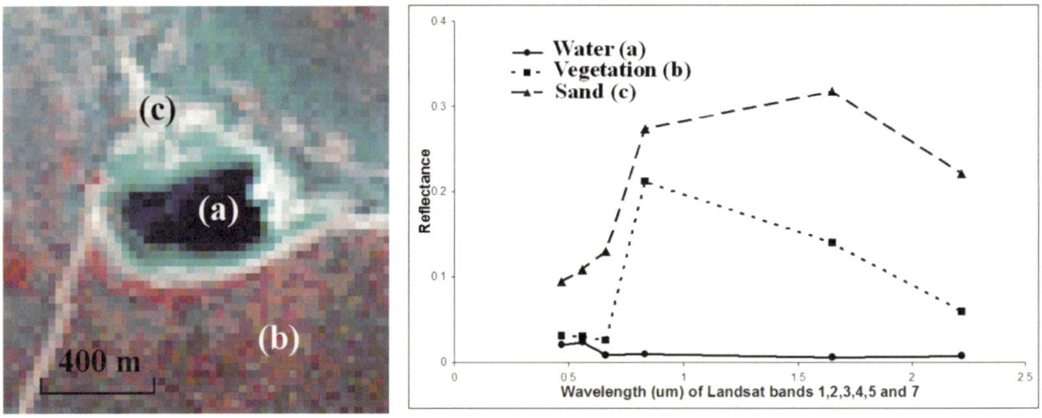

(a) (b)

Fig. 8. Reflectance values samples in a Landsat sub-scene for the visible, near and mid-infrared bands (a) Image section in false color, (b) graph of reflectance values for water, dry savanna vegetation and sandy deposits.

(less border pixels) than disadvantage (spectral variation). This special context led us to stipulate that the lake edge pixels can be subdivided into proportions of water and water edge using a weighted interpolation method. Amongst the various interpolation methods we opted for the minimum curvature interpolation (a variation of bi-cubic spline) with tension as described in Smith & Wessel (1990). This interpolation method has the advantage of being able to generate a smooth surface without generating undesirable fluctuations (artifact peaks or dips) by using a tension parameter. This interpolation proved better than "inverse distance weighted" that tends to produce artifact dips between sampling points (Maune et al., 2001). The minimum curvature worked well and fast and generated smooth ramps while keeping a sharp water-land edge. Figure 9 illustrates the effect of interpolating the Landsat data to 5 m on the lake extraction processing.

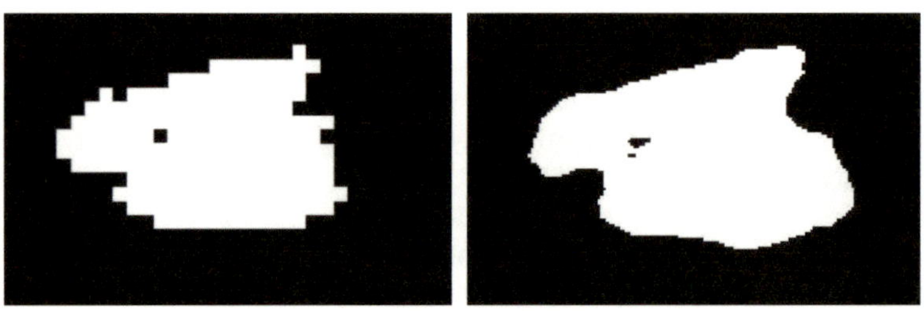

(a) 30 m (b) 5 m

Fig. 9. Comparison of the lake extraction methods using the original 30 m Landsat data (a) and the 5 m interpolated data (b).

3.1.4 Classification

Because the classification was binary in nature (water *vs* non-water), a supervised pixel-based approach was chosen to yield maximum control. Classification approaches such as maximum likelihood can produce posterior probability maps to which can be applied a threshold (hardened). This approach has the advantage to require training data only for the object of interest whereas classical classification procedures require all classes to have been defined using training data. In this case the posterior probability is simply the Gaussian probability density of the "water" class. In simple nominal classification, a pixel can be classified as pertaining to a particular class even if its probability is low, as long as it is higher than for all the other classes. By using a high threshold value (i.e. $> 90\%$) to attribute a water label to a pixel, we are able to use but a single class and avoid having to gather training data for other objects or surfaces.

3.1.5 Validation

Two validation data sets were used for testing the performance of the extraction of the lake contours from the interpolated Landsat data which also involved our definition of the "water-land" edge. First, the contours from the dry season image of 2006 were compared against the contours extracted from a pan-sharpened Ikonos image (1 m) five days apart from the Landsat image. Secondly, the four lakes of the VPSP (data from the larger lake outside the park could not be acquired) were surveyed using a geodetic GPS in kinetic mode to be compared with the contour from the Landsat image (with a five days difference). Coordinates of the lake contour were acquired at an interval of 15 meters with an approximate precision of 10 cm.

The validation was done by two complementary methods: 1) by expressing the difference between the areas as a proportion of the validated area ($1 - \frac{A_{real} - A_{observed}}{A_{real}} \times 100$); and 2) by overlapping the two contours (interpolated Landsat and validation data) and dividing the overlap area (intersection) by the merged areas (union) of both contours as illustrated in Figure 10. The latter accounts for errors of registration and edge definition of the lakes.

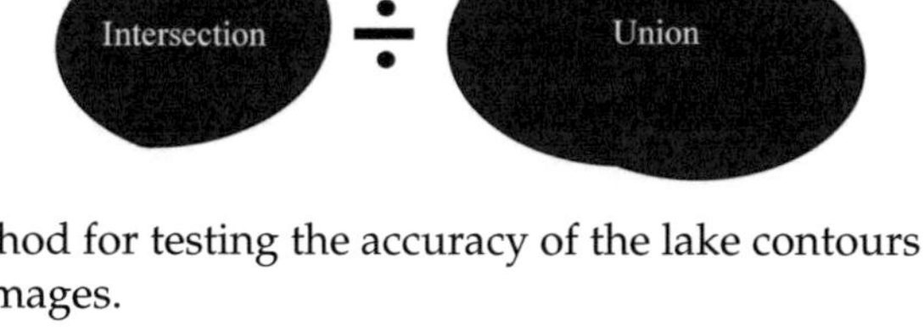

Fig. 10. Validation method for testing the accuracy of the lake contours extracted from the interpolated Landsat images.

3.1.6 Correlation between the lake areas and the water balance

If the behavior of the area of the lakes can be related to a local climate change, then the water budget should be the best indicator of such relationship. Even though the response of the water level to a change in the water budget is not spontaneous, the trend should still be statistically perceptible. Because the areas of the lakes are not normally distributed, a regression was not recommended. Spearman's correlation does not assume a normal distribution of the dependant variable and was chosen instead. The correlation was also computed between the area of the lakes themselves as a means to infer a generalized trend.

3.2 Results

Figure 11 shows the annual budget averaged every five years for the period along with the average budget for the whole period (black line). Apart from the two first periods (1984-1989 and 1990-1994) which appear as exceptionally high and exceptionally low respectively, the other periods do not show any trend towards an increase or a decrease.

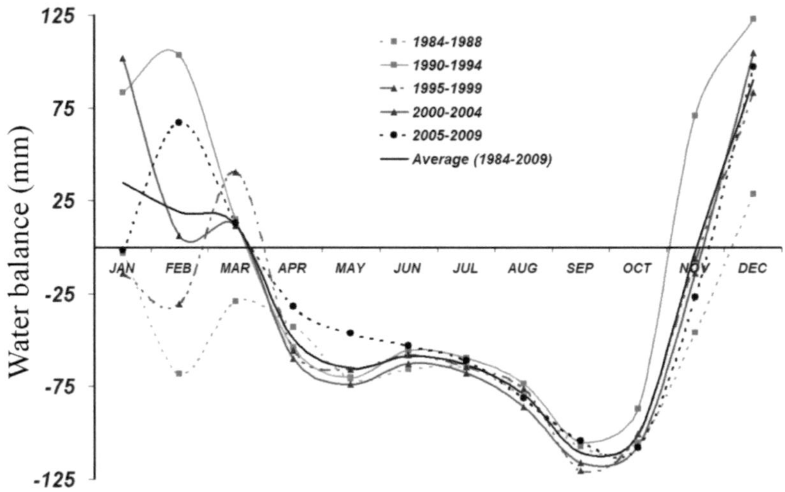

Fig. 11. Water balance over the region averaged for every five years between 1984 and 2009 and overall average (black continuous line).

3.2.1 Extraction and validation of the lake contours

All 51 selected Landsat images were geometrically rectified, registered to a UTM grid, corrected for atmospheric interferences (using Chavez's DOS method) and transformed in reflectance values. The images were then interpolated to a 5 m resolution using a minimum curvature algorithm. Apart from a few exceptions, the multi-temporal dataset shows an almost constant shrinking of the lake surfaces areas and the disappearance of one small water body (Lake Four). Figure 12 shows the 1984 and 2009 image sections side by side to illustrate the shrinking of all six lakes. The triangular area at the bottom of the 1984 image, was part of a eucalyptus plantation and is now naturally regenerating into *cerrado* vegetation.

Extraction of the lake surface area of open water using the posterior probability of the maximum likelihood classification yielded good visual results in all images. This was evaluated by looking at the spatial consistency of the results. Validation of the 2010 classification results confirmed the appropriateness of the methodology. By using the posterior probability of a single water class, we found that there was always an easily identifiable break between the water and non-water classes that made the selection of a threshold very easy. The threshold was applied to all 51 images and the area of all six

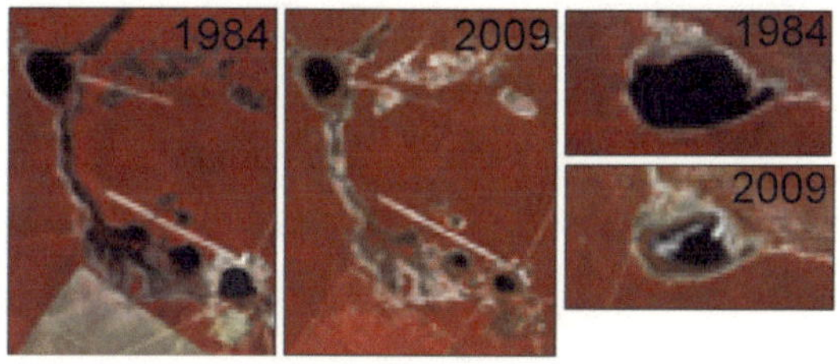

Fig. 12. Comparison of the lakes between October 1984 and September 2009.

lakes computed for every date. The graph in Figure 13 shows how these areas have changes between 1984 and 2009. Table 4 gives an over view of the shrinking of the six lakes. The lake areas of 1990 are also indicated (in bold) for being the record size for all lakes. While Lake Four has completely disappeared since 2000, four other lakes have lost between 59 and 80% of their area. Lake Azul has somewhat retained much more of its original area (loss of 29%) and it is also the only lake surrounded by hydromorphic gley soil with a higher clay content.

Areas	Lakes					
m^2	Four	Three	Two	One	Azul	Formosa
1990	**4962**	**28778**	**37413**	**56402**	**105389**	**296237**
1984	375	14795	32471	39030	92670	291502
2009	0	2928	7228	12243	65829	170409
% loss	100%	80,2%	77,7%	68,6%	29,0%	58,5%

Table 4. Comparison of the areas of all six lakes between 1984 and 2009 with the shrinking expressed in percentage (1990 was the record year for all lakes).

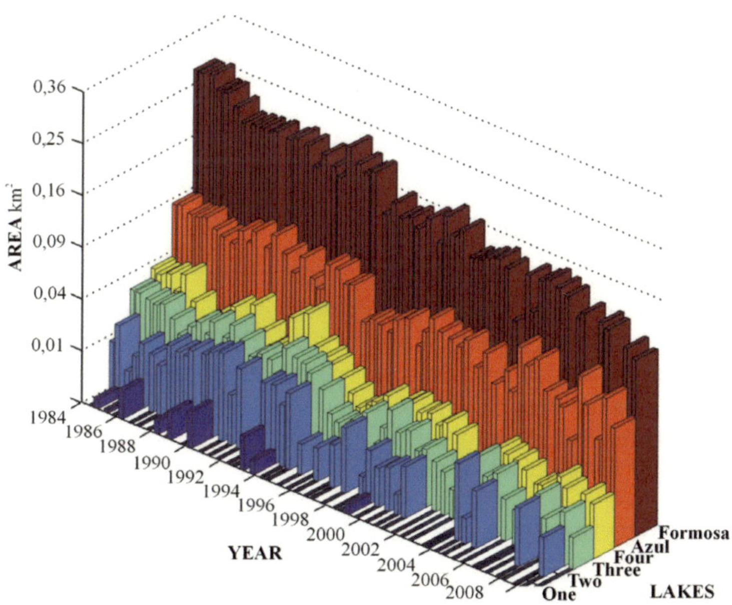

Fig. 13. Graph showing the evolution of the area of all six lakes for the period 1984-2009.

Lakes	Area Comparison		Intersection/Union×100	
	GPS	Ikonos	GPS	Ikonos
Three	94,54%	n/a	81.05%	n/a
Two	93,34%	86,01%	91,53%	71,04%
One	89,50%	94,41%	89,16%	83,85%
Azul	94,18%	96,36%	92,13%	93,20%
Formosa	n/a	95,08%	n/a	92.66%

Table 5. Validation of the lake contour extraction using the GPS survey and the Ikonos scene. Column 2 and 3 show the results for the area comparison; column 4 and 5 show the accuracy obtained with the $\frac{intersection}{union} \times 100$ approach.

Since we did not have precise elevation data, the water surfaces areas could not be associated with precise altimetric level measurements. We used the digital elevation surface (DES) from the ASTER sensor (ground resolution of 30 m) to overlay the contours of the lakes to estimate the height of the water level for the 1984-2009 period. Our analysis shows that Lake Azul has lowered by about 1 meter whereas lakes One, Two, Three and Formosa appear to have lowered by slightly more than 2 meters. Figure 14 shows the 1984 and 2009 levels on the ASTER DES profile for Lake Formosa (we did not, however have access to bathymetric data and the depth of the lake is unknown).

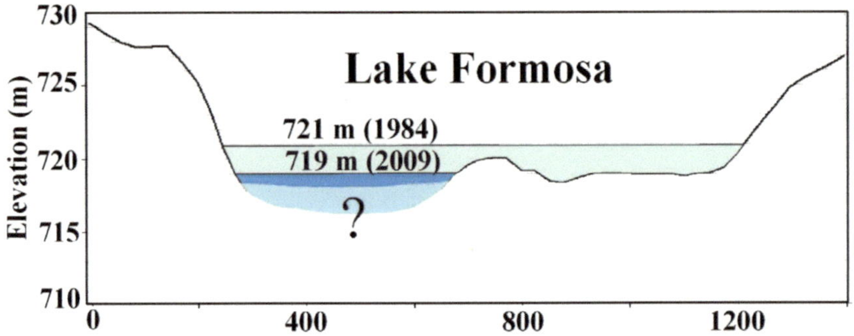

Fig. 14. Water level of Lake Formosa in 1984 and 2009 on an ASTER DES profile.

The validation of the data was done using the approach described in section 3.1.5. Table 5 shows the validation obtained with both control datasets (GPS and Ikonos image) and with the two methods of comparison (simple comparison of areas and "intersection ÷ union" approach). As expected, the accuracies with the latter method are slightly lower but since all accuracies but one are well above 80%, we conclude that both our extraction method and our geometric correction are within very acceptable boundaries. Figure 15 shows the contours extracted from the Landsat image of 2010 and the GPS survey contours for three of the lakes.

3.2.2 Statistical testing

Spearman's correlation test was applied to the area series of all lakes along with the AW data for the same period. The results are presented in Table 6. The only correlation between the areas of the lakes and the AW is Lake Four which has dried up since 2000 and the level of significance is p=0.05. Conversely, all the lakes are strongly related among themselves with a significance of 0.01. This confirms that the trend is statistically significant and that we can

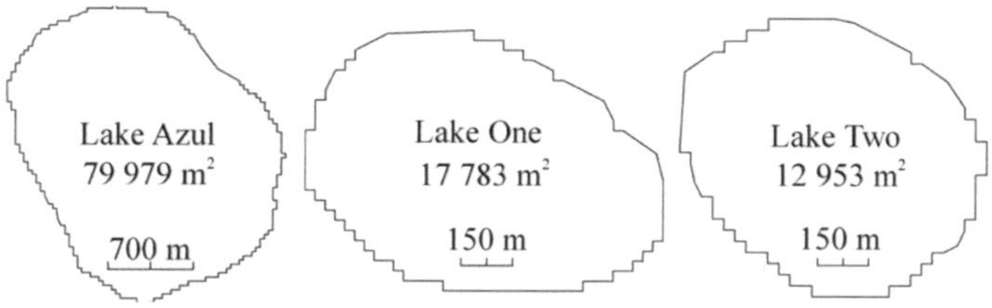

(a) Landsat lake contours (2010)

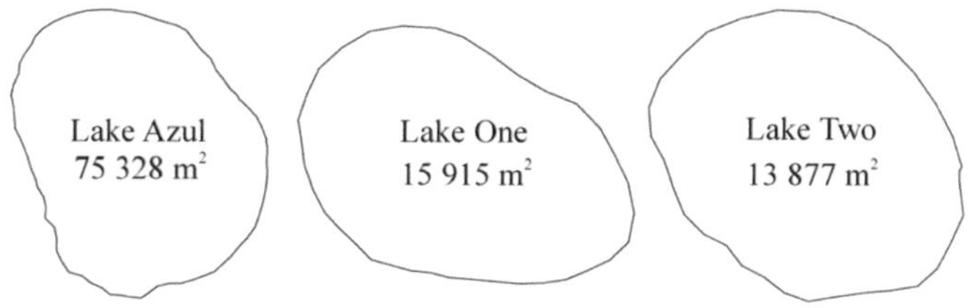

(b) Geodetic GPS lake contours

Fig. 15. Comparison of the contours of three of the six lakes using the interpolated Landsat data (top) and the geodetic GPS survey data (bottom) made five days after image acquisition.

infer that the lakes are rapidly shrinking. Even Lake Azul which has kept a much more constant surface area is strongly correlated with all the other lakes (0.601 to 0.871). Since the AW cannot be said to be correlated with the shrinking of the lakes, the meteorological explanation becomes much less plausible and the human pressure on the watershed can more easily be pinpointed as responsible.

	AW			**Lakes**		
Lakes		Four	Three	Two	One	Azul
Four	*0.329					
Three	0.209	**0.455				
Two	0.209	**0.611	**0.834			
One	0.075	**0.566	**0.735	**0.957		
Azul	0.259	**0.601	**0.871	**0.866	**0.789	
Formosa	0.068	**0.524	**0.674	**0.899	**0.897	**0.730

* Significant at 0.05 ** Significant at 0.01

Table 6. Results of the Spearman's correlation tests.

3.2.3 Discussion and future research

In this study we proposed an innovative approach for monitoring small lakes using medium resolution Landsat data. The approach uses minimum curvature interpolation to artificially improve the resolution of the image data and produce a much cleaner lake contour that matches the actual measured contour with a high success rate (15 validation out of 16 with better than 80% and 10 better than 90%). Using posterior probability of a maximum likelihood classifier, we were able to systematically extract contours from six lakes for 50 different dates with ease and good matching of control data. Even though we did not have bathymetric data or even precise elevation data of the surroundings of the lakes, the digital elevation surface produced from ASTER data with a resolution of 30 m, made it possible to estimate that the aquifer lowered, during the 25 year period (1984-2009), by up to two meters. The water balance using the Thornthwaite approach is well suited for area with limited climatological information and provides valuable insight on the climatological condition ruling water availability. In the present case, the water balance could not be statistically correlated (Spearman's correlation) to the shrinking of six small lakes in Northern Minas Gerais, Brazil. It became clear that, if the present situation continues, these small lakes (and the nearby palm swamps) will disappear with drastic consequences for the populations of humans and animals.

Future studies will concentrate on matching the lake size with precise elevation data and piezometric measurements. Although Landsat data proved most useful for extracting the open water surface, we plan to shift towards more precise satellite data such as RapidEye for which the Minas Gerais Government is acquiring on a regular base (twice a year) for the whole state. Future research will also explore more thoroughly the possibilities of artificially increasing resolution through interpolation. More interpolations methods need to be tested and compared with various situations. With the recent installation of a nearby weather station, precise local data will yield better control on monitoring the water budget throughout the year.

4. Overall conclusions

Multi-temporal remote sensing offers countless opportunities for monitoring past and present changes in land cover and land use. By monitoring the size and shape of water bodies, we can infer on human pressure and climate change. Small water bodies are especially fragile areas with a very high ecological value (the value of the services provided by lakes and wetlands has been considered as high as 8.498 and 14.785 $\$ha^{-1}yr^{-1}$ respectively according to Costanza et al., 1997) that are very sensitive to changes in temperature or the equilibrium of nutrients input (Mitsch & Gosselink, 2000).

In this chapter, two new approaches for monitoring small lakes and wetlands were used. First by using a region-based unsupervised classification based on an hybrid implementation (watershed and Markov random fields) we ensured a non-arbitrary systematic approach that did not rely on training samples or a subjective threshold. The MAGIC program proved very reliable for processing a large number of scenes while maintaining a very stable and predictable behavior. Although it turned out to work better by choosing a larger number of classes than actually needed, finding the water class was always easy and could easily be automated in certain cases like this one (for example by ordering the signatures through their mean). Future work will concentrate on determining the parameters that govern the precise amount of water within a pixel for it to fall in the water class.

Secondly, an interpolation method was used to artificially increase the resolution (from 30 m to 5 m) of a series of Landsat images to improve the contour definition of a set of very small lakes and to characterize their dynamic throughout a 25 years period. Much care was taken to validate the methodology by using two distinct methods of validation to account for all type of errors. The validation yielded a precision between 80% and 93% in all cases except one. Future work will concentrate on having this approach improve by using precise elevation data to associate an actual water level with the size of the lakes.

The use of historical satellite data is often made difficult by the absence of validation data and one must generally rely of sparse observations to corroborate results. One solution lies on validating the methodology using recent data and then to apply it to the historical data. Landsat has been an invaluable source of data since the 80's (Thematic Mapper) and even the 70's (Multi Spectral Scanner) by systematically acquiring data at regular predictable intervals over the same region. The newer generations of satellites platforms work mostly on a "per demand" scheme and require more carefully planned logistics of image acquisition. It is also likely that future post-Landsat multi-temporal studies will have to deal with data from different sensors with different resolutions and even different spectral specifications. This will bring new challenges to multi-temporal studies for which much research is still needed.

5. Acknowledgements

The authors are thankful to the Forestry Institute of Minas Gerais (IEF-MG) for providing the Ikonos and RapidEye data and field support. We are most thankful to Thaís Amaral Moreira for her hard work in mapping and statistics.

6. References

Barbosa, I. & Maillard, P. (2010). Mapping a wetland complex in the Brazilian savannah using an Ikonos image: assessing the potential of a new region-based classifier, *Canadian Journal of Remote Sensing* 36(Suppl. 1): S231–S242.

Bonn, F. & Rochon, G. (1992). *Précis de Télédétection: Principes et Méthodes*, Vol. 1, Presses de lt'Université du Québec. 485 p.

Bryant, R. G. & Rainey, M. (2002). Investigation of flood inundation on playas within the zone of chotts, using a time-series of AVHRR, *Remote Sensing of Environment* (3): 360–375,.

Chavez Jr., P. S. (1988). An improved dark-object subtraction technique for atmospheric scattering correction of multispetral data, *Remote Sensing of Environment* 24(2): 459–479.

Clausi, D. A., Qin, K., Chowdhury, M. S., Yu, P. & Maillard, P. (2010). Magic: Map-guided ice classification, *Canadian Journal of Remote Sensing* 36(Suppl. 1): S13–S25.

Costanza, R., d'Arge, R., de Groot, R., Farber, S., Grasso, M., Hannon, B., Limburg, K., Naeem, S., O'Neill, R. V., Partuelo, J., Raskin, R. G., Sutton, P. & van den Belt, M. (1997). The value of the world's ecosystems services and natural capital, *Nature* 387(15 May): 253–260.

Deng, H. & Clausi, D. A. (2005). Unsupervised segmentation of synthetic aperture radar sea ice imagery using a novel Markov random field models, *IEEE Trans. on Geoscience and Remote Sensing* 43(3): 528–538.

Esteves, F. A. (1988). *Fundamentos de limnologia*, 2nd edn, Interciência, Rio de Janeiro.

Geman, D., Geman, S., Graffigne, C. & Dong, P. (1990). Boundary detection by constrained optimization, *IEEE Trans. Pattern Anal. Machine Intell.* 12(7): 609–628.

Gilhuis, J. P. (1986). *Vegetation survey of the Parque Florestal Estadual do Rio Doce, MG, Brasil*, Master's thesis, Universidade Federal de Viçosa, Brazil.

Ikusuma, I. & Gentil, J. G. (1985). *Limnological Studies in Central Brazil, Rio Doce Valley Lakes and Pantanal Wetland (1st Report)*, Water Reseach Institute, Nagoya University, chapter Macrophyte and its environment in four lakes in Rio Doce Valley.

Jain, S. K., Singh, R. D., Jain, M. K. & Lohani, A. K. (2005). Delineation of flood-prone areas using remote sensing technique, *Water Resources Management* 19(4): 337–347.

Jensen, J. R. (2005). *Introductory Digital Image Processing*, 3rd edn, Pearson Prentice Hall, New Jersey. 526 p.

Ji, L., Zhang, L. & Wylie, B. (2009). Analysis of dynamic thresholds for the normalized difference water index, *Photogrammetric Engineering and Remote Sensing* 75(11): 1307–1317.

Junk, W. J. & Silva, C. J. (1999). O conceito do pulso de inundação e suas implicações para o pantanal de Mato Grosso, *Anais do II Simpósio sobre Recursos Naturais e Sócio-Econômicos do Pantanal*, EMBRAPA-DDT, pp. 17–28.

Li, S. Z. (1995). *Markov Random Field Modeling in Computer Vision*, Springer-Verlag, New York, NY, USA.

Maillard, P., Alencar-Silva, T. & Clausi, D. A. (2008). An evaluation of radarsat-1 and aster data for mapping *veredas* (palm swamps), *Sensors (MDPI)* 8: 6055–6076.

Maillard, P., Augustin, C. H. R. R. & Fernandes, G. W. (2009). *Arid Environments and Wind Erosion*, Novascience Publisher, chapter Brazil's Semiarid Cerrado: A Remote Sensing Perspective.

Markham, B. L. & Townshend, J. R. G. (1981). Land cover calssification accuracy as a function of sensor spatial resolution, *Proceedings of the 15th Int. Symp. on Remote Sensing of the Environment*, Ann Arbor, MI, pp. 1075–1090.

Maune, D. F., Kopp, S. M., A., C., Crawford & Zerdas, C. E. (2001). *Digital Elevation Model Technologies and Applications*, 1 edn, American Society for Photogrammetry and Remote Sensing, Bethesda, MD, chapter Introduction, pp. 1–34.

McCoy, R. (2005). *Field Methods in Remote Sensing*, The Guildford Press, 159 p., New York, NY.

Meis(de), M. R. M. & Tundisi, J. (1986). Geomorphological and limnological processes as a basis for lake typology: the middle Rio Doce lake system, *Anais da Academia Brasileira de Ciências* 58(1): 103–120.

Mitsch, W. & Gosselink, J. (2000). *Wetlands*, John Wiley and Sons, New York, NY, USA.

Pivari, M. O. D., Pott, V. J. & Pott, A. (2008). Macrófitas aquáticas de ilhas flutuantes (baceiros) nas sub-regiões do Abobral e Miranda, Pantanal, MS, Brasil, *Acta Botanica Brasilica* 22(2): 559–567.

Pott, V. J. & Pott, A. (2003). *Ecologia e manejo de macrófitas aquáticas*, Editora da Universidade Estadual de Maringá, chapter Dinâmica da vegetação aquática do Pantanal, pp. 145–162.

Smith, W. H. F. & Wessel, P. (1990). Gridding with a continuous curvature surface in tension, *Geophysics* 55: 293–305.

Thornthwaite, C. W. & Mather, J. R. (1955). *The water balance*, Publications in Climatology, Drexel Institute of Technology, New Jersey.

Tso, B. & Mather, P. (2001). *Classification Methods for Remotely Sensed Data*, Taylor and Francis, London, England.

Tundisi, J. G., Matsumura-Tundisi, T., Pontes, M. & Gentil, J. (1981). Limnological studies at quaternary lakes in eastern Brazil, *Revista Brasileira de Botânica* 4: 5–14.

Veloso, H. P., Rangel Filho, A. R. L. & Lima, J. C. A. (1991). *Classificação da Vegetação Brasileira Adaptada a um Sistema Universal*, Rio de Janeiro: FIBGE.

Vincent, L. & Soille, P. (1991). Watersheds in digital spaces: an efficient algorithm based on immersion simulations, *IEEE Transactions on Pattern Analysis and Machine Intelligence* 13(6): 583Ű598.

Xu, H. (2006). Modification of normalised difference water index (NDWI) to enhance open water features in remotely sensed imagery, *International Journal of Remote Sensing* 27(14).

Yu, Q. & Clausi, D. A. (2008). IRGS: image segmentation using edge penalties and region growing, *IEEE Trans. Pattern Analysis and Machine Intelligence* p. paper accepted for publication.

3

Satellite-Based Snow Cover Analysis and the Snow Water Equivalent Retrieval Perspective over China

Yubao Qiu[1]*, Huadong Guo[1], Jiancheng Shi[2] and Juha Lemmetyinen[3]
[1]*Center for Earth Observation and Digital Earth, Chinese Academy of Sciences, Beijing*
[2]*Institute for Computational Earth System Science, University of California, Santa Barbara*
[3]*Finnish Meteorological Institute (FMI), Arctic Research Centre, Sodänkylä*
[1]*China*
[2]*USA*
[3]*Finland*

1. Introduction

Global changing is a great challenge that affects the nowadays world, even arises and becomes kinds of the political issues. The changing of the snow is not only a sensitive factor act as a driving force but can be influenced much in the global temperature variation, especially for the seasonal snow cover which is vastly distributed over the northern hemisphere. Snow cover influences the atmosphere and ocean, and therefore the climate system, through both direct and indirect effects (Judah, 1991). In the climate regime, the snow cover alters the surface energy and water circle in a global scale in the climate processing (Fg.1). From the IPCC (2001), the recent and anticipated reductions in snow cover due to future greenhouse warming are an important topic for the global change community. Large seasonal variations in snow cover are of importance on continental to hemispheric scales induces to investigate its natural variability in the climate-system forcing of such trends, versus possible anthropogenic influences (Roger, 2002). So, understanding the spatial pattern in the temporal variability of snow cover increase the current understanding of global climate change and provide a mechanism for exploring future trends (Steve Vavrus, 2007) . As such, snow cover is an appropriate indicator of climate perturbations and may be a suitable surrogate for investigations of climate change (Serreze , 2000; IPCC, 2001; Roger, 2002; Wulder, 2007; IPCC AR4, 2007).

Recent research result over China area revealed that the long time series snow trend is not suit for the whole trend over northern hemisphere and regional northern American (Qin, 2006; Xu, 2007; Wang, 2008;). From Qin's research (2006) of snow cover for the period of 1951

*Corresponding Author

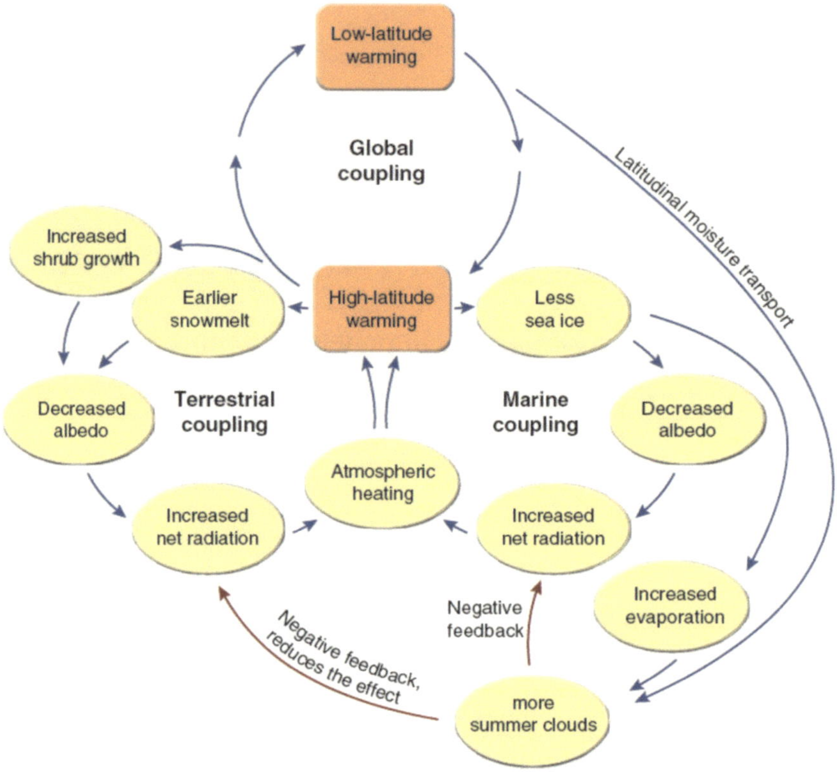

Fig. 1. This conceptual diagram illustrates the connectivity of the positive ice/snow albedo feedback, terrestrial snow and vegetation feedbacks and the negative cloud/radiation feedback. (Source: Chapin III, 2005)

and 1997, the results show that western China did not experience a continual decrease in snow cover during the great warming periods of the 1980s and 1990s. The positive trend of snow cover in western China snow cover is consistent with increasing snowfall, but is in contradiction to regional warming. Xu's result (2007) also show that the SCA of the entire *Tarim* basin in Xinjiang Province revealed a slowly increasing trend from 1958 to 2002, the SCA change in the cold season was positively correlated with the contemporary precipitation change. Wang (2008) reported an inconsistent tend with a reported Northern Hemisphere increasing trend based on limited in situ observations in Xinjiang Province which is a western province in China. While over the area located in the southern parts of the high land of *Tibet Plateau*, China, some investigators explore that annual snow cover has declined by -16% per decade between 1990 and 2001, which is explained due to the contribution of enhanced Indian black carbon (Menon et al., 2010) and the additional absorption of solar radiation by soot on snow cover area (Chand, 2009). Over *Tibetan Plateau* area, Pu's study (2007) indicated that a decreasing trend of snow cover fraction using snow data of 2000-2006 from the Moderate Resolution Imaging Spectroradiometer (MODIS) data is -0.34% per year. In their study, the meteorological station data (Xu, 2007) and the satellite sensor (Scanning Multichannel Microwave Radiometer, SMMR) observed snow depth (SD), NOAA snow cover area data and MODIS snow cover fraction products are used. When concerning the climate change impact on snow cover, the variability of snow cover area is negatively associated with to air temperature(Wang, 2008), and positive trend of the snow cover area is connected with the increasing precipitation records (Qin, 2006;) in western China. While over *Tibet Plateau* area, China, it is difficult to analyze the long time series trend for its highly rugged mountain, west-east variation and sparse meteorological stations. The data used in these studies are mostly based on the some single satellite products and meteorological station records which are sparse over the high altitude area of *Tibet Plateau*. Furthermore, the meteorological stations are affected by station location, observing practices and land covers, and are not uniformly distributed. Therefore, it is important to evaluate the gross representative satellite data in a large scale area for more than twenty years and try to deliberate the climate impact on the snow behaviors over *Tibet Plateau* area in mid-latitude.

According to the importance of the snow and the climate singularity aspects, in this work, we used the available snow cover area (Snow Cover Area), snow depth (Snow Water Equivalent, SWE) products to examine the climatological characteristics and time series analysis over *Tibetan Plateau* area and study the new snow-retrieval algorithm over China area which often experiences the shallow snow situation. This chapter includes two parts, the first is to analyse the snow products, include the near-time optical and passive microwave remote sensing and the blended SCA and SWE products, the second is to analyse the perspective view of the shallow snow retrieval analysis based on the passive microwave high frequency.

2. Climatology analyses of the satellite-based snow parameters over China

2.1 Introduction

The climatology features for a long time series of snow parameters over land could provide the signature of climate changes across the globe. According to the IPCC AR4 report, the snow extent is sharply decreasing over Northern Hemisphere from the prediction of the nine General Circulation Models since 2000. This part provides a climatology analysis of the SCA and SWE over China area and *Tibetan Plateau* from the satellite observation. The data set includes snow extent and snow water equivalence. Snow extent products are 24 km daily Northern Hemisphere snow and ice coverage from the NOAA/NESDIS Interactive Multi-sensor Snow and Ice Mapping System (IMS), Near-Real-Time SSM/I-SSMIS EASE-Grid Daily Global Ice Concentration (NISE) and Snow Extent and the Moderate-resolution Imaging Spectroradiometer (MODIS, TERRA/AQUA) snow cover fraction (SCF) products from 1999 to now, and the SWE products include Global Monthly EASE-Grid Snow Water Equivalent Climatology from 1978 to 2007, and the Advanced Microwave Scanning Radiometer for EOS (AMSR-E) from 2002 to now. The SCF (MODIS) and SWE (AMSR-E) are employed to analyse the ten years' time series over *Tibetan Plateau* (the area is defined by the area where the atmosphere pressure is less than 700 hPa).

2.2 Satellite–based snow products and processing method

2.2.1 Snow extent and snow cover fraction products

a. IMS Daily Northern Hemisphere Snow and Ice Analysis at 24 km Resolution

This data is 24 km daily Northern Hemisphere snow and ice coverage by the NOAA/NESDIS Interactive Multi-sensor Snow and Ice Mapping System (IMS) (National Ice Center, 2008). The key parameters for this type of data are listed below:

- Time Span: 1997~2011
- Polar Stereographic Projection
- 1024*1024 grid
- Spatial Resolution :~24km
- Time frequencies: Daily
- Four types parameters: Ocean\Land\Sea ice\Snow
- Optical satellite and other sources (environmental satellite imagery)
- Distorted much in China Area

From the sample data map in fig.2, we can find that the SCA data is distorted over China area. Over the northern hemisphere, the China area is not a dominant domain in the continent analysis. It is fit for the onset, duration and end of the snow for its daily resolution.

b. Near-Real-Time SSM/I-SSMIS EASE-Grid Daily Global Ice Concentration and Snow Extent

The Near-Real-Time SSM/I-SSMIS EASE-Grid Daily Global Ice Concentration and Snow Extent product (Near-real-time Ice and Snow Extent, NISE) provides daily, global near-real-time maps of sea ice concentrations and snow extent. They are derived from the passive microwave data from the Special Sensor Microwave Imager/Sounder (SSMIS) on board the Defense Meteorological Satellite Program (DMSP) F17 satellite (Nolin, 1998).

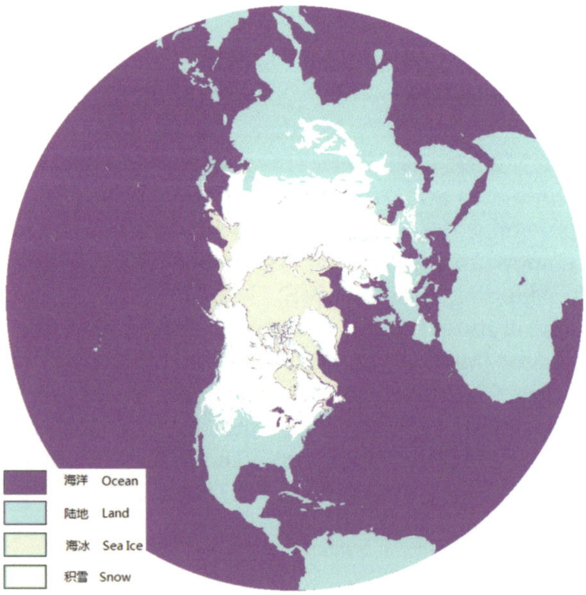

Fig. 2. The sample product from the 24km IMS SCA products, the SCA over China is obvious for its Plateau shape in the upper part of the map

- Time Span: 1995.05~2011.08
- EASE-Grid Projection.
- 721*721 grid
- Spatial Resolution :~25km

- Parameters : Snow extent, Sea ice concentration
- Time frequencies: daily
- NISE Product Source: passive microwave remote sensing data
- Distorted much in China Area

The snow cover over China is showed in the right part of the EASE-GRID projection image (Fig. 3), which is also distorted for some extent. The data could be used to evaluation the onset, duration and end of the snow appearance.

c. MODIS/Aqua Snow Cover 8-Day L3 Global 0.05Deg Climate Modeling Grid (CMG)

The MODIS Snow Cover 8-Day L3 Global 0.05Deg CMG (Fig. 4.) is a global map of snow cover expressed as a percentage of land, i.e. snow cover fraction, in each CMG cell for an eight-day period, which are derived from the Normalized Difference Snow Index (NDSI) of MODIS spectro-radiometer data (Hall, 1995). The percentage of snow-covered land is based on the clear-sky view of land in the CMG cell, and count the number of snow observation over land. So the amount of snow observed in a CMG cell is based on the cloud-free observations mapped into the CMG grid cell for all land in that cell (Hall, 2007). Compared with the daily snow-cover products, the eight-day SCFs products greatly reduce the percent of cloud obscured or masked pixels from near half to less than 7% over *Tibet Plateau* (Riggs, 2003), which is more suitable to analyse the trend for at a long time span.

- Time span: 2002 to 2010
- Latitude/longitude projection
- Grid resolution is 0.05 degrees
- Parameters: Snow cover fraction
- Time frequencies: eight days
- Source: MODIS optical remote sensing under cloud-free condition
- Suit for the CMG projection

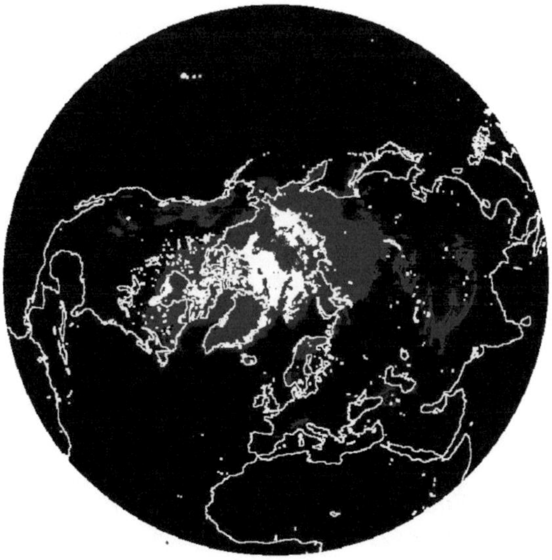

Fig. 3. The sample product from the NISE products, the SCA over China is also obvious for its Plateau shape in the right part of the map (EASE-GRID)

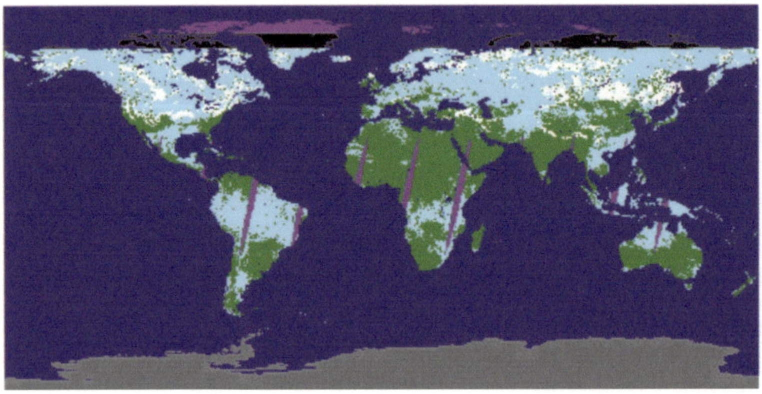

Fig. 4. Sample image derived from MODIS/Aqua Snow Cover Daily L3 Global 0.05Deg CMG data set 08 February 2004 (cited from http://nsidc.org/data/modis/data_summaries/cmg_sample.html)

2.2.2 Snow water equivalent products

a. Global Monthly EASE-Grid Snow Water Equivalent Climatology

This data set comprises global, monthly SWE from November 1978 to 2007, with periodic updates released as resources permit. Global data is gridded to the Northern and Southern 25 km Equal-Area Scalable Earth Grids (EASE-Grids) (Fig.5).

- Time Span: 1978 - 2007
- EASE-Grid Projection.
- 721*721grid
- Spatial Resolution :~25km
- Time frequencies: Monthly
- Parameters : Snow water equivalent and Snow cover frequency of occurrence
- Source: Scanning Multichannel Microwave Radiometer (SMMR) and selected Special Sensor Microwave/Imagers (SSM/I) and Visible snow parameters as a factor
- Distorted much in China Area

b. AMSR-E/Aqua L3 Global Snow Water Equivalent EASE-Grids

We also use the AMSR-E snow products to check the SWE variation and climatology over *Tibet Plateau*. The SWE_Northern daily data (Tedesco, 2004) is used in the next process. The data characteristics are listed.

- Time span: 2002~2010
- EASE-Grid Projection.
- 721*721grid
- Spatial Resolution :~25km
- Time frequencies: Daily
- Parameters: Snow Water Equivalent (mm)
- Source: AMSR-E passive microwave remote sensing
- Distorted much in China Area

This data are processed into the maximum and average SWE for the *Tibet Plateau*.

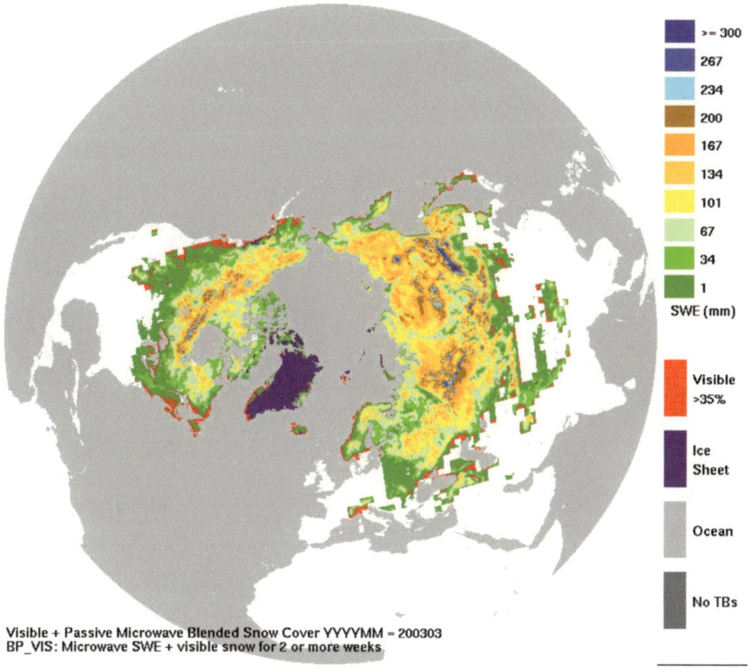

Fig. 5. Northern Hemisphere average snow water equivalent (mm) from passive microwave, with additional area indicated as snow by Northern Hemisphere EASE-Grid weekly snow cover in red, March, 2003.(cited: http://nsidc.org/data/docs/daac/nsidc0271_ease_grid_swe_climatology/NL200303.NSIDC8.BP_VIS35.png)

2.2.3 Multisource satellite data processing method

According to the data characteristics mentioned above, the different projections and resolutions data need to be projected in the same project that could provide a same base for the later analysis. We select the equal latitude and longitude project to provide a more effective understanding for the China mid-latitude area. A tool has been developed to processing the EASE_Grid, Polar Stereographic Projections into the 0.05 degree latitude and longitude map. Fig. 6 shows the transform scheme from the multi-projection to the equal latitude and longitude.

When all of these data products are resampled, we analyse the onset and duration of the data from the SCA products (named: IMS and NISE) using the accumulating, the first and the end day of the snow. The monthly SWE products are used to calculate the climatological characteristics over China by the averaging method.

2.2.4 Onset, duration of the snow cover over China

After all of the data mentioned above is projected into the same equal latitude and longitude grid. The IMS and NISE daily snow cover data are processed to the onset, duration and the end time map, the base-time for IMS product is 31/May, and the day of the year 183 (almost 31/May) for NISE products. The Global Monthly EASE-Grid Snow Water Equivalent

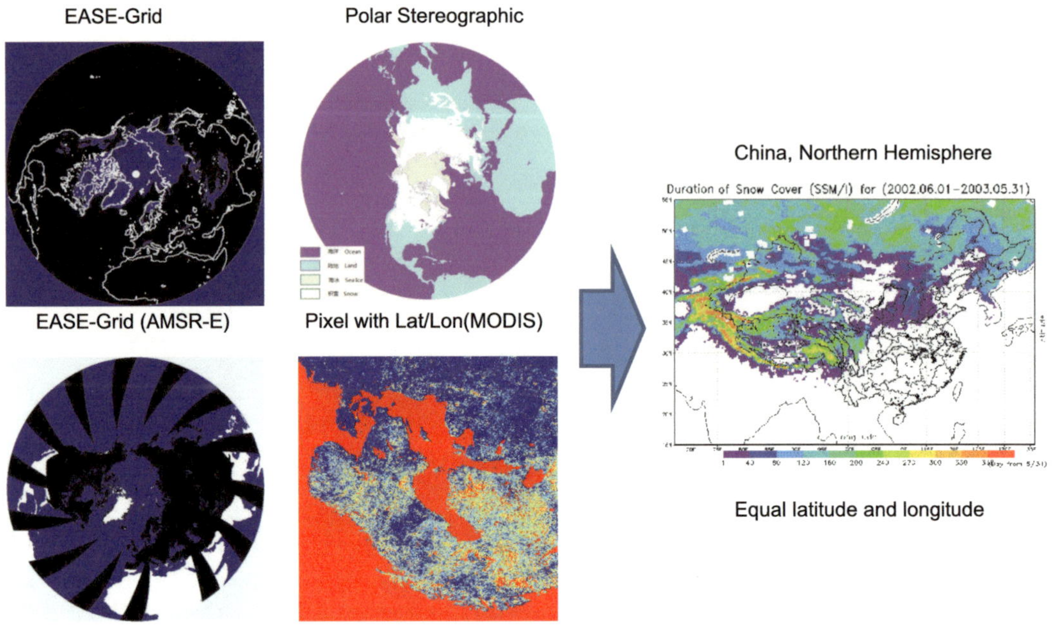

Fig. 6. The resampling processing in the transition process (multi-projection to equal latitude and longitude)

Climatology data is reprocessed to analyse the averaged monthly climatologic characteristics. The data quality control has been done to make sure that the representativeness suit for statistical analysis. The China area is defined as 15^0N-56^0N, and 67^0E-136^0E, includes all of the Chinese land area, part of the center-Asia, Mongolia, and part of the southern Russia, where the snow often appear.

a. Onset of the snow cover over China

The onset of the snow cover is plotted using the data from IMS and NISE products for fourteen (1997~2011) and sixteen (1995~2011) at whole year respectively (Fig.7 just shows the corresponding 4 years of these two dataset). From fig.7, the snow cover over Tibet High Mountain and the Centre Asia Mountain is always influenced much by the mountain glaciers, the mostly early snow are showed in the northern part and *Tibet Plateau* area of high mountains marked the permanent snow area (the onset data value is 1). Along with the latitude which changes from south to north, the snow appearance shows its latitude dependency over land area, the high latitude experience early snow cover compared to the low latitude area. The NISE and IMS onset of the snow cover all show postpone in the first snow occurrence, while the IMS records give an explicit result.

These two products show the same regime of the onset of the snow cover but they have explicit difference when compare together (compare these two column in fig.7). The data from NISE take larger area as blank or snow-free area, such as the Yellow River area at Central Plains China. There is more snow record at the beginning of the 31/May at the south margin of the Tibet Plateau that is not suit for the rain forest area in the Northern Indian Mountains. Over the Khrebet Kropotkina area and the northern glacier rich areas the NISE products show the early records about the snow appearance. Overall, the NISE

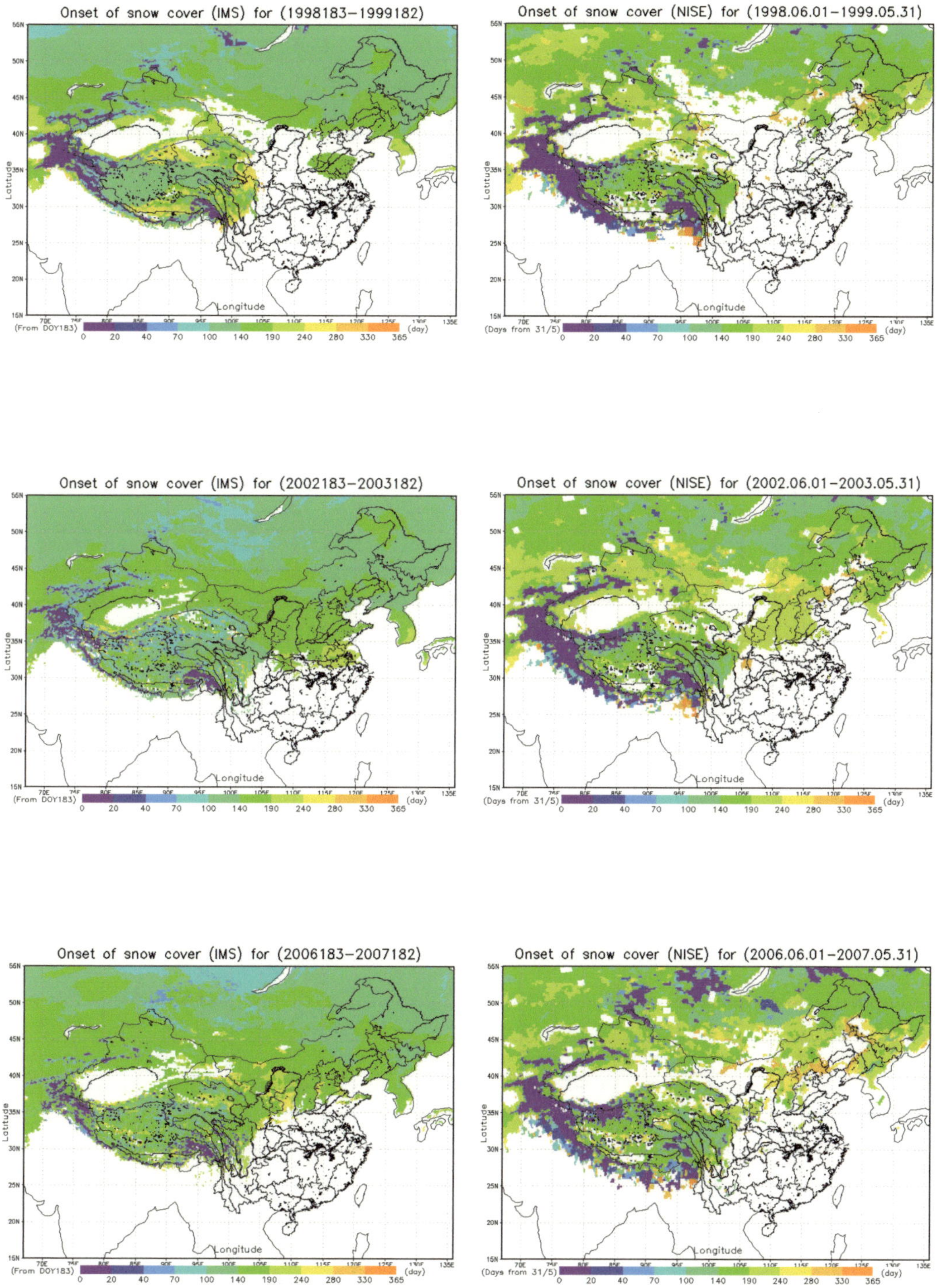
Onset of snow cover (IMS) for (1998183–1999182)
Onset of snow cover (NISE) for (1998.06.01–1999.05.31)
Onset of snow cover (IMS) for (2002183–2003182)
Onset of snow cover (NISE) for (2002.06.01–2003.05.31)
Onset of snow cover (IMS) for (2006183–2007182)
Onset of snow cover (NISE) for (2006.06.01–2007.05.31)
Latitude
Longitude
(From DOY183)
(Days from 31/5)
0 20 40 70 100 140 190 240 280 330 365
(day)

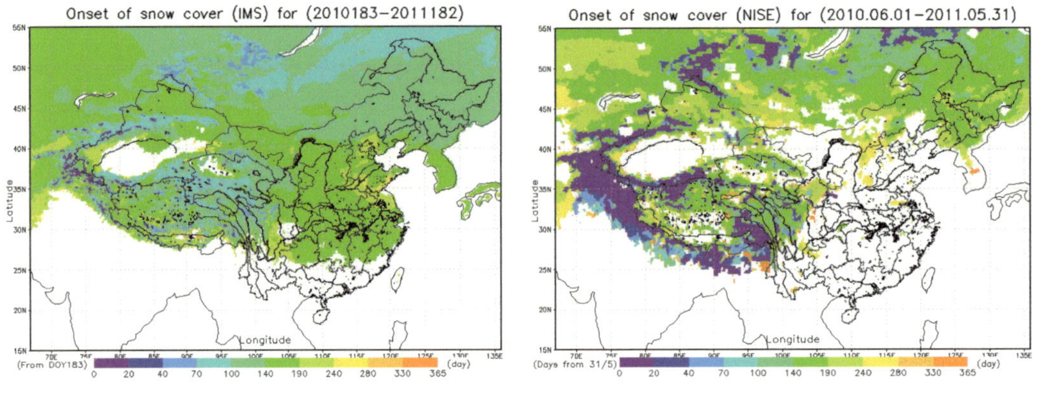

Fig. 7. The onset time of the snow appearance over China, left column is from the IMS products, and the right column is from NISE products. We just give the winter of 4 years, 1998-1999, 2002-2003, 2006-2007 and 2010-2011.

products give out a relative late onset time than that of IMS over flat area which could be attributed to the rugged mountain's influence in the microwave signal in NISE. While for the IMS products, the coverage area are larger than that from NISE (SSM/I) products, and show the early first snow occurrences. The IMS spatial distribution of the snow are possibly more accuracy than that of NISE, such as the Korea Island and the southern China where there is snowy in the January or February.

b. Duration of Snow cover over China

The duration of snow cover for a region is also a sign of climate condition. The duration of the snow is derived from the snow products, IMS and NISE. From fig.8, the duration distribution of the snow is inhomogeneous. The high land area experiences the longest time of the snow cover, such as the expected *Tibet Plateau* and the northern glacier rich area. The duration of the snow have direct relationship with the latitude (higher latitude, longer snow duration), and the southeastern China has the least time of snow cover where the climate is temperate continental climate.

The snow duration data from these two dataset are similar distributed but the NISE (i.e. SSM/I) product shows the longer time when compare the same region in the lower latitude at high land, for example, the snow over Tibet Plateau. Over the high latitude area, the time-span of the snow existence from IMS is somewhat longer than that from NISE. These aspects reveal that the satellite snow products of optical and microwave estimation are different in northern part of China and high land of Tibet Plateau, which is similar with the finding of Wang (2007).

From fig.8, the time series of the snow cover duration is increasing over the patchy snow cover areas, such as the low land of the China area, e.g. south-east of China and the Yellow river area. It seems that there is somewhat a little bit of longer and longer duration of the

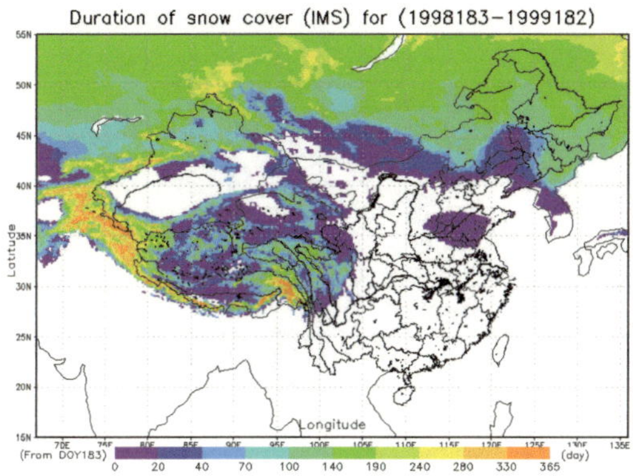
Duration of snow cover (IMS) for (1998183–1999182)
Latitude
Longitude
(From DOY183)
0 20 40 70 100 140 190 240 280 330 365
(day)

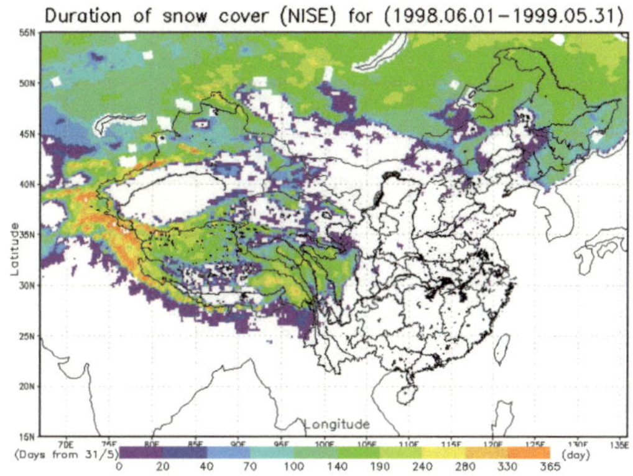
Duration of snow cover (NISE) for (1998.06.01–1999.05.31)
Latitude
Longitude
(Days from 31/5)
0 20 40 70 100 140 190 240 280 330 365
(day)

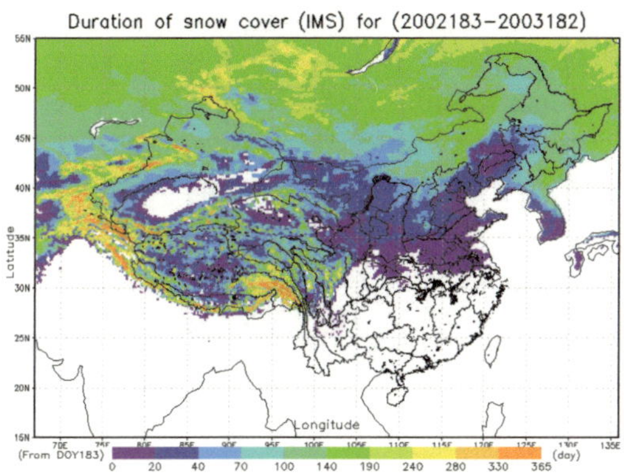
Duration of snow cover (IMS) for (2002183–2003182)
Latitude
Longitude
(From DOY183)
0 20 40 70 100 140 190 240 280 330 365
(day)

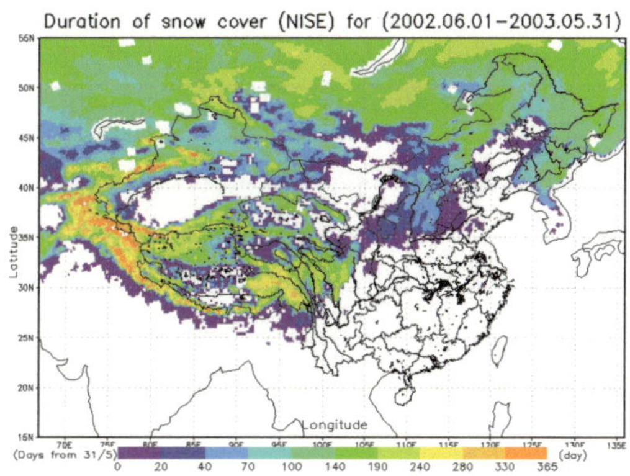
Duration of snow cover (NISE) for (2002.06.01–2003.05.31)
Latitude
Longitude
(Days from 31/5)
0 20 40 70 100 140 190 240 280 330 365
(day)

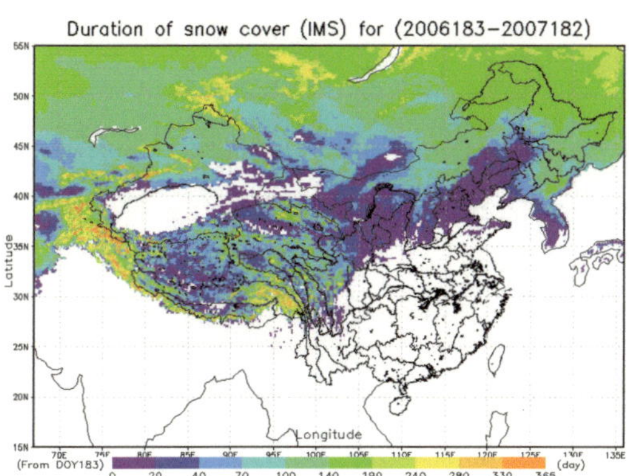
Duration of snow cover (IMS) for (2006183–2007182)
Latitude
Longitude
(From DOY183)
0 20 40 70 100 140 190 240 280 330 365
(day)

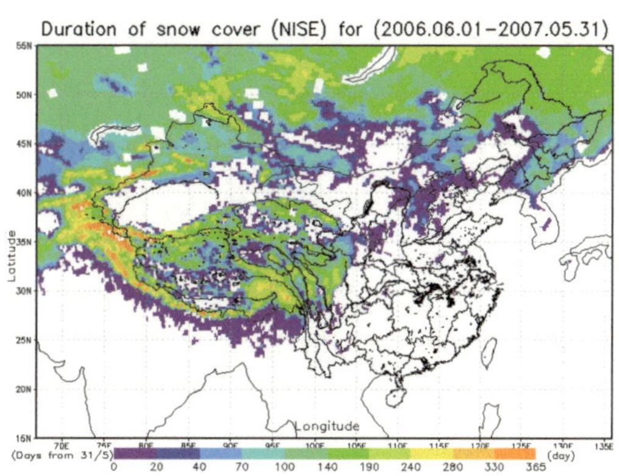
Duration of snow cover (NISE) for (2006.06.01–2007.05.31)
Latitude
Longitude
(Days from 31/5)
0 20 40 70 100 140 190 240 280 330 365
(day)

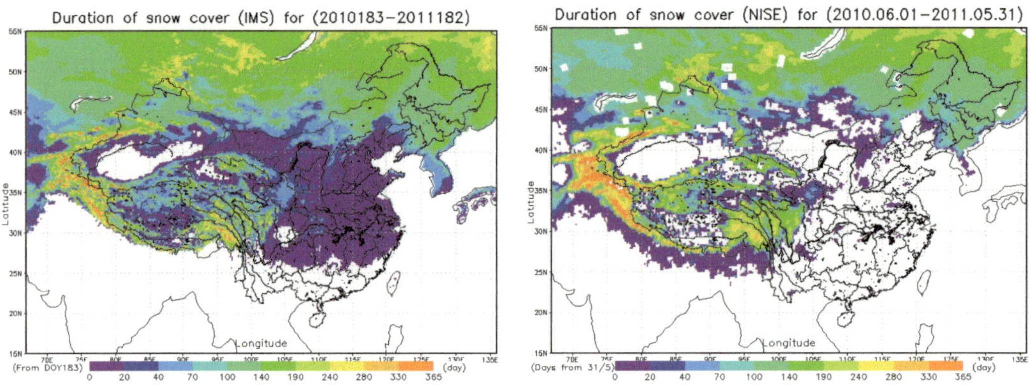

Fig. 8. The duration of the snow appearance over China, left column is from the IMS products, and the right column is from NISE products

snow over Tibet Plateau area from IMS records for fourteen years, but an ambiguous trend is for the 14 or 16 years. Over the southeastern China, the snow obviously exits in every winter time, but the NISE products does not record for its empirical ancillary data in the algorithm, we could get that the data from IMS is more reliable for the situation over China than that from NISE (SSM/I).

c. The monthly climatologic characteristics over China

The monthly snow climatology map is derived from the EASE-Grid Snow Water Equivalent (SWE) for about 30 years' satellite records. From fig.9, the seasonal snow change is obvious in the most area of China. The winter and early spring time from December to the March of next year is the snowiest over the northern China. The maximum snow cover area is in January. From May to September, the snow cover became less and less except the high altitude of *Tibet Plateau* area. The minimum snow cover area is in August. While the maximum SWE and snow cover area of northern China and *Tibet Plateau* area is quite different, the SWE (mm) reach its peak in November over *Tibet Plateau*, and the northeastern China suffered its maximum SWE (mm) in February. The snow cover area in Qinghai-Xizang (Tibet) experiences the largest snow cover in January, which is consistent with Qin's result (2006), while the western area (Xinjiang province) of China reaches its maximum snow in February along with the maximum SWE (mm) which is earlier than that of Qin's (2006). The southeastern China is almost snow-free for the all year time.

The climatological characteristic of *Tibet plateau* area is different than that of the low land area of China, especially the northern part of China. The latitude dependency is obvious in the northern China. Another control factor is the altitude, especially over the Northern Mongolia when compared with the ASTER Global Digital Elevation Map (http://asterweb.jpl.nasa.gov/images/GDEM-10km-colorized.png).

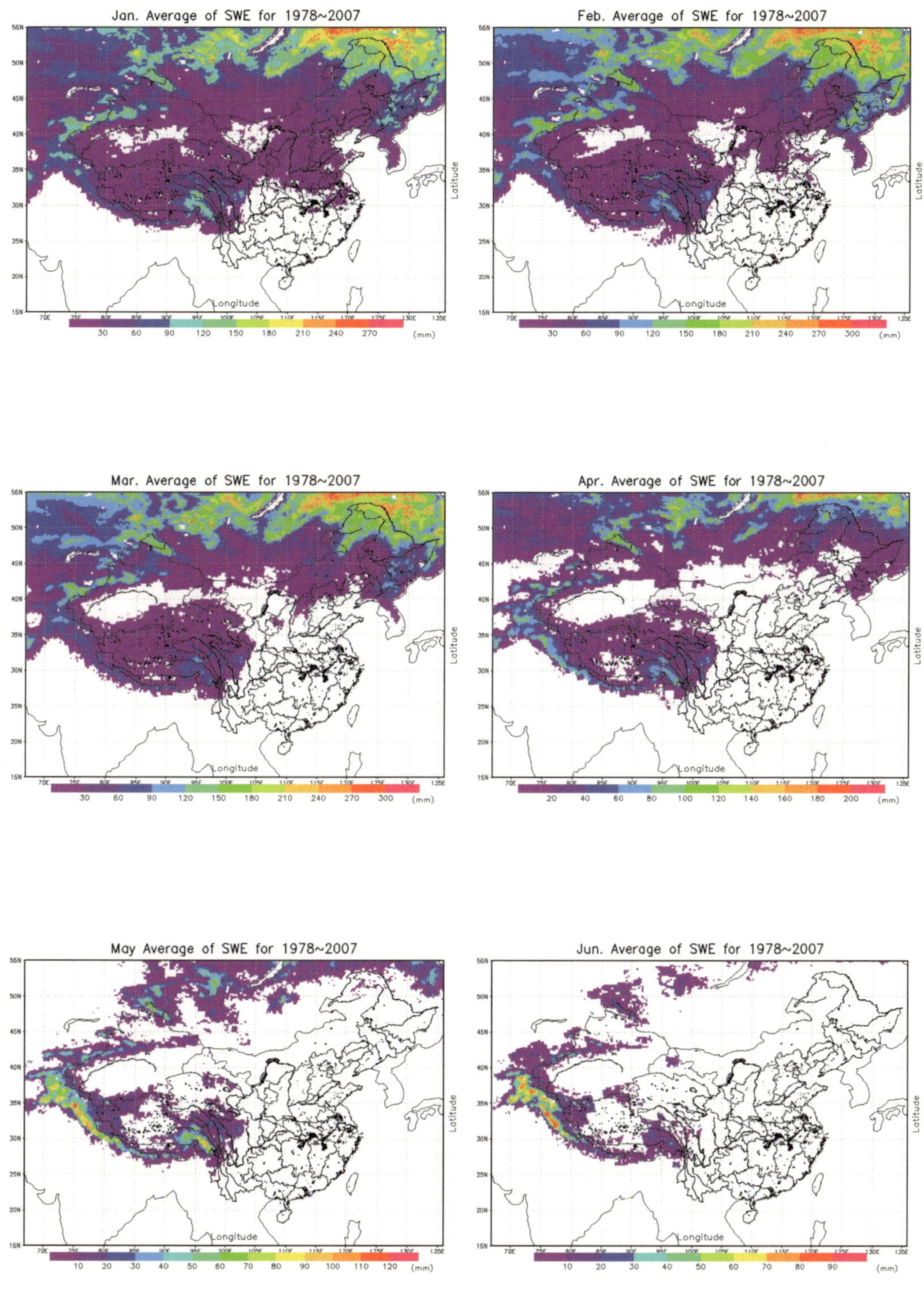
Jan. Average of SWE for 1978~2007
Longitude
Latitude
30 60 90 120 150 180 210 240 270
(mm)
Feb. Average of SWE for 1978~2007
Longitude
Latitude
30 60 90 120 150 180 210 240 270 300
(mm)
Mar. Average of SWE for 1978~2007
Longitude
Latitude
30 60 90 120 150 180 210 240 270 300
(mm)
Apr. Average of SWE for 1978~2007
Longitude
Latitude
20 40 60 80 100 120 140 160 180 200
(mm)
May Average of SWE for 1978~2007
Longitude
Latitude
10 20 30 40 50 60 70 80 90 100 110 120
(mm)
Jun. Average of SWE for 1978~2007
Longitude
Latitude
10 20 30 40 50 60 70 80 90
(mm)

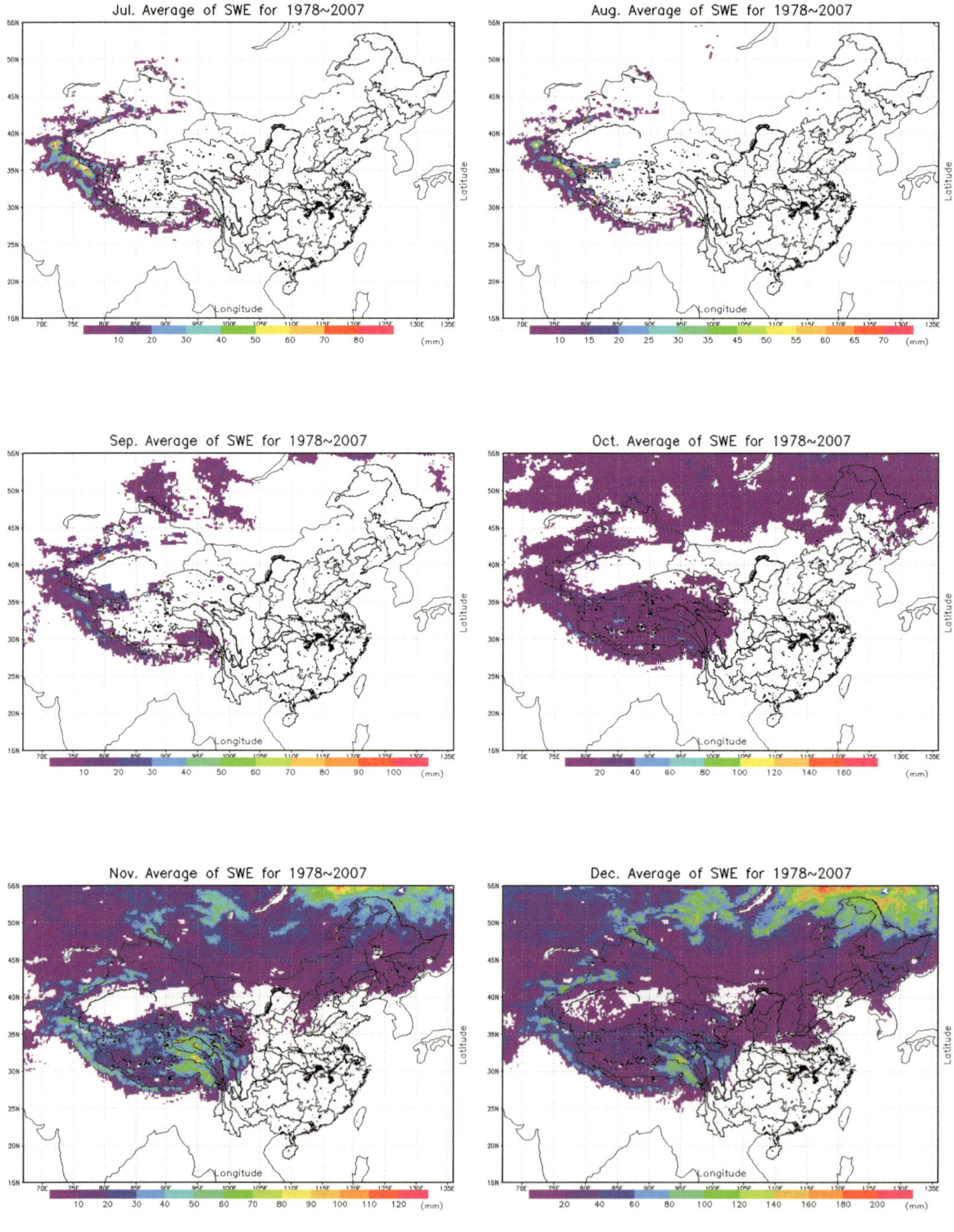

Fig. 9. The monthly averaged SWE (mm) of the snow appearance over China, data is from Global Monthly EASE-Grid Snow Water Equivalent Climatology for 1978-2007

2.2.5 SWE from AMSR-E/Aqua and SCA MODIS/Terra (Aqua) over Tibetan Plateau for the last ten years

From the above analysis, the Tibet Plateau area is quite special in the seasonal snow cover not only for the SCA (Squa. km) but also for the SWE (mm). We consider the high land of *Tibet Plateau* as one whole area by filtering the atmosphere pressure that is lower than 700 hPa, which includes all of the Tibet, China, part of the Qinghai province and the Center Asia mountain areas (see Fig.10). The AMSR-E/Aqua L3 Global Snow Water Equivalent EASE-Grids and the MODIS/Aqua Snow Cover 8-Day L3 Global 0.05Deg Climate Modeling Grid (CMG) data are employed to analyse the snow time series trend over the Tibetan Plateau area.

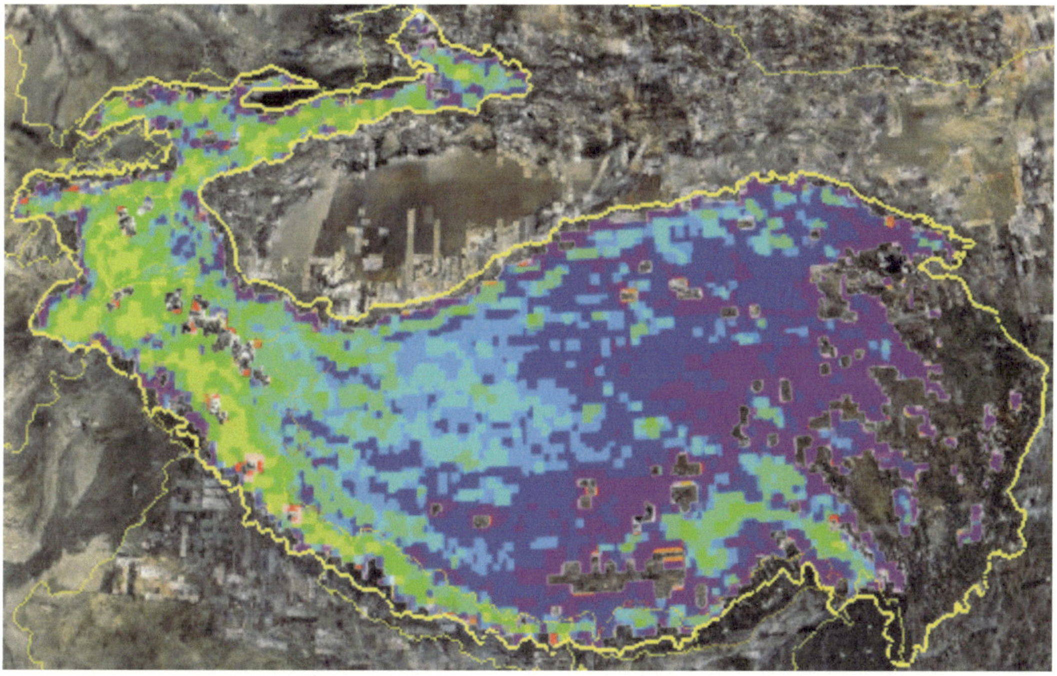

Fig. 10. Definition of *Tibet Plateau* area- according to the air pressure (when < 700hpa)

a. Time-series climatological analysis

The AMSR-E/Aqua provides 8 years' monthly average SWE for the study area, and the total area of the pixels covered by snow is also presented monthly. The time series analysis is in Fig.11, which give a slightly increasing trend for 8 years from 2002 (launch time) to summer, 2010. From fig.11, the average monthly SWE (mm) reach the max value in February (2002/2003, 2003/2004, 2005/2006, 2009/2010) or March (2006/2007, 2007/2008, 2008/2009), and the minimum value appear in August except the summer in 2005, which is quite similar with the section in 2.2.4 c. When we check the SCA from AMSR-E/Aqua, the SCA (Squa. km) reach its maximum extent in January except the winter of 2005/2006, the minimum extent is in July (2002, 2005, 2008) or August (2003, 2004, 2006, 2007, 2009). These tells a positive trend of SCA and SWE over the high altitude region (<700hpa) of *Tibet Plateau* area.

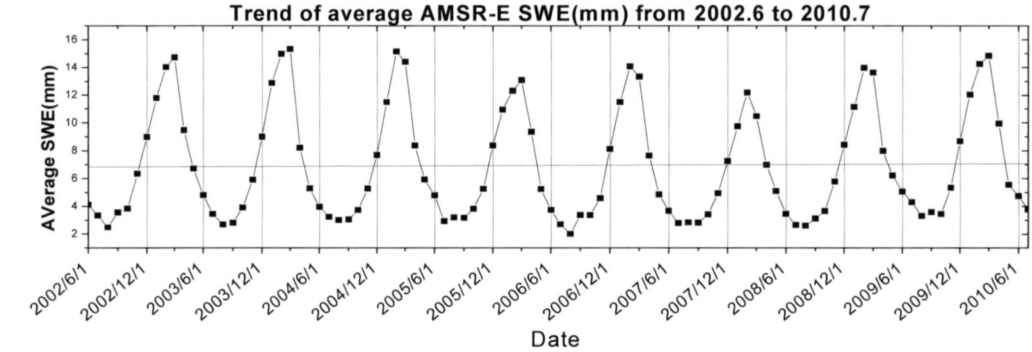

Fig. 11. Time series of the averaged AMSR-E Snow Water Equivalence (SWE) (Average snow cover area, the SCA plot is not showed here).

b. The monthly averaged SWE (mm) and SCA(Squa. km) from 2002.6 to 2010.7

The time series analysis for the averaged SWE (mm) is presented in Fig.12, the fluctuation for each month in the near eight years is small but can find that the slightly trend (see Table 1). The trend analysis shows that the SWE (mm) experience a slightly increasing in this eight years from June to September, which is almost in the summer and autumn time of one year in China, while other time (winter and spring) are decreasing in the averaged SWE (mm). The SCA parameter of the study area shows the same trend as the averaged SWE (see table1 at right column). This climatological characteristic is fit for the Warming and Wetting of the *Tibet Plateau* (Bao, Q., 2010) for the increasing precipitation in the summer time, while the increasing precipitation could not influence the winter and spring snow-rich situation.

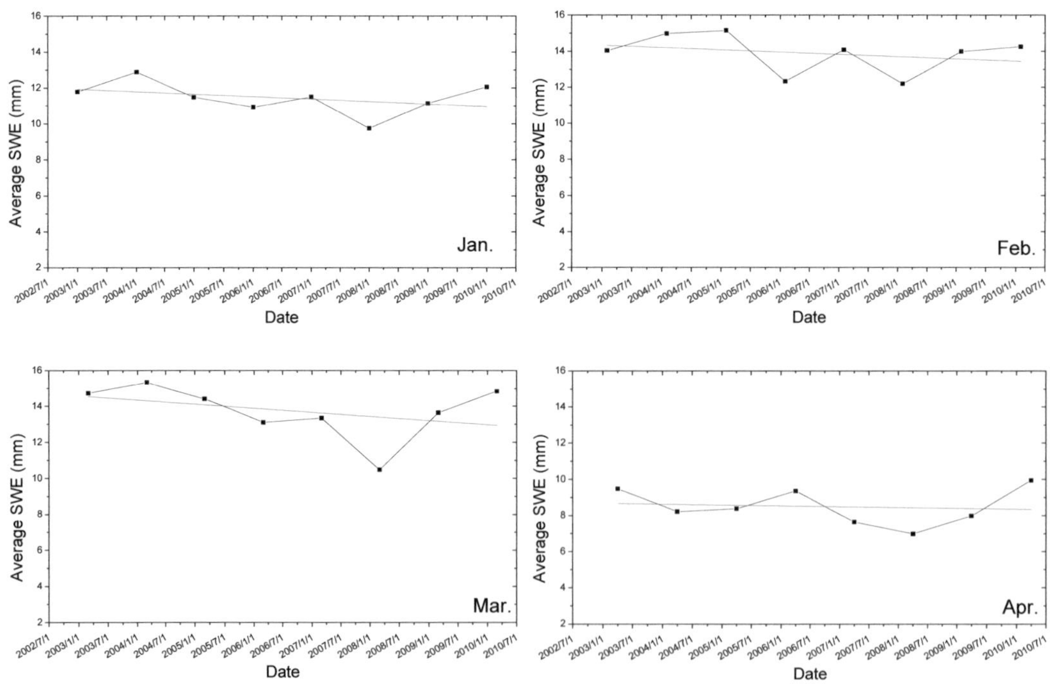

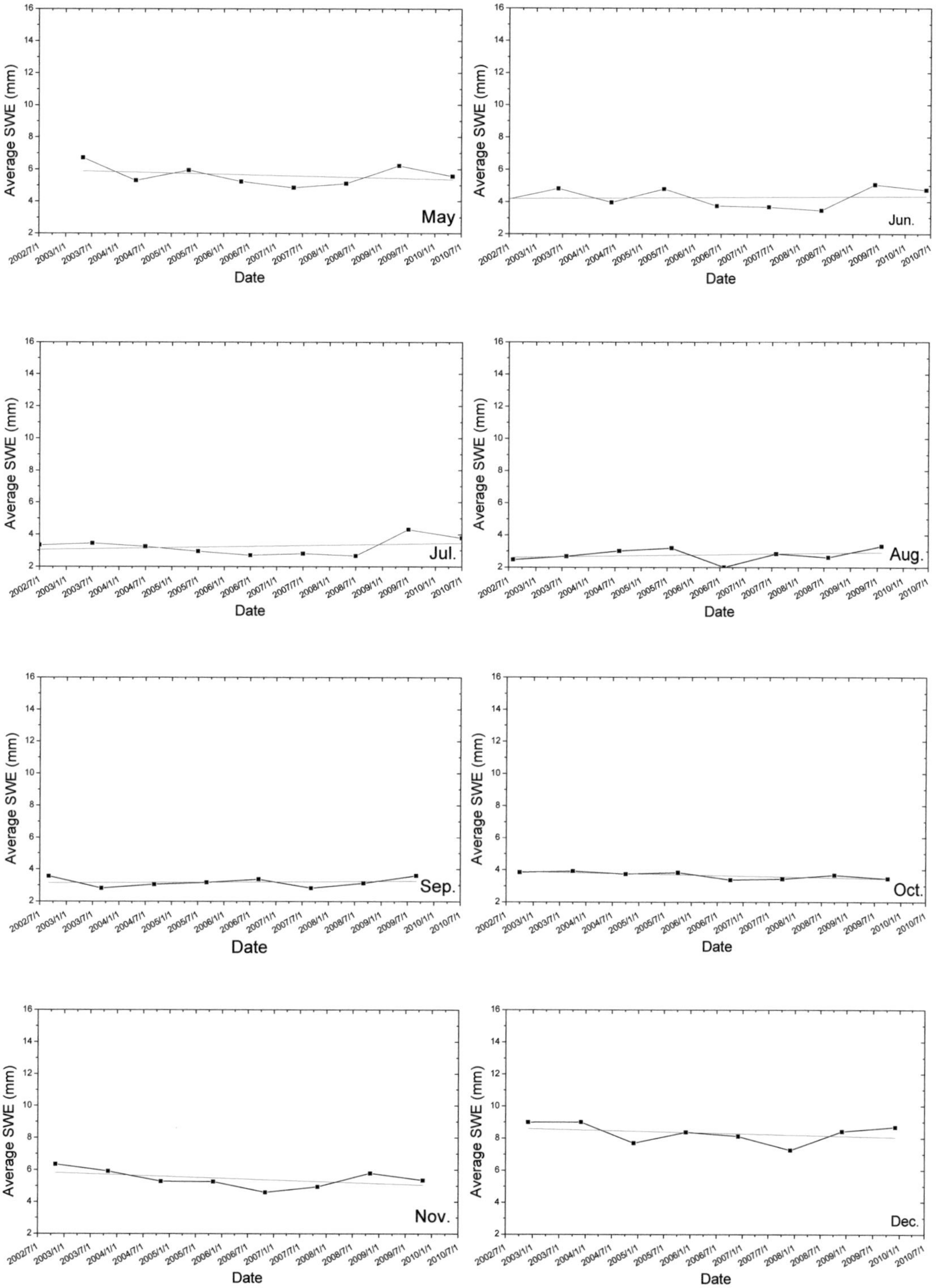

Fig. 12. The time series of average SWE (mm) for twelve months in one year

Month	Rate of Changing Averaging SWE (mm)	Rate of Changing SCA (Squa. KM)
Jan.	-0.00037	-7.36287
Feb.	-0.00035	-17.9450
Mar.	-0.00062	-17.4998
Apr.	-0.00012	-29.1789
May	-0.00021	-89.2610
Jun.	0.000049	26.7051
Jul.	0.00014	23.9970
Aug.	0.00012	20.2665
Sep.	0.00004	-48.7194
Oct.	-0.00017	-14.1196
Nov.	-0.00030	21.7514
Dec.	-0.00021	-2.1661

Table 1. The trend slope for the average SWE (mm) and SCA (Squa. km) from AMSR-E/Aqua eight years records

c. The time series of the monthly snow cover fraction

For the snow cover fraction area (SFC) statistic for *Tibet Plateau* study area in Fig.13. The snow data from MODIS/Aqua (Terra) can provide the snow cover fractional distribution in different time at the same day (morning and afternoon). In Fig. 13, the time series of the SCFs are plotted for different span (0-10%, 10-20%, 20-30%, 30-40%, 40-50%, 50-60%, 60-70%, 70-80%, 80-90%, 90-100% and 100%) for two satellites (MODIS/Terra and Aqua). Compared these two figures, the SCF area from Aqua satellite is general larger than that of Aqua with the same seasonal characteristic possibly for its different overpass time (morning and afternoon) at mid-latitude area. The summer time (almost in later August or early September) has the least area for the SCF which is greater than 20%, while the winter time (especially in the February) has the maximum area. When focus on the SCF less than 20%, the situation is a different result than that more than 10%, the summer time has the greater area than that in winter time for these two satellites, due to the summer patchy snow fractional pixels influence the satellite estimation. The time series analysis trends for these different SCF's range are showed in Table.2. The changing rate indicates a positive trend for the last ten year, especially for the large SFC which almost distribute in the high altitude mountain area. The largest increasing rate is the SFC between 90% and 100% which indicate the high mountain area suffering an increasing snow cover because the full cover areas are mostly in the high elevation mountain area. Another aspect is that the changing rate for MODIS/Terra record is larger than MODIS/Aqua's, but the reason has not discovered in this study.

	0-10%	10-20%	10-20%	30-40%	40-50%	50-60%	60-70%	70-80%	80-90%	90-100%	100%
MODIS/Aqua	-0.05629	1.82353	1.62721	1.77621	1.51221	1.54443	2.00536	2.14856	3.00974	7.61426	2.35225
MODIS/Terra	1.52948	1.63037	1.14803	1.2696	1.29663	1.36101	1.52654	1.59566	2.20187	8.62898	4.85512

Table 2. The slope coefficients for different SCF range during ten year (Terra) and eight year (Aqua) running

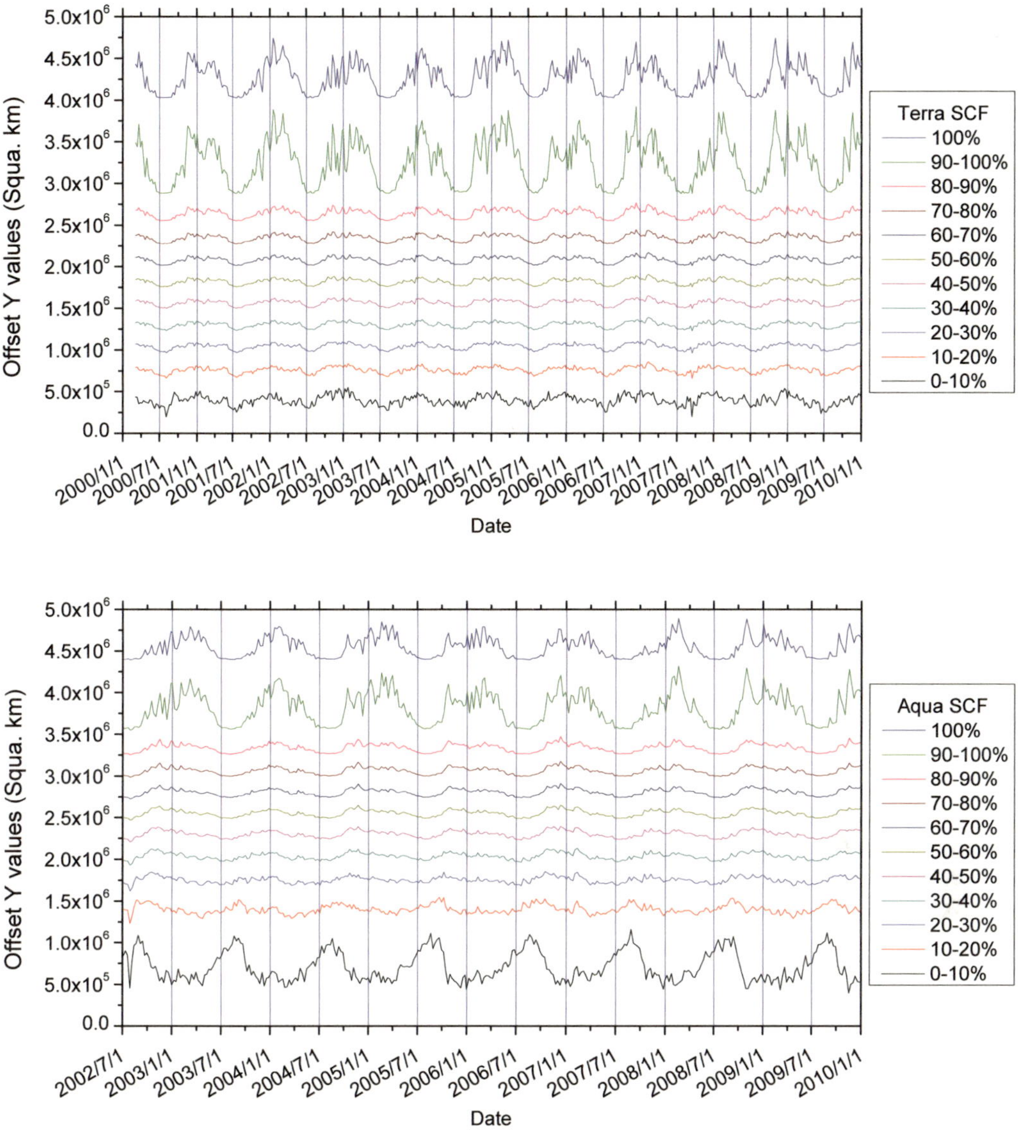

Fig. 13. The time series analysis for different snow cover fractions derived from MODIS/terra and MODIS/Aqua

2.3 Conclusions

From what we have analyzed, the climatological characteristics show that the onset time of snow over China area are slightly postponed, while the duration is undecided by the satellite record of NISE (SSM/I) and IMS, the monthly climatology analysis reveals that the snow distribution is quite different in the altitude and latitude, the Tibet Plateau area experiences the maximum SWE in November. The northern China and lower land reach the maximum area in December and January.

The products of SCA and SWE could provide a long time series data and derived snow climatological analysis, when compared the optical and microwave remote sensing products of snow, IMS SCA and NISE SCA show difference each other, the blank area in Tibet and northwestern China could not enough to provide analytical result, though these are some clues on it. The snow product of IMS seems provide more reliable results over China area, and it is recommended that a new snow algorithm from satellite is needed for the accuracy assessment.

In the traditional view, the satellite data could provide more reliable large-scale snow parameters than the local observational station, the trend from several snow products provides the same continental regime over Northern American, it looks like snow cover gets a negative response to the global warming, while, a near local look over the Tibet Plateau, the result shows that the snow cover area appears a positive trend with snow equivalent water from PSW dataset, and the situation is also same over the China West Area.

From the monthly snow water equivalent (mm) which is recorded from the AMSR-E/Aqua, two snow parameters are derived, one is the averaged SWE monthly and another is the snow cover area (squa. km). The result reveals the positive trend of the averaged SWE (mm) and snow cover area (squa. km) over the Tibet Plateau area, which is the same situation with the result of western China (Qin, 2006 and Xu, 2007). While the monthly trend for more than ten years, we can find some interest results (see b part in 2.2.5). The averaged SWE and Snow cover area experience slightly increasing trend in the summer and autumn time (June, July, August and September), while in the winter and spring time (from October to next May), these parameters shows its negative trend.

From the MODIS SCF time series analysis according to the different percentage pixels, we can find that the SCF less than 20% are quite variable with more pixels in summer time than that in the winter time, while all of the pixels that contain more snow indicate a similar positive trend, and less pixels in summer time than that in winter time. The higher of the SCF, the higher trend value for the line. The data from the MODIS/Aqua show very similar result as that of MODIS/Terra, but larger area than Terra's.

It is hoped that China mainland area whose cryosphere is a major element in the climate now undertake national programs designed to address questions of global environment change.

3. Analysis between AMSR-E brightness temperature and ground snow depth over Tibet Plateau, China

3.1 Introduction

Over the Tibet Plateau (Western China), snow cover is presented only for a few months per year, except mountainous areas. However, it highly influences the energy flux, atmosphere dynamics and surface water reservoirs. Recently, much effort has been put into developing region-specific retrieval algorithms for snow parameter retrieval from passive microwave measurements. Automatic station observations of snow cover are essential factors in the development of these retrieval algorithms, but they cannot provide comprehensive information on the snow cover distribution. The recent study has improved the snow depth accuracy for some extent, but the method highly depend on the method training for the artificial neural network methodology (Yungang Cao, 2008) without more physical explanation. From the Fig.14, the distribution of the meteorological stations in the Tibet Plateau can be seen to be very sparse, especially over the main part of the Plateau. Furthermore, many of them are near areas of human activity, and provide few measurements for a long time span with very shallow snow depth values (see Fig.15 example for DanXung Station) (Che, 2004). Armstrong (2001) notes that passive microwave remote sensing tends to underestimate the snow in the fall and early winter due to the weak signal of thin snow with the 36.5GHz and 18.7GHz (Armstrong, 2001), while the situation is the opposite over Tibet Plateau. Matthew H. Savoie (2009) improved the accuracy of the snow measurement by considering the atmospheric influence to some extent; Qiu etc. (2009) paid attention to the atmosphere influence via the experiment and model simulation.

Due to the thin snow (snow occurrence) is often seen over western China, especially over the Tibet Plateau, more comprehensive analysis is urgent with the station observation data and microwave Tbs. In this work, we consider the shallow snow situation, and try to explain the discrepancy between the in situ time series measurement of snow (snow depth, SD) and the values retrieved from passive microwave remote sensing with the traditional difference between the brightness temperature at 36.5GHz and 18.7GHz, and that from 89.0GHz-18.7GHz and 18.7GHz-10.7GHz. Then, we analyze the ability of the higher frequencies in snow parameter retrieval over the Tibet Plateau (e.g. 89.0GHz at AMSR-E) using the time series data comparison.

3.2 Snow depth and AMSR-E brightness temperature

3.2.1 Snow depth data

We selected the snow depth measurements at the NamCo station over 4700m in altitude, which is located beside the NamCo Lake and the Mt. Nyainqenttanglha (Fig.14, circle). The Institute of Tibetan Plateau Research, Chinese Academy of Sciences, operates a station in the

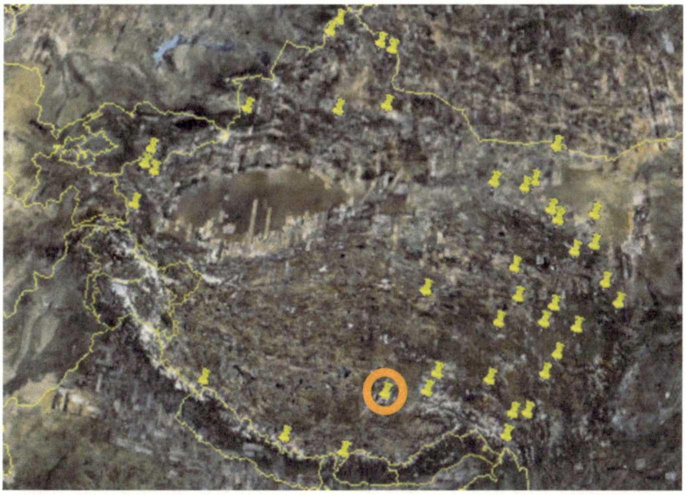

Fig. 14. The distribution of the selected meteorological station over Tibet Plateau and western China (from China Meteorological Data Sharing Server System, data used in this work) and the geographic location of the Namco station site (30^{0}46.44'N, 90^{0}59.31'E).

area. A snow campaign covering the whole winter offseason between 2006.10~2007.2 was conducted. SD records are acquired over three sites around the Namco station. Compared to the AMSR-E/Aqua satellite footprint, these sites are regarded as one site and represent the general situation of the whole area in this work, though this is a fairly inaccurate estimation in mountainous areas. Other time-series SD data in this work is from the winter-time observation (stations at Fig.14) in 2009~2010, when northern China were suffered from vast snowfall.

Fig 15 (left) shows a time series of the measured in situ snow depth values. From 24/10/2006, snow depth increases from 23cm to about 45cm on 8/11/2006, after which the depth decreased to 17cm on 28/1/2007. In this time span, several snowfall events happened on 12/11/2006, 14/11/2006 and 16/1/2007, with 2 cm of new snow in the last case. A relatively large shift appeared on 14/12/2006 because of the change of the observation sites for the surface wind.

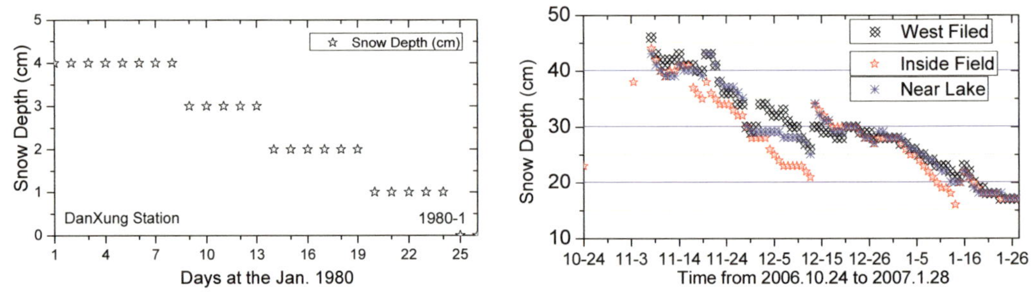

Fig. 15. The SDs from the Danxung station (No.55493) (Left) and the SD (cm) field campaign near NamCo station (Right)

3.2.2 AMSR-E L2A swath dataset and processing

We selected the AMSR-E daily L2A swath brightness temperature (ascending and descending pass, A/D, http://nsidc.org/data/docs/daac/ae_l2a_tbs.gd.html) over the experiment site and other western stations in China according to the geographic coordinate, which means that the extracted swath Tbs are in the area of 10km^2 around the site. We chose the Tb difference between 89.0/36.5/18.7GHz and 10.7GHz channels for the gradient time series comparison with station snow depth (cm).

3.3 Comparison result at Nam Co experiment site

3.3.1 The AMSR-E swath L2A Tb gradient time series

We plotted the Tb gradient between 89.0/36.5GHz and 18.7GHz with different resolutions corresponding to the snow measurement time at Nam Co in Fig. 15. Compared to the snow depth (the solid lines), the brightness temperature gradient (traditional algorithm prototype) at Fig. 15 shows a good relationship for the snow depth decreasing period (24/11/2006~26/1/2007) at 89.0GHz (named high frequency) gradient and 36.5GHz (named low frequency) gradient. For this period of time (snow depth are less than 30 cm), we can understand that the high frequency are more sensitive to the snow evolution than the low

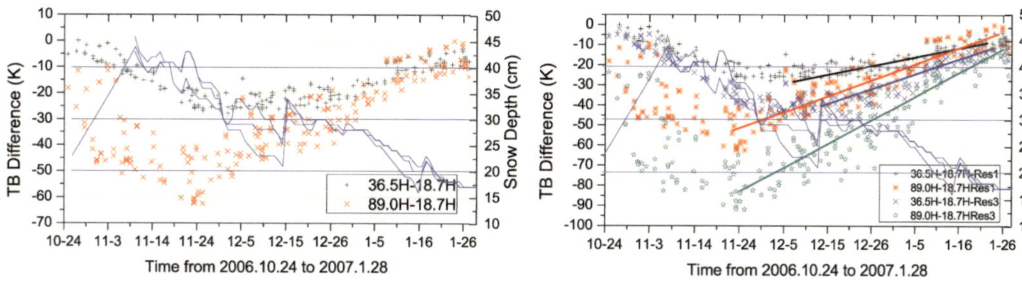

Fig. 16. The brightness temperature gradient between 89.0/36.5GHz and 18.7GHz with different resolution corresponding to the snow measurement time. The solid lines stand for the ground snow depth.

frequency, and the bottom panel in Fig.16 indicates the high resolution (L2A.res3) resample is more sensitive to the snow evolution than that of low resolution (L2A.res1).

Fig.17 shows a sample, located beside (20km away from the field measurement site, the time stamp starts from October 1st), it indicates the same trend as the previous Fig.15. During the first 15days, the 89.0GHz shows a good sensitivity to the fresh snow. During the later succeeding 17days (7/11/2006~24/11/2006), the gradient becomes more and more large, but the snow depth decreases (compare to the Fig.15). This can be explained preliminarily by the evolution of the snow grain size and density, which typically increases with time. A more physical explanation needs e.g. the model simulation work on the snow emission of snow grain size, snow depth and snow density, in order to decide which part plays a dominant role (Pulliainen, 1999).

3.3.2 Comparison to the AMSR-E daily SWE product

Another comparison (Fig.17) has been done by using the AMSR-E daily snow products, which are EASE-Grid with coarse resolution (25km) and the snow depths at Fig.15. From these figures, we can find that the maximum estimated AMSR-E SWE(mm) (using 36.5GHz gradient) over NamCo station at the winter of 2006~2007 occurs around 26/12/2006, which are do not match the snow measurement, with a delay of almost one month.

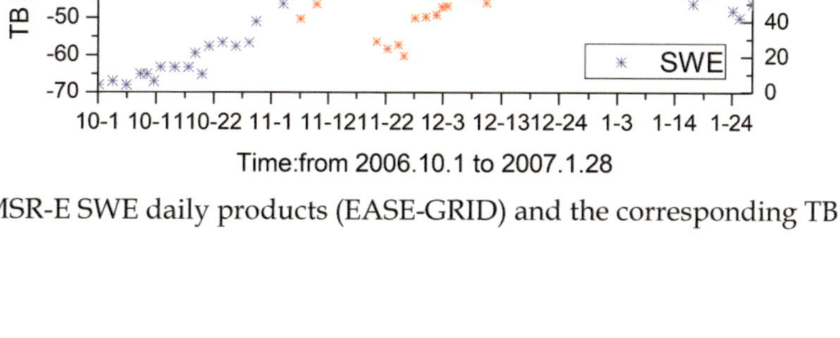

Fig. 17. The AMSR-E SWE daily products (EASE-GRID) and the corresponding TB difference.

Compared with the gradient figure, from Fig.16, we can find that the 36.5GHz gradient shows a relatively stable value from 1/12/2006~26/12/2006. The discrepancy between the SWE and TB gradients is probably due to the response of 37 GHz saturating for SWE values over 120-140mm, or the mixed pixel by the lake. This requires a more extensive field dataset to acquire the explanation.

If we consider the typical snow density over Tibet area to be approximately 0.239g/cm^3, we get a maximum snow depth value of about 75 cm from the AMSR-E observation. This is quite larger compared to the in situ measurements, an indication that the AMSR-E SWE value is overestimated, which is consistent with the result in paper(Pulliainen, 1999).

3.4 Time series analysis between Tb and snow depth

We selected several observations over western China (Fig.14) for the qualitative analysis, which include the Xinjiang and Neimenggu deep snow and Gansu, Qinghai and Tibet shallow snow depth situations.

All of the figures in Fig.18 are plotted with the three gradients (Tb difference at 89.0-18.7, 36.5-18.7 and 18.7-10.7) and the corresponding snow depths (see Fig.18).

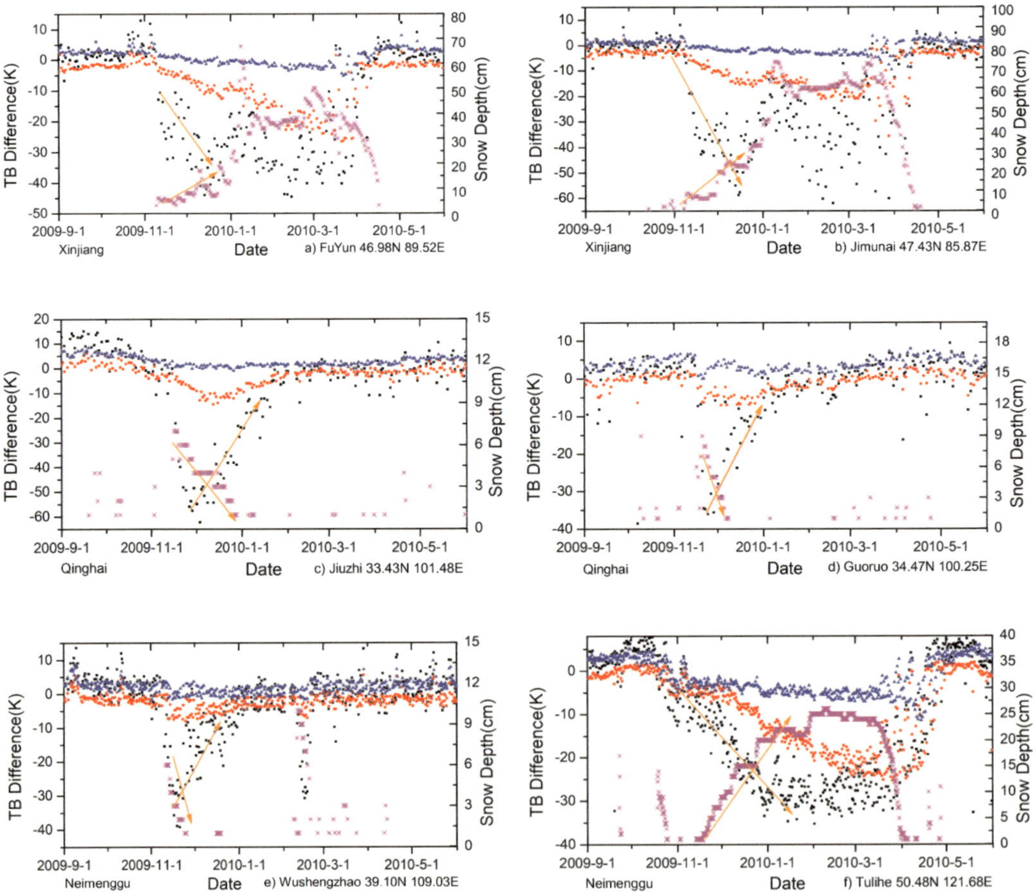

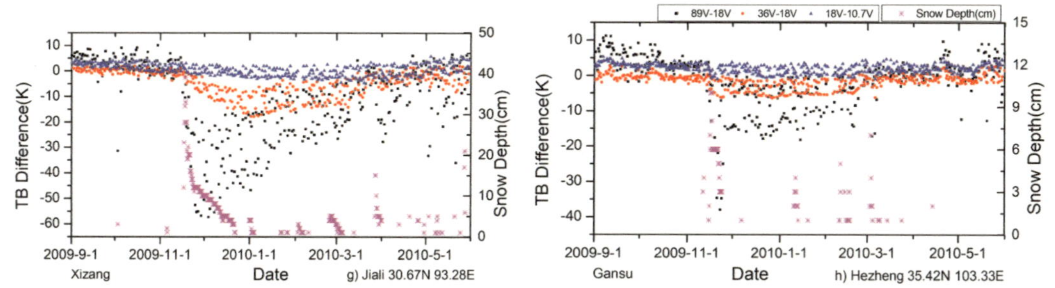

Fig. 18. The time series Tb difference (89/36.5-18.7GHz and 18.7-10.7GHz) and the ground snow depth (cm). The first 4 figures are plotted only ascending Tb and the corresponding ground measurements (local morning time), the last four figures are plotted with all of the Tb (A/D) and snow depth at any available time.

We get the preliminary analysis result, the Tbs at18.7-10.7GHz are insensible to the snow evaluation except the deep snow depth (a, b and f), although the depression in f is obvious due to the local vegetation influence. Over deep snow (a, b, f and g, continuous accumulation > 20cm), the Tbs at 36.5-18.7GHz are more reliable than that of high frequency, while over the shallow snow (c, d, e and h, discontinuous snow occurrence, < 15cm), the pair 36.5/18.7 is insensitive, but the high frequency pair (89.0/18.7) shows its distinct response. The pair 89.0/18.7 shows its shallow snow retrieval ability in a, b, c, d, e and when the snow depth over 20cm, the signal is more variable and suspect. The pair 89.0/18.7 indicates its sensitive response to the quick presence of the snowfall, and keeps turbulence when the snow depths are unchanged due to the temporal snow physical characteristics and climate factors. The last four figures show that the A/D Tbs act the similar behaviors with difference correlation intensity.

3.5 Conclusions

From what we have shown above, it can be argued that the high frequency (89.0GHz) shows its sensitive to the relative shallow snow pack, which suggests that we can develop the shallow snow depth retrieval via the good Tb pair and ground snow depths over the western China. Model simulation work is needed to explain the discrepancy of the snow evolution and brightness temperature gradient at high frequencies, and we should enhance the following aspects, the possible mixed pixel effect, the atmosphere effect elimination, and the vegetation effect removal.

4. Acknowledgement

This work is now supported by the Chinese "973" Program "Earth Observation for Sensitive Factors of Global Change: Mechanism and Methodologies" (NO. 2009CB723906), the Director Foundation of Center for Earth Observation and Digital Earth Chinese Academy of Sciences, the National Natural Science Funds (Grant: 40901175), and the "Open Foundation" of Institute of Plateau Meteorology.

5. References

[1] Armstrong, R. L., & Brodzik, M. J. (2001). Recent Northern Hemisphere snow extent: A comparison of data derived from visible and microwave sensors. Geophysical Research Letters, 28(19), 3673–3676.

[2] Armstrong, R. L., M. J. Brodzik, K. Knowles, and M. Savoie. 2007. Global Monthly EASE-Grid Snow Water Equivalent Climatology. Boulder, Colorado USA: National Snow and Ice Data Center. Digital media.

[3] Basist, A., Garrett, D., Ferraro, R., Grody, N. C., &Mitchell, K. (1996). A comparison between snow cover products derived from visible and microwave satellite observations.Journal of Applied Meteorology, 35(2), 163−177.

[4] Bao, Q., Y. M. Liu, J. C. Shi, and G. X. Wu. 2010. Comparisons of soil moisture datasets over the Tibetan Plateau and application to the simulation of Asian summer monsoon onset. Advances in Atmospheric Sciences 27:303–314.

[5] Cao Yungang, Xiuchun Yang and Xiaohua Zhu, Retrieval snow depth by artificial neural network methodology from integrated AMSR-E and in-situ data—A case study in Qinghai-Tibet Plateau, Chinese Geographical Science, Volume 18, Number 4, 356-360, 2008, DOI: 10.1007/s11769-008-0356-2

[6] Chand, D., Wood, R., Satheesh, S. K., Charlson, R. J., Anderson, T. L.," Satellite-derived direct radiative effect of aerosols dependent on cloud cover", Nat. Geosci., 2,181–184(2009) [doi:10.1038/ngeo437(2009)]

[7] Chapin III, F.S., Sturm, M., Serreze, M.C., McFadden, J.P., Key, J.R., Lloyd, A.H., McGuire, A.D., Rupp, T.S., Lynch, A.H., Schimel, J.P., Beringer, J., Chapman, W.L., Epstein, H.E., Euskirchen, E.S., Hinzman, L.D., Jia, G., Ping, C.L., Tape, K.D., Thompson, C.D.C., Walker, D.A. and Welker, J.M. (2005). Role of land surface changes in Arctic summer warming. Science, 310, 657-660

[8] Che Tao, Li Xin, Geo Feng, Estimation of the snow water equivalent in the Tibet Plateau using passive microwave remote sensing data (SSM/I). Journal of glaciology and Geocryology, vol. 26, no 3, pp.363-368. 2004 (in Chinese)

[9] Cohen, J. and D. Entekhabi. "The influence of snow cover on Northern Hemisphere climate variability," Atmosphere-Ocean, Vol. 39, 2001, pp. 35-53.

[10] Gong G, Entekhabi D, Cohen J (2003) Relative impacts of Siberian and North American snow anomalies on the winter Arctic Oscillation. Geophysical Research Letters 30(16):1848–1851

[11] Groisman, P. Ya. and T. D. Davies. "Snow cover and the climate system," in: H. G. Jones, J. W. Pomeroy, D. A. Walker, and R. W. Hoham, eds., The Ecology of Snow.Cambridge, UK: Cambridge University Press, 2000, pp. 1-44.

[12] Hall, D. K., Riggs, G. A., Salomonson, V. V.," Development of methods for mapping global snow cover using moderate resolution imaging spectroradiometer data", Remote Sens. Environ., 54(2), 127–140(1955)[doi:10.1016/0034-4257(95)00137-P, 1995.]

[13] Hall, Dorothy K., George A. Riggs, and Vincent V. Salomonson. 2007, updated daily. MODIS/Aqua Snow Cover 8-Day L3 Global 0.05deg CMG V005, [list the dates of the data used]. Boulder, Colorado USA: National Snow and Ice Data Center. Digital media.

[14] IPCC (2001) Climate Change 2001: The scientific basis. Contribution of working group I to the third assessment report ofthe Intergovernmental Panel on Climate Change. WMO/ UNEP. Cambridge University Press, Cambridge, 944 pp

[15] Judah cohen, David Rind ,the effect of snow cover on the climate, Journal of climate, July, 1991, PP:689-706

[16] Kripalani, R.H. and A. Kulkarni 1999: Climatology and variability of historical Soviet snow depth data: some new perspectives in snow-Indian monsoon teleconnections. Clim. Dyn., 15, 475-489.

[17] Matthew H. Savoie, Richard L. Armstronga, Mary J. Brodzika and James R. Wang, Atmospheric corrections for improved satellite passive microwave snow cover retrievals over the Tibet Plateau, Remote Sensing of Environment,vol.113, no. 15, PP.2661-2669,2009

[18] Menon, S., Koch, D., Beig, G., Sahu, S., Fasullo, J., Orlikowski, D.,"Black carbon aerosols and the third polar ice cap", Atmos. Chem. Phys., 10, 4559–4571(2010) [doi:10.5194/acp-10-4559-2010, 2010.]

[19] National Ice Center. 2008, updated daily. IMS daily Northern Hemisphere snow and ice analysis at 4 km and 24 km resolution. Boulder, CO: National Snow and Ice Data Center. Digital media.

[20] Nolin, A., R. L. Armstrong, and J. Maslanik. 1998, updated daily. Near-Real-Time SSM/I-SSMIS EASE-Grid Daily Global Ice Concentration and Snow Extent, [list the dates of the data used]. Boulder, Colorado USA: National Snow and Ice Data Center. Digital media.

[21] Pulliainen, J., Grandell, J., Hallikainen, M., "HUT snow emission model and its applicability to snow water equivalent retrieval". IEEE Transactions on Geoscience and Remote Sensing, vol. 37, no. 3, pp. 1378-1390, 1999.

[22] Qin Dahe, Liu Shiyin, Li Peiji, 2006: Snow Cover Distribution, Variability, and Response to Climate Change in Western China. J. Climate, 19, 1820–1833

[23] Qiu Yubao, Jiancheng Shi, Juha Lemmetyinen, Anna Kontu, Jouni Pulliainen, Huadong Guo,Lingmei Jiang, James R. Wang, Martti Hallikainen, Li Zhang , the atmosphere influence to AMSR-E measurements over snow-covered areas: simulation and experiments, Proceedings of IGARSS'09, 13-17 July, Cape Town, Africa, 2009

[24] Riggs GA, Hall DK, Salomonson VV (2003) MODIS snow products user guide for collection 4 data products. Available at http://modis-snow-ice.gsfc.nasa.gov

[25] Roger G.Barry, The Role of Snow and Ice in the Global Climate System: A Review, Polar Geography, Volume 26, Number 3, 1 July 2002 , pp. 235-246(12)

[26] Serreze M, Walsh J, Chapin III F, Osterkamp T, Dyurgerov M, Romanosky V, Oechel W, Morison J, Zhang T, Barry R (2000) Observational evidence of recent change in the northern high-latitude environment. Climatic Change 46:159–207.

[27] Steve Vavrus, The role of terrestrial snow cover in the climate system, Clim Dyn (2007) 29:73–88, DOI 10.1007/s00382-007-0226-0

[28] Tedesco, Marco, Richard E. J. Kelly, James L. Foster, and Alfred T. C. Chang. 2004, updated daily. AMSR-E/Aqua Daily L3 Global Snow Water Equivalent EASE-Grids V002, 2002~2010. Boulder, Colorado USA: National Snow and Ice Data Center. Digital media.

[29] Walsh, J. E. "Large-scale effects of seasonal snow cover," in B. E. Goodison, R. G. Barry, and J. Dozier, eds., Large Scale Effects of Seasonal Snow Cover. Wallingford, UK: IAHS Press, IAHS Publ. No. 166, 1987, pp. 3-14. Wang, X., Xie, H., Liang, T., "Evaluation of MODIS snow cover and cloud mask and its application in Northern Xinjiang, China", Remote Sens. Environ., 112, 1497–1513(2008) [doi:10.1016/j.rse.2007.05.016, 2008.]

[30] Wulder M A, Nelson T A, Derksen C, Seemann D, Snow cover variability across central Canada (1978-2002) derived from satellite passive microwave data, Climatic Change (2007), Volume: 82, Issue: 1-2, Publisher: Springer, Pages: 113-130

[31] Xu Changchun; Yaning, Chen; Weihong, Li; Yapeng, Chen; Hongtao, Ge, Potential impact of climate change on snow cover area in the Tarim River basin, Environmental Geology, 2007,Volume 53, Issue 7, pp.1465-1474

4

Seagrass Distribution in China with Satellite Remote Sensing

Yang Dingtian and Yang Chaoyu
State Key Laboratory of Tropical Oceanography,
South China Sea Institute of Oceanology,
Chinese Academy of Sciences, Guangzhou
China

1. Introduction

In nowadays, seagrass has been regarded as one of the healthy indexes for costal ecosystem, for it can provide shelter for fish living and laying egg, and also provide food for fish, tortoise, Dugong and seabirds. Management and preservation of coastal marine resources is a formidable challenge given the rapid pace of change affecting coastal environments. Fast, accurate, and quantitative tools are needed for detecting change in coastal ecosystems. Traditional in-situ surveys are time and labor intensive, generally lack the spatial resolution and precision required to detect subtle changes before they become catastrophic, and can be difficult to maintain from year to year (Orth & Moore, 1983, Peterson & Fourqurean 2001). In recent times satellite technology has played a vital role in seagrass monitoring. Remote sensing was a useful method for detection of land use change and seagrass. Satellite remote detecting of seagrass was different from that of terrestrial vegetation for water absorbing greatly at red and infer-red spectrum. When seagrass distributed underwater, visible spectrum was often used to detect the density and living state of seagrass. Lennon introduced the advantage of satellite remote sensing on detection of seagrass in 1989 and regarded red, blue and green as the most useful channel for detecting seagrass distribution (Lennon, 1989). Dahdouh-Guebas (1999) also used channel of blue, red and green to map the distribution of seagrass in Kenyan coast. Understanding of light scattering by plant canopies is crucial for remote sensing quantification of vegetation abundance and distribution (Jacquemoud et al. 1996). Hyperspectral data is very useful for assessing seagrass resources as it contains plentiful information. High turbidity is one of the important reasons for seagrass decline and usually was a problem for detection of seagrass with remote sensing. Phinn (2005) retrieved the seagrass along the coast of Moreton Bay, Australia and found that seagrass in turbid water was relatively difficult to detect. After studied variation of seagrass distribution and species affected by land use change, Batish (2002) concluded that hurricane and strong rainfall was the main factors for mud losing and sediment resuspension, which increased the water turbidity. The need for precise detection of living status and distribution of seagrass led some researchers to use high resolution remote sensing data. Among them SPOT data was very useful, for it had spatial resolution of 2.5m, 5m, 10m and four bands (visible and inferred, Bands B1: 0.50–0.59μm; B2: 0.61–0.68μm; B3: 0.78–0.89μm; B4 : 1.58–1.75 μm with a resolution of 20 m). Pasqualini (2005) used SPOT 5 data to map the distribution of Posidonia oceanica along Zakinthos Island in Greece with Principal Component Analysis, obtained good results and the accuracy was between 73%-96%. However, SPOT data only have two visible bands, which limited the use of image for retrieval of substrate information. IKONOS was another satellite data with high resolutions. Some researchers used the IKONOS data to retrieve the types of sea bottom and seagrass distribution. Andrefouet (2005) used IKONOS data to classify tropical coral reef environment, overall accuracy was 77% for 4-5 classes, 71% for 7-8 classes, 65% in 9-11 classes, and 53% for more than 13 classes. Compared with SPOT data, IKONOS and Quickbird have their advantage on more bands cover visible spectrum. Compared with IKONOS data, Quickbird is better for higher spatial resolution. Some researchers compared Quickbird data with Landsat-5 and CASI data, the conclusion was reached that Quickbird was better for mapping of seagrass cover, species and biomass to high accuracy levels (> 80%) (Phinn et al. 2008).

New technology provided by ocean color remote sensing provides high spatial and temporal resolution of the benthos, but the application of ocean color data in shallow waters is still in its early stage. Unverified classification techniques neglect the confounding effects of reflectance from benthos and spectral shifts, and this can lead to considerable errors. In optically shallow water, the radiance can be modified due to spectral scattering and absorption by phytoplankton, suspended organic and inorganic matter and dissolved organic substances (Dekker et al., 1992). Therefore, more efforts should be paid in accuracy of classification of seagrass on water column correction. A foremost problem for mapping seagrass by analysis of remote sensing data is water column effect. Most photons are absorbed or scattered by all particulates in optically shallow water (Morel et al.1977, Gordon et al., 1983). So water depth and water column inherent optical properties must be measured for mapping seagrass (Holden et al., 2001). While most water column correction procedures may not be appropriate for mapping or deriving quantitative information of seagrass. For instance, simply subtracting a deep-water remote sensing reflectance from each pixel is based on the assumption that energy traveling through a water column behaves the same way regardless of substrate type and water depth. In fact, when light penetrates water its intensity decreases exponentially with increasing depth. This process is known as attenuation and it exerts a profound effect on remotely sensed data of aquatic environment. The severity of attenuation differs with the wavelength of electromagnetic radiation. The single/quasi-single scattering theory and numerical simulations are often used to estimate water column effects (Liang, 2007). In these approaches a parameterized forward model for reflectance such as Hydrolight (for aquatic applications) takes a series of parameters describing the optical properties of participating media or canopy structure. Image analysis then uses a search algorithm to find the parameter space location which minimizes the distance of the corresponding spectral space location from the image pixel reflectance (Goodman & Ustin, 2007). With the approach, the model may take substantial computational effort. Often the model is simplified or approximated to facilitate inversion. So, model accuracy must be compromised. Conger et al. (2006) applied the approach separating determinations of the spectral albedos of typical materials covering the floor (Morel, 1993) to develop a simple technique to merge SHOALS (Scanning Hydrographic Operational Airborne Lidar Survey) LIDAR bathymetry data with Quickbird data. The remote sensing data with respect to depth was linearized in the model by subtracting an optically deep water value from the entire waveband under consideration and taking the natural logarithm of the result. Subtracting a deep-water remote sensing reflectance from each pixel (Cannizzaro & Carder, 2006) or utilizing the water optical properties which are derived from adjacent deep waters, were used to correct water column effects. Modeled shallow water reflectance typically has an inverse exponential relationship with depth (Mobley, 1994), so a regular subdivision of a depth parameter would over-sample the deep water spectral space region with similar reflectance spectra produced from unnecessary forward model runs. At the same time the shallow water spectral space region would be relatively under-sampled with a higher discretization error in the tabulated reflectance (Froidefond & Ouillon, 2005). The interaction between multiple parameters (e.g. depth vs. water clarity) makes the general problem of efficient and accurate LUT (look-up tables) construction very hard to tackle by analytical means (Hedley et al., 2009). Routine or large-scale operational image analysis by physics-based methods therefore demands the development of efficient approaches for both modeling and inversion. Often the model is simplified or includes too many parameters which cannot be measured directly (i.e. the attenuation coefficients for the upward streams originating from the water column and from the bottom) (Sathe & Sathyendranath, 1992; Nichols & Kyrala, 1992). In our investigation an optical model of in-coming solar radiation transfer was developed, in which multi-layer water was considered. Implementation of the method was found to be effective for improving the accuracy of coastal habitat maps and essential for deriving empirical relationships between remotely sensed data and interested features in the marine environment. Retrieved bottom reflectance was then used to study the relationship between reflectance and the LAI (leaf area index) of seagrass. In addition, it was found that the peak location of retrieved bottom reflectance was highly correlated to LAI of seagrass measured in Sanya Bay, South China Sea. Then, this paper studied the seagrass distribution along the northeast coast of Xincun Bay and Sanya Bay. The first aim was to give a regional bio-optical model of seagrass for detecting seagrass distribution with high accuracy. The second was to provide the information of seagrass distribution and living state for government and people to understand how to preserve and protect seagrass.

2. Water column model

In our investigation an optical model of in-coming solar radiation transfer was adopted, in which multi-layer water was considered (Yang *et al.* 2010). This algorithm which had been

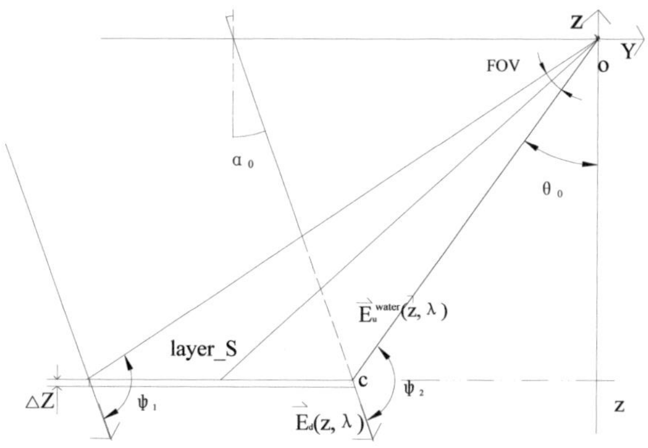

Fig. 1. The optical model of light propagation in optically shallow water

used to successfully extract seagrass and coral reefs information in Sanya Bay enables us to simulate radiation fields in a wide range of optical characteristics of these layers to analyse the mechanisms of the formation of the radiation characteristics inside and outside the layers, and to estimate any contribution of each region. Differences in water column properties would only modify the input to the equation for retrieving bottom reflectance but not the equation.

At the core of the inversion method by Yang *et al.* (2010) lies an analytical expression for $R_{rs}^{water}\left(0^{-},\lambda\right)$, subsurface remote sensing reflectance of the water column, for an optical shallow water body:

$$R_{rs}^{water}\left(0^{-},\lambda\right)=Q\cdot\left(\frac{\exp\left(-2k(\lambda)\left(z-z_{surf}\right)\right)-1}{-2k}\right)\cdot \tag{1}$$
$$\left[-2\pi\beta_{w}\left(90^{\circ},\lambda_{0}\right)\cdot\left(\frac{\lambda_{0}}{\lambda}\right)^{4.32}\cdot\left(\cos\Psi+\frac{0.835}{3}\cos^{3}\Psi\right)-\frac{b_{p}}{2g}\cdot\frac{1-g^{2}}{\left(1+g^{2}-2g\cos\Psi\right)^{1/2}}\right]_{\psi_1}^{\psi_2}$$

Where Q is the radio of the subsurface upward irradiance to radiance conversion factor, and has a value of 3.25 (Morel & Gentili 1993)；λ is the wavelength; k is a unique attenuation coefficient which is invariable with respect to depth; z is water depth measured downward from the detector; z_{surf} is the distance between ocean surface and the detector; $\beta_w(\psi,\lambda)$ is the total volume scattering function (VSF) for pure seawater (Morel 1977); Morel gave the values of β (90°, λ_0) at the reference wavelengths of 350 and 600nm, and λ_0 is the wavelength value selected from the reference wavelength table; The wavelength dependence of $\lambda^{-4.32}$ results from the wavelength dependence of the index of refraction; The factor 0.835 is attributed to the anisotropic properties of the water molecules (Chami *et al.* 2006); b_p is the particle scattering coefficient; g is a parameter that can be adjusted to control the relative

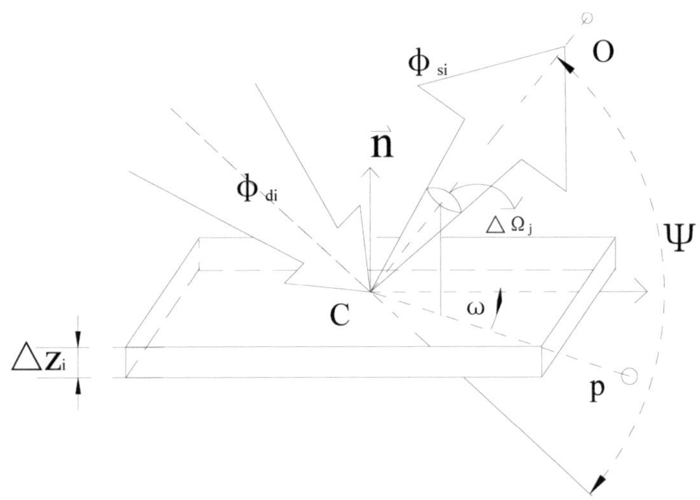

Fig. 2. The fraction of incident power scattered out of the beam through an angle ψ into a solid angle $\Delta\Omega_j$.

amounts of forward and backward scattering in β_{HG}, Henyey-Greenstein phase function (Henyey-Greenstein, 1941); R (0^-, λ) is the irradiance reflectance just beneath the sea surface.

In addition, ψ is the angle of reflection of the incident beam; ψ_1, and ψ_2 can be expressed as:

$$\Psi_1 = -\left(\alpha_0 + \theta_0 + \frac{FOV}{2}\right) + \pi \tag{2}$$

$$\Psi_2 = -\left(\alpha_0 + \theta_0 - \frac{FOV}{2}\right) + \pi \tag{3}$$

Where, FOV is the field of view of the sensor, α_0 is the solar attitude, and θ_0 is the view angle. In addition, the final signal detected by the receiver results from the flux reflected by the bottom (as if the bottom were black) and the flux reflected by the bottom (when it is not black) (Maritorena *et al.*, 1994). For simplification, reflectance, mentioned below, represents the remote sensing reflectance. Finally, we can get the bottom reflectance, $R^b{}_{rs}$:

$$R_{rs}^b = R_{rs}(0^-,\lambda) - R_{rs}^{water}\left(0^-,\lambda\right) \tag{4}$$

Here R is the total irradiance radio just below the surface.

3. Results and analysis

3.1 The difference between uncorrected and corrected reflectance

Fig. 3 shows the retrieved bottom reflectance was lower than the in situ measured reflectance and the change between the corrected and uncorrected reflectance did exist (Fig. 4) (Yang et al. 2010). The range of the variation varied widely, and the difference was much larger at 400-450 nm. The variation in spectral reflectance was determined mainly by absorption and scattering properties of shallow water.

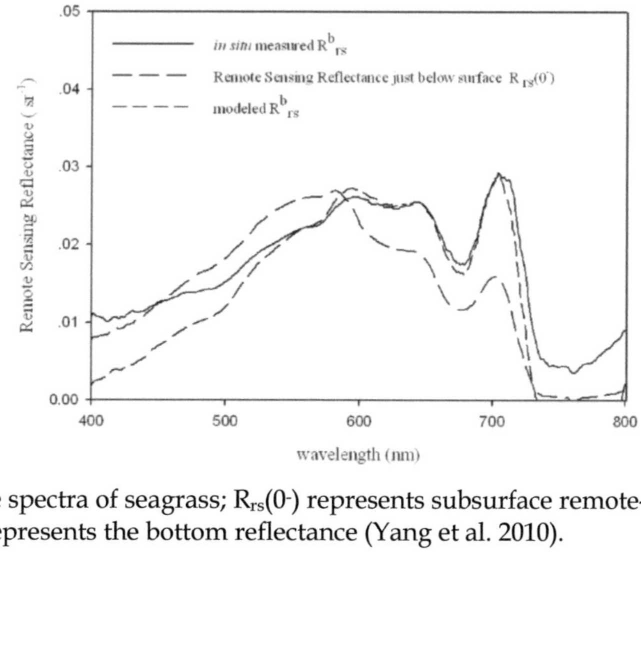

Fig. 3. Reflectance spectra of seagrass; $R_{rs}(0^-)$ represents subsurface remote-sensing reflectance; $R_{rs}{}^b$ represents the bottom reflectance (Yang et al. 2010).

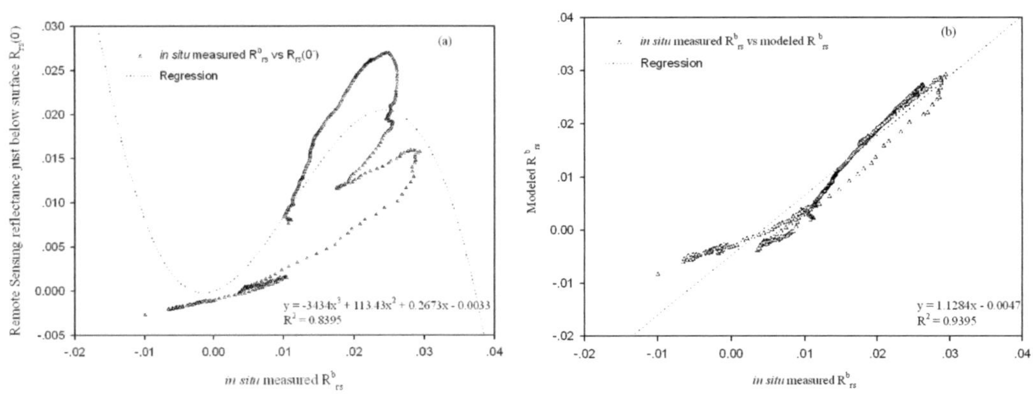

Fig. 4. (a) Comparison of the in situ measured bottom reflectance with subsurface remote-sensing reflectance; (b) Comparison of the in situ measured bottom reflectance with the retrieved bottom reflectance from the optical model; $R_{rs}(0^-)$ represents subsurface remote-sensing reflectance; $R_{rs}{}^b$ represents the bottom reflectance (Yang et al. 2010).

3.2 Spectral characteristics of seagrass

Fig. 5 shows that the reflectance of Thalassia increased between 518 and 532 nm, which might indicate changes in xanthophyl-cycle pigmentation. In South China Sea the leaves of Thalassia display olive-drab color, and this just coincides with the relevant spectral features of Thalassia detected around 550 nm. The reflectance differences between 600 and 650 nm can mainly be attributed to different proportions of red, orange, yellow and brown carotenoids. Between 650 and 680 nm reflectance decreased, and that suggests reduced absorption of light by chlorophyll, which may be resulted from reduction in chlorophyll. The reflectance overlapped around 720nm may be relative to the total effect whereby the relationship between light harvesting efficiency and chlorophyll content is non-linear due to pigment self-shading among thyllakoid layers. The particularly strong package e ffect observed in seagrass was largely attributed to restriction of chloroplasts to the leaf epidermis (Cummings *et al.* 2003, Enriquez 2005). Zone between 800 and 840 nm

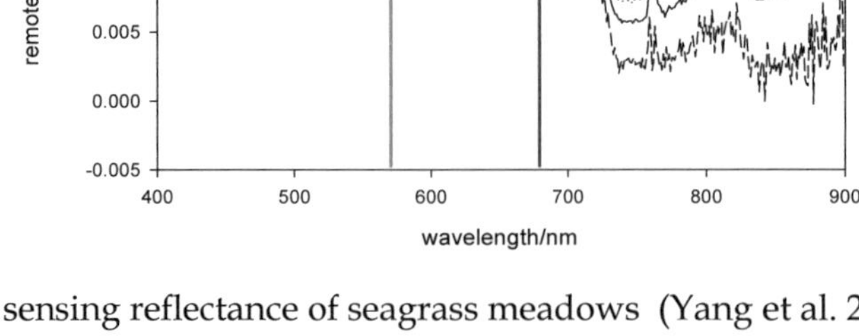

Fig. 5. The remote sensing reflectance of seagrass meadows (Yang et al. 2010).

encompasses a spectral region of maximum reflectance, although a weak water absorption feature occurs near the 812 nm (Becker *et al.* 2005). In addition, the seagrass can also spectrally be identified with leaf samples coated by epibionts including a diverse array of microalgae, bacteria, juvenile macroalgae and sessile invertebrates such as tubeworms and bryozoans. It was found that seagrass reflectance reduced drastically at the green peak without having a noticeable effect on the chlorophyll absorbance trough in infrared. However, biliproteins of algal epibionts were responsible for the increased reflectance peaks observed between 560-670 nm.

3.3 LAI and the spectral response

The spectral reflectance of seagrass measured at different station changed regularly with LAI. The bands with good relationship with LAI were 555,635, 650 and 675 nm (Yang and Yang 2009), for absorption and reflectance of seagrass photosynthetic and accessory pigment (Fig. 6), and correlation coefficient of quadratic equation was greater than 0.79. These results can be explained as lower leaf area index the exposed area of sediment was greater, and water leaving radiance from seagrass bed is weaker; with higher seagrass LAI, water leaving radiance from bottom is dominant by seagrass and increased with increasing LAI. Previous study showed that the bands with good relationship with LAI were 555,635, 650 and 675 nm, for absorption and reflectance of seagrass photosynthetic and accessory

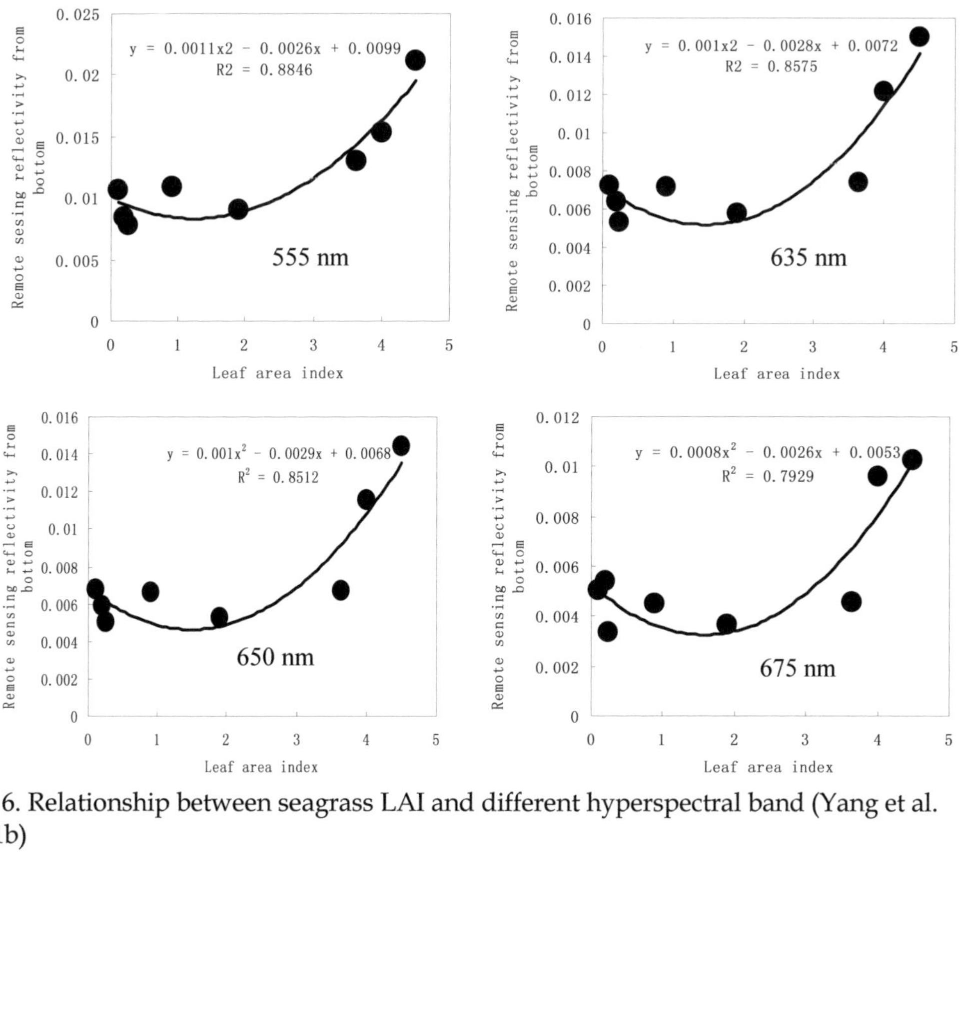

Fig. 6. Relationship between seagrass LAI and different hyperspectral band (Yang et al. 2011b)

pigment, the results in the paper are also confident with the results of previous study. Compared with Quickbird bands (Band1 (0.45-0.52μm), 2(0.52-0.59μm), 3(0.63-0.69μm) and 4(0.77-0.89μm)). Band 2 and band 3 of Quickbird can be well used for retrieving seagrass distribution.

3.4 The peak location and LAI

Red edge refers to the region of rapid change in reflectance of chlorophyll in the near infrared range. Vegetation absorbs most of the light in the visible part of the spectrum but is strongly reflective at wavelengths greater than 700 nm. The change can be from 5% to 50% reflectance between 680 nm to 730 nm. The phenomenon accounts for the brightness of foliage in infrared photography. It is used in remote sensing to monitor plant activity and could be useful to detect light-harvesting organisms on distant planets. Fig.7 also shows the typical reflectance characteristics of seagrass with an obvious spectral peak at red edge, and the peak at the wavelength of 695-710 nm range shifted to the red with increasing of leaf areas. Fig.7 indicates a strong relationship between the peak location and LAI with a coefficient of correlation of 0.7263 in the near infrared range.

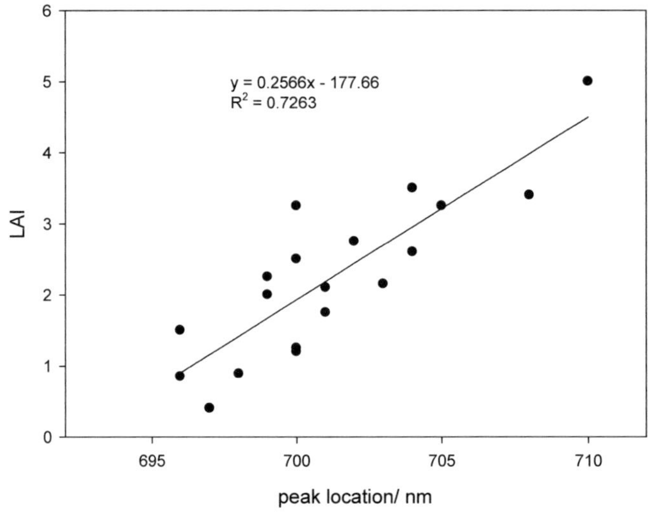

Fig. 7. Relationship between the peak location and LAI (Yang et al. 2010).

3.5 Relationship between seagrass LAI, NDVI and Hyperspectral Bands

The Normalized Difference Vegetation Index (NDVI) is a simple numerical indicator that can be used to analyze remote sensing measurements. In order to fully explore the useful information in hyperspectra, the red band reflectance of NDVI was replaced by the green and blue band reflectance. The equations are listed as follows:

$$\text{Red NDVI} \quad RNDVI=(NIR-Red)/(NIR+Red) \tag{5}$$

$$\text{Green NDVI} \quad GNDVI=(NIR-Green)/(NIR+Green) \tag{6}$$

$$\text{Blue NDVI} \quad BNDVI=(NIR-Blue)/(NIR+Blue) \tag{7}$$

where, NIR is near infrared reflectance. In the paper, VNDVI is taken as the general name of BNDVI, GNDVI and RNDVI. Hyperspectra, leaf area index and NDVI (Normalized Difference Vegetation Index) of seagrass were measured and calculated with equations (5, 6, 7), and the relationship between them was obtained.

As Fig.8 indicates, the spectral reflectance of seagrass measured at different stations changed regularly with LAI. The bands with a good correspondence with LAI were 555, 635, 650 and 675 nm, for absorption and reflectance of seagrass photosynthetic and accessory pigment. However, at the band around 400 and 720 nm, a relatively poor relationship with LAI was found (Table 1). The peaks and troughs on the reflectance spectra were also affected by factors such as reflectance and transmission of single leaves, types of background, leaf angle, the geometry of sun and sensor angles, etc, and these effects can be reduced to great extent when hyperspectral measurement are taken *in situ*.

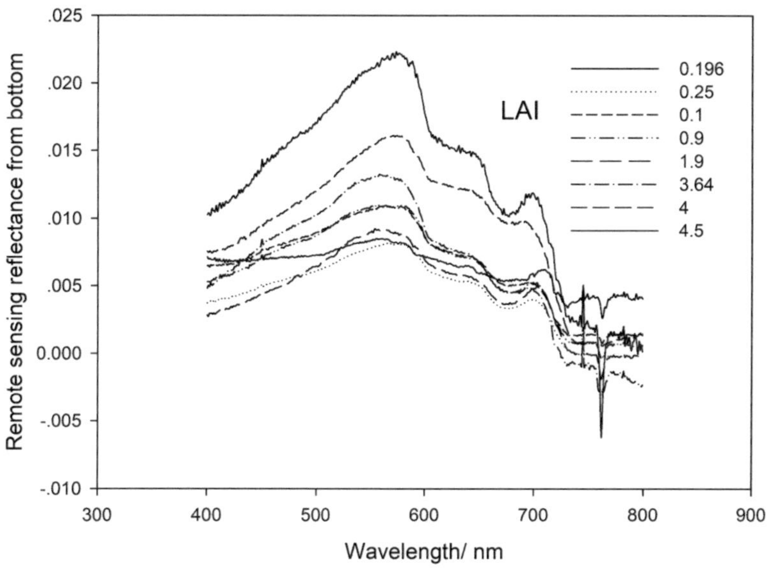

Fig. 8. Relationship between spectral bands and seagrass LAI (Yang et al. 2009).

550 nm	$y = 0.0011x^2 - 0.0026x + 0.0099$	$R^2 = 0.8846$
635 nm	$y = 0.001x^2 - 0.0028x + 0.0072$	$R^2 = 0.8575$
650 nm	$y = 0.001x^2 - 0.0029x + 0.0068$	$R^2 = 0.8512$
675 nm	$y = 0.0008x^2 - 0.0026x + 0.0053$	$R^2 = 0.7929$
400 nm	$y = 0.0011x^2 - 0.0041x + 0.0068$	$R^2 = 0.7728$
700 nm	$y = 0.0008x^2 - 0.0024x + 0.0057$	$R^2 = 0.7604$

Table 1. Relationship between LAI and different hyper spectral band. In Table 1, y is remote sensing reflectivity from bottom, x is LAI.

In order to monitor seagrass with satellite remote sensing, the relationship between Leaf Area Index (LAI) and the Normalized Difference Vegetation Index (NDVI) was correlated. Analysis indicated that a good relationship existed between NDVI and LAI. Relationships between every VNDVI (RNDVI, GNDVI and BNDVI) and LAI were studied. Fig.9 shows that the VNDVI increased with the increase of LAI. The correlation coefficient between G-NDVI and LAI (0.7357) was better than the RNDVI (0.6705) and BNDVI (0.6729), which means that GNDVI is more sensitive than RNDVI and BNDVI when applied to seagrass remote sensing. However, when LAI is less than 1.5, the correlation coefficient between VNDVI and LAI is relatively low. A0s the band set in the visible and infrared of the satellite remote sensing data (CBERS, Landsat TM, and QuickBird) is blue (0.45-0.52μm), green (0.52-0.59μm), red (0.63-0.69μm) and infra-red (0.77-0.89μm), NDVI was retrieved with blue, green and red also, which is compatible with the satellite data band set.

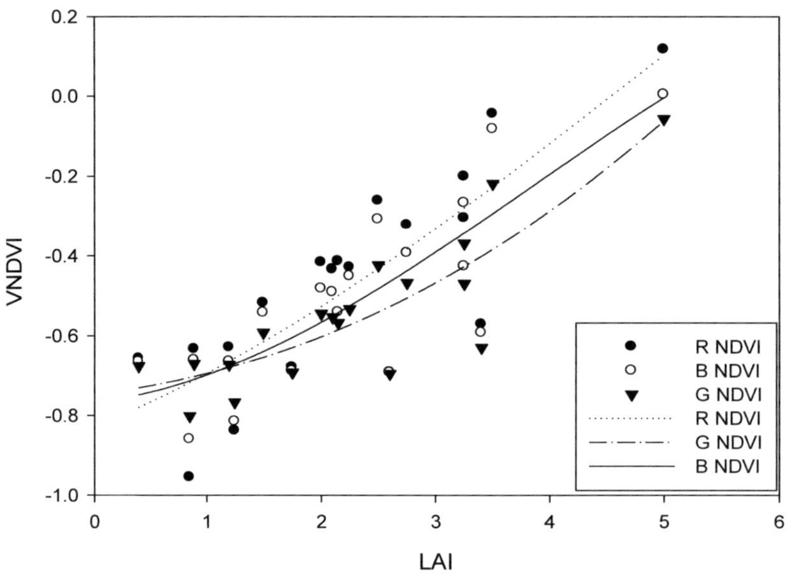

Fig. 9. Relationship between LAI and NDVI (Yang et al. 2009).

3.6 Seagrass detection with satellite remote sensing

QuickBird data was used for correcting seagrass density and detailed distribution retrieved with Landsat and CBERS data, for detailed seagrass distribution was very important for us to compare seagrass distribution changes (Yang and Yang 2009). With the water body correction, sun glint correction and computation of bands, seagrass distribution along the south coast of Xincun Bay was retrieved and showed in Fig 10 (Yang 2008). From the Quickbird image, substrate types, such as sand, seagrass can be detected clearly, and the profiles from bank to the center of Xincun Bay were sea pond, sand, seagrass and optically deep water. In shallow water near bank and relatively deeper water, no seagrass can be detected. The pattern of seagrass distribution can be detected clearly, and seagrass mainly distributed in the pattern of cluster, with tens meter distance away from the bank.

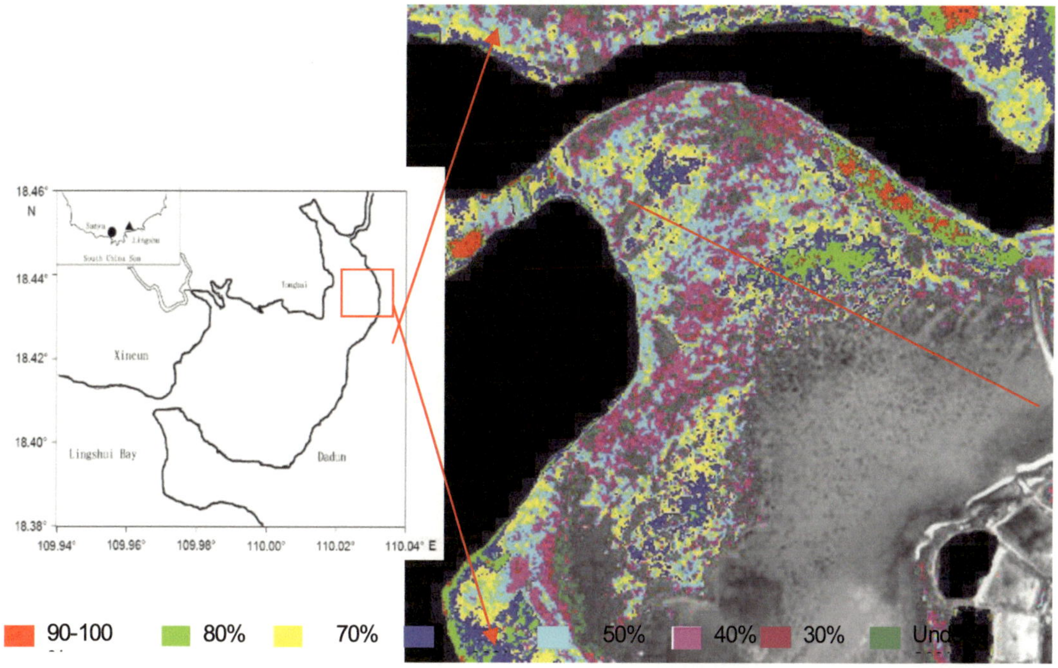

Fig. 10. Seagrass distribution in the northwest of Xincun Bay with Quickbirds (Yang et al. 2011b).

The pattern of seagrass distribution can also be clearly classified, and seagrass was mainly distributed in a stripe pattern, some tens of meters away from the coastline (YANG and HUANG 2011a). Seagrass density is regular in the main seagrass bed. From the outside to the center of the main seagrass bed, seagrass distribution coverage was under 20%, 20-40%, 40-60%, 60-80% and greater than 80%. Among them, the area of seagrass coverage greater than 80% accounted for more than 30% of the total seagrass bed (Fig. 11). Seagrass species in

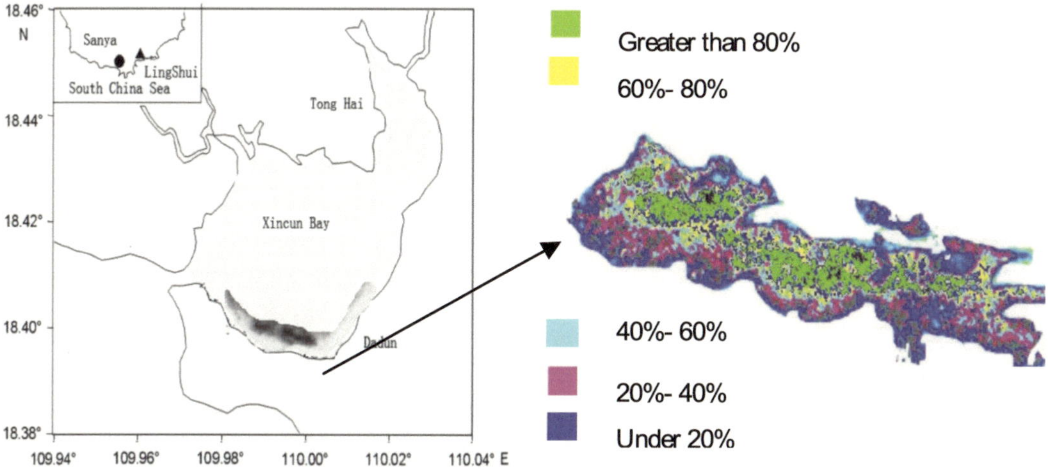

Fig. 11. Seagrass density retrieved with QuickBird (Yang et al. 2009).

Xincun Bay are mainly *Enhalus acoroides, Thalassia hemprichii, Cymodocea rotundata and Halodule uninervis;* however, we could not differentiate one species from another with QuickBird data. Detection accuracy of seagrass with Quickbird data was mainly by comparing pixels of satellite remote sensing with *in situ* observations. In this paper, the accuracy was more than 80% for seagrass coverage greater than 20% when compared with the *in situ* observation results. Detailed information on seagrass distribution was very important for us to know the density of seagrass distribution and can be used as basis for comparing seagrass distribution changes. In order to further study seagrass distribution changes in Xincun Bay, Landsat TM and CBERS data was used (Yang et al. 2009). Compared with QuickBird, seagrass detected with Landsat TM and CBERS had fewer classes, which only showed the distribution range, for pattern and species cannot be clearly obtained. However, the distribution contour can be detected clearly, which was enough for comparison of seagrass distribution changes. Landsat data was usually used for detecting seagrass distribution for cost-effective and relatively higher revisit frequency, and visible band was regarded as the most useful. Lennon introduced the advantage of satellite remote sensing with Landsat TM data on detection of seagrass in 1989 and regarded red, blue and green as the most useful channel for detecting seagrass distribution (Lennon, 1989). Dahdouh-Guebas (1999) also used channel of blue, red and green of Landsat TM data to map the distribution of seagrass in Kenyan coast with good results.

The imperfect was that band spectrum coverage was relatively wider and the pixel sizes (20–30 m) are of a similar magnitude to the size of the habitat patches. So it was problematical when applied Landsat data on detecting seagrass in small area and distinguishing seagrass species. In 1970s, Yang et al (1979) investigated the seagrass distribution in the Chinese costal water, and distribution of seagrass in coastal water of Hainan province showed in the references. However, seagrass in the west coast of Hainan province disappeared in recent years. Some researchers regarded that it is mainly caused by aquaculture. Seagrass distributed in the east coast of Hainan province only confined within a few bays. Because different species of seagrass live in different environment, distribution of seagrass is confined by its growth habitat. Generally speaking, vertical distribution of seagrass from coastal to relative deep water around the coast of Hainan Province were *Halophila, Cymodocea, Syringgodium, Ehaus, Thalassia, Ha-lophila*. Substrates of sand, seagrass and coral were differentiated with different reflective characteristics. Based on the results, we can find that area of seagrass distribution in Sanya bay decreased, from distributed in the whole south coast in 1991 to less than 1 hectare in southwest coast of Sanya bay in 1999 (Fig 12).

Seagrass distribution in the south coast of Xincun Bay was mainly studied (Yang et al. 2009). In order to compare seagrass distribution in detail, we divided seagrass distribution regions as A, B and C (Fig. 12). Seagrass distribution in region A and region B was connected as one big seagrass bed in 1991, however, the two region seagrass separated gradually and they were only connected with a line of seagrass in 1999. Finally, complete separation was observed in 2001. Seagrass distribution in region C was relatively large with an elliptical shape in 1991; however, the shape of seagrass bed became thinner by 1999 and became a line in 2001. In 2006, seagrass at region C could be detected with satellite remote sensing.

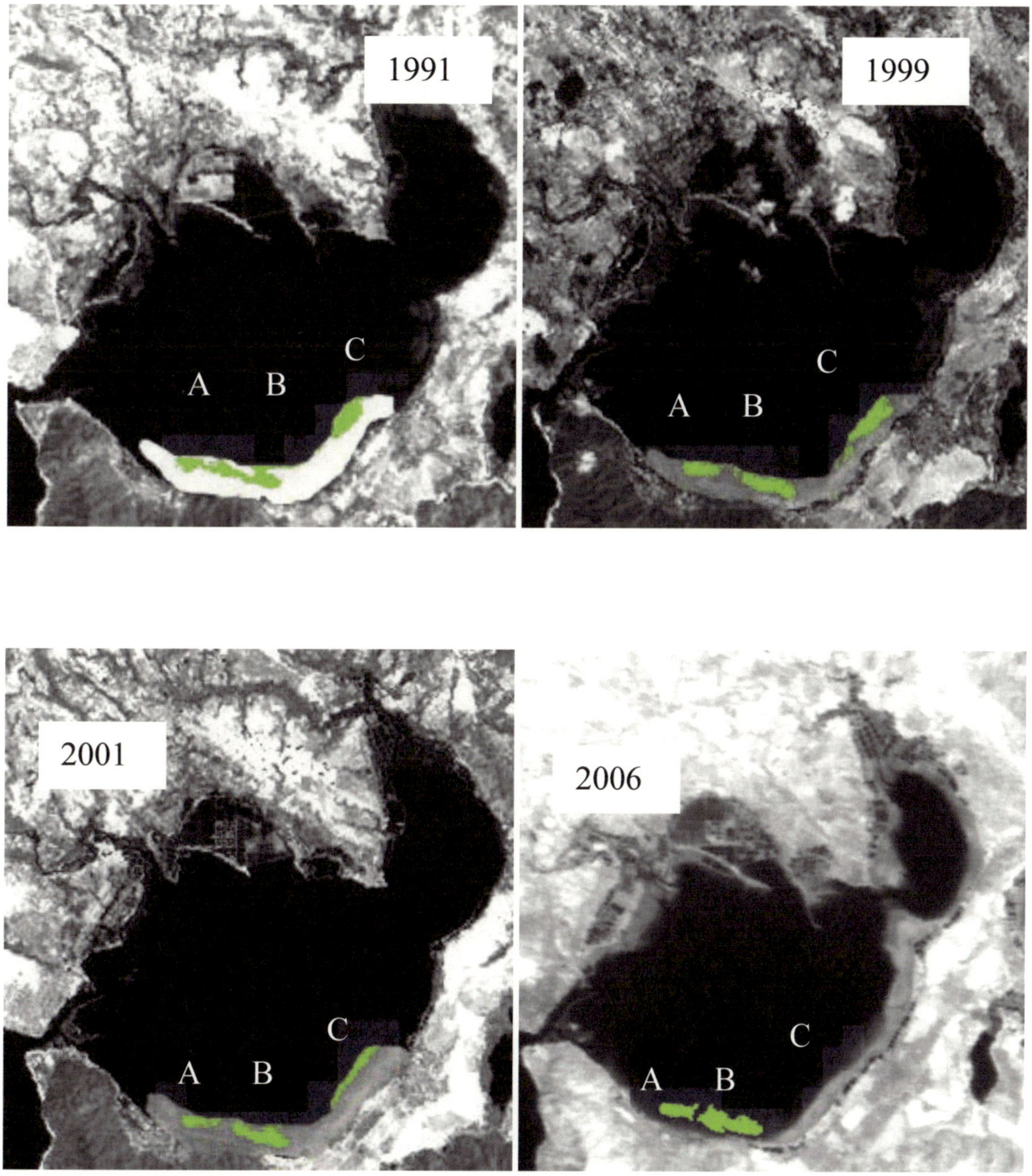

Fig. 12. Seagrass distribution change from 1991 to 2006 (Yang et al. 2009).

4. Impacts of environmental factors for seagrass distribution

4.1 Effluent diffusion in Sanya Bay

As water with high concentration of nutrients in Sanya River flow into Sanya Bay, it diffused by current and wave. The effluent plume streaming out of the Sanya river into the Sanya Bay flows straight for only a short distance before it is washed westward by the longshore current (Yang 2008). Sanya Bay is an open bay, the flows-in water only stay in a short time. Landsat TM data of 1991 was used to detect the diffuse pattern in Sanya Bay. From the satellite images, the water plume of Sanya River mainly distributed along the south of Sanya Bay. The reason is when tidal water rushed into Sanya Bay it splits and flows back from the south of Sanya Bay. In Sanya Bay, Seagrass mainly distributed along the north coast of Lu Huitou peninsular, the plume of Sanya River affected seagrass distribution in great extent.

Fig. 13. The diffuse pattern of flows-in water in Sanya Bay (Yang 2008).

4.2 Land use change

Satellite Landsat data was used to retrieve land use change around Sanya bay. Sanya River and the island between the two rivers was chose as land use change indexes (Yang 2008). From the satellite images, area of eastern and western Sanya Rivers reduced in 1999 compared with that in 1991. However the area of island between the two rivers enlarged more than 30%. The decreased river area, correspondently decreased the ecological wet land along the Sanya River, reduced the area for waste water cleaning. Seashore land use change was also retrieved with remote sensing data. From remote sensing data, the shape of costal line at the northeast of Lu Huitou peninsular changed greatly. In situ observation proved that buildings were constructed just along the coastal line. In situ observation also found that coast at the middle north of Lu Huitou peninsular was constructed as sea bath and sea diving area.

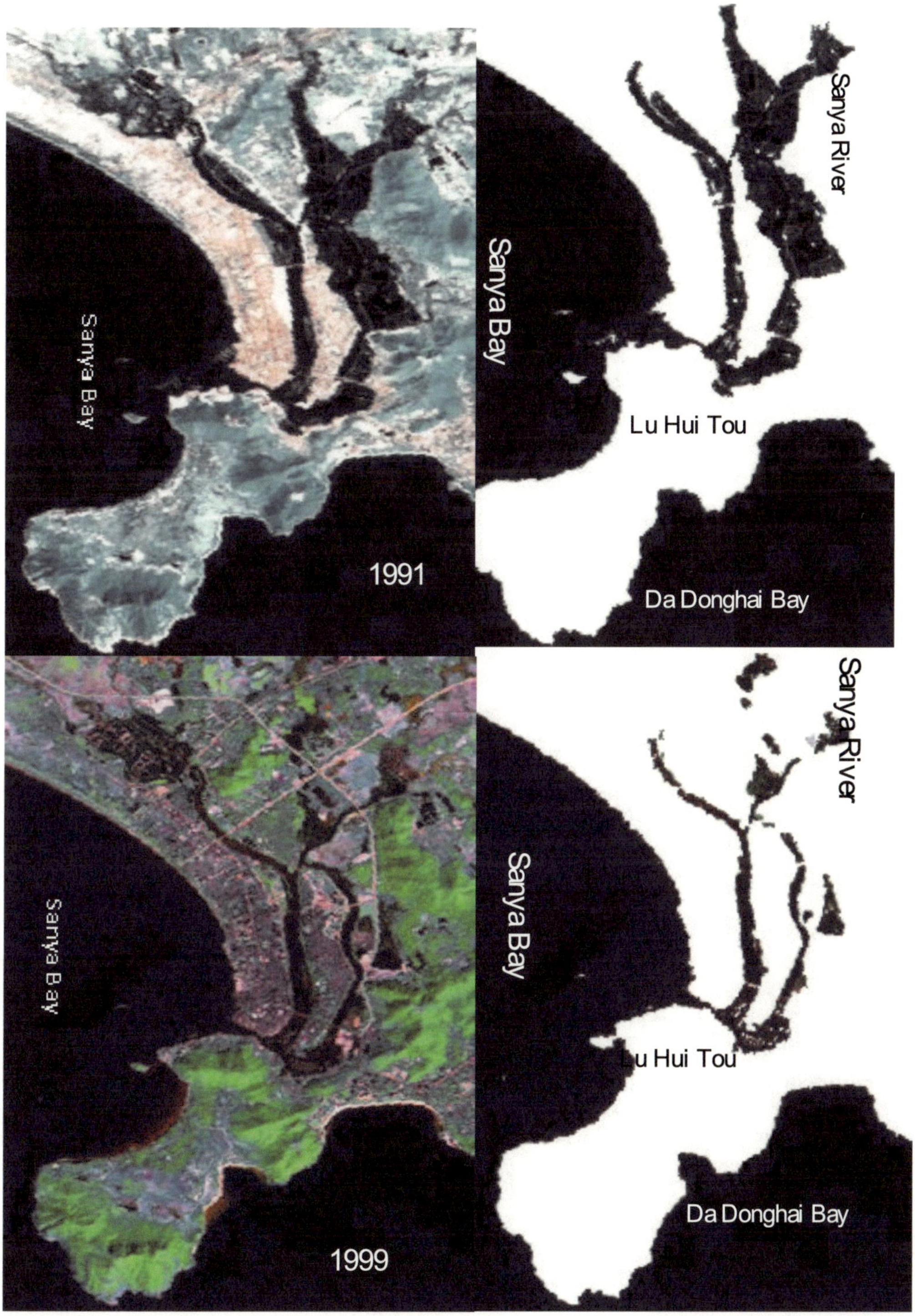

Fig. 14. Landsat TM data of Land use change around Sanya Bay in 1999 compared with that of in 1991 (Yang 2008).

4.3 Tendency of water chemical indexes by land use change in Sanya Bay

Water quality of Sanya River and Sanya Bay, such as water transparency and water quality indexes, was provided by Sanya ecological field station and references. In situ water chemical data of inorganic nitrogen, phosphorus and chlorophyll a concentration was used for validation and correction of the results from satellite remote sensing data. Results showed that water quality of Sanya River degraded in 2002 compared with that in 1991 (Fig.15) ; water quality in Sanya bay near the Sanya River mouth was also degraded. However, water quality in other part of Sanya bay changed little. Perhaps Sanya bay is an open bay, where water exchange rate is relatively high.

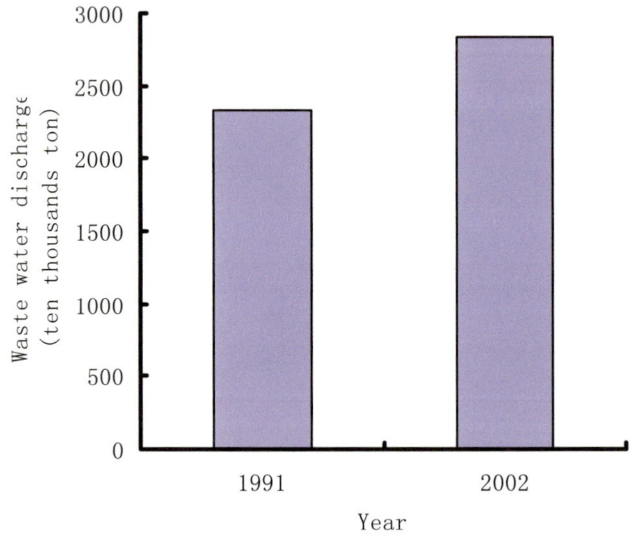

Fig. 15. Waste water discharge in Sanya Bay in 1991 and 2002 (Yang 2008).

4.4 Relationship between land use change and seagrass distribution

Seagrass distribution in Sanya area conversely correlated with land use change, the more area of land use change the less coverage of seagrass distribution. This mainly because distribution of seagrass was confined by the following factors: (1) Sediments, in the area close to the bank, were sand and frequently affected by hydrodynamics, on which seagrass cannot grow flourish. (2) Human activities, such as aquaculture, digging clam worm and ship sailing, also affected seagrass growth in shallow waters; (3)Transparency was also an important factor for seagrass growth. Clear water mainly distributed in the southwest of Sanya Bay, which provided the suitable condition for large area continuous seagrass distributed in the area.

5. Discussions and conclusions

As described in our investigation, much is known about the photophysiology of seagrass, while much is still required for us to effectively manage this important yet diminishing resource. New coastal ocean remote sensing techniques permit benthic habitats to be explored with higher resolution than ever before, however, the application of ocean color remote sensing to quantitative mapping of sparse seagrass species is still in its early development.

An optical model was proposed to simulate the radiation transfer in multi-layer, non-homogenous, heterogeneous, natural media. This algorithm enables us to simulate radiation fields in a wide range of variations of optical characteristics of these layers and to analyze the mechanisms of the formation of the radiation characteristics inside and outside the layers, as well as to estimate any contribution of each region. Based on the algorithm, we appropriately removed the distorting influence of the water column on the remotely sensed signal to retrieve an estimate of the reflectance of seagrass. Implementation of the method was found to be effective for improving the accuracy of coastal habitat maps and essential for deriving empirical relationships between remotely sensed data and features of interest in the marine environment. Retrieved bottom reflectance was then used to study the optical characteristics of seagrass. Through spectrum analysis it was found that the wavelengths for the discrimination and mapping of seagrass meadows of Sanya Bay, South China Sea lay between 500-630 nm as well as 680-710 nm. An appropriate hyper spectral band set for the remote sensing of seagrass should include narrow bands (maximum 5-10 nm bandwidth) centered around 555, 650, 675 and 700 nm. If satellite images were used, the effect of atmosphere should be taken into account. Though the blue band is more easily affected by atmosphere, the accurate surface reflectance could be acquired with the development of the theory and models in atmosphere correction. The relationship between seagrass leaf area index (LAI) and hyperspectra is very important when satellite remote sensing data is applied for detecting seagrass distribution.

Seagrass distribution in Xincun Bay spanning 15 years (1991-2006) was retrieved with satellite remote sensing. From the seagrass detection results, the resolution of satellite remote sensing image is very important for seagrass detection, so QuickBird data was more suitable for seagrass detection than Landsat TM and CBERS, especially when the seagrass distribution area was relatively small. Results in the paper proved that five classes can be classified clearly with QuickBird; however, only seagrass distribution contours can be detected with Landsat TM and CBERS data.

Though the accuracy of seagrass detection with satellite remote sensing can be affected by many factors, seagrass in Xincun Bay can be detected clearly for the sediment there was sand. Compared with satellite remote sensing data in 1991, the seagrass distribution area was reduced gradually and large areas of seagrass had disappeared by 2006. Human activities and extreme natural disasters were the main reasons for seagrass reduction, especially land use changes in recent years. The effect of land use change on seagrass distribution can be concluded as following: seagrass distribution in Sanya area conversely correlated with land use change, the more area of land use change the less coverage of seagrass distribution. Mainly because of land use change changed the water quality and sediment type.

Except for hydrodynamic effect, distribution of seagrass was also affected by the following factors: (1) Sediments, in the area close to the bank, were sand and frequently affected by hydrodynamics, on which seagrass cannot grow flourish. (2) Human activities, such as aquaculture, digging clam worm and ship sailing, also affected seagrass growth in shallow waters; (3)Transparency was also an important factor for seagrass growth. Clear water distributed in the northeast of Xincun Bay, which provided the suitable condition for large area continuous seagrass distributed in the area.

Human activities, such as construction of shrimp ponds, aquaculture, fishing with standing net, clam digging, boat sailing, capturing prawns and fishes with blasting and trawling, affected seagrass growth in shallow waters. The area dedicated to shrimp ponds increased greatly in recent years, which had great negative effects on seagrass distribution.

6. Acknowledgment

The National Basic Research Program of China (973 Program) under grant No.2010CB951203; the National Natural Sciences Foundation of China under grant No. 41176161 and No. 40876092; the National Natural Sciences Foundation of Guangdong Province under grant No.8351030101000002.

7. References

Andrefouet, S., & Guzman, H. M. (2005). Coral reef distribution, status and geomorphology-biodiversity relationship in Kuna Yala (San Blas) archipelago, Caribbean Panama. *Coral Reefs*, Vol.24, No.1, pp. 31-42, ISSN. 0722-4028.

Batish, S. (2002). Demonstration of the Impact of Land Use on the Benthic Habitats of the U. S.Virgin Islands with LandSat Satellite Imagery. *Littoral 2002, The Changing Coast. EUROCOAST / EUCC, Porto – Portugal Ed. EUROCOAST – Portugal, ISBN 972-8558-09-0,103.*

Becker, B. L., Lusch, D. P., & Qi, J. G. (2005). Identifying optimal spectral bands from in situ measurements of Great Lakes coastal wetlands using second-derivative analysis. *Remote Sensing of Environment*, Vol.97, No.2, pp. 238-248, ISSN. 0034-4257.

Cannizzaro, J. P., & Carder, K. L. (2006). Estimating chlorophyll a concentrations from remote-sensing reflectance in optically shallow waters. *Remote Sensing of Environment*, Vol.101, No.1, pp. 13-24, ISSN. 0034-4257.

Chami, M., Mckee, D., Leymarie, E., & Khomenko, G. (2006). Influence of the angular shape of the volume-scattering function and multiple scattering on remote sensing reflectance. *Applied Optics*, Vol.45, No.36, pp. 9210-9220, ISSN. 1539-4522.

Conger, C. L., Hochberg, E. J., Fletcher, C. H., & Atkinson, M. J. (2006). Decorrelating remote sensing color bands from bathymetry in optically shallow waters. *Ieee Transactions on Geoscience and Remote Sensing*, Vol.44, No.6, pp. 1655-1660, ISSN. 0196-2892.

Cummings, M. E., & Zimmerman, R. C. (2003). Light harvesting and the package effect in the seagrasses Thalassia testudinum Banks ex Konig and Zostera marina L.: optical constraints on photoacclimation. *Aquatic Botany*, Vol.75, No.3, pp. 261-274, ISSN. 0304-3770.

Dahdouh-Guebas, F., Coppejans, E., & Van Speybroeck, D. (1999). Remote sensing and zonation of seagrasses and algae along the Kenyan coast. *Hydrobiologia*, Vol.400, pp. 63-73, ISSN. 0018-8158.

Dekker, A. G., Malthus, T. J., Wijnen, M. M., & Seyhan, E. (1992). Remote-Sensing as a Tool for Assessing Water-Quality in Loosdrecht Lakes. *Hydrobiologia*, Vol.233, No.1-3, pp. 137-159, ISSN. 0018-8158.

Enriquez, S. (2005). Light absorption efficiency and the package effect in the leaves of the seagrass Thalassia testudinum. *Marine Ecology Progress Series*, Vol.289, pp. 141-150, ISSN. 0722-4028.

Froidefond, J. M., & Ouillon, S. (2005). Introducing a mini-catamaran to perform reflectance measurements above and below the water surface. *Optics Express*, Vol.13, No.3, pp. 926-936, ISSN. 1094-4087.

Goodman, J. A., & Ustin, S. L. (2007). Classification of benthic composition in a coral reef environment using spectral unmixing. *Journal of Applied Remote Sensing*, Vol.1, pp. -, ISSN. 1931-3195.

Gordon, H. R., & Morel, A. Y. (1983). Remote assessment of ocean color for interpretation of satellite visible imagery: A review. ISBN 3-540-90923-0.

Hedley, J., Roelfsema, C., & Phinn, S. R. (2009). Efficient radiative transfer model inversion for remote sensing applications. *Remote Sensing of Environment*, Vol.113, No.11, pp. 2527-2532, ISSN. 0034-4257.

Henyey, L. G., & Greenstein, J. L. (1941). Diffuse radiation in the galaxy. *The Astrophysical Journal*, Vol.93, pp. 70-83, ISSN. 0004-637X.

Herbert, D. A., Perry, W. B., Cosby, B. J., & Fourqurean, J. W. (2011). Projected Reorganization of Florida Bay Seagrass Communities in Response to the Increased Freshwater Inflow of Everglades Restoration. *Estuaries and Coasts*, Vol.34, No.5, pp. 973-992, ISSN. 1559-2723.

Holden, H., & Ledrew, E. (2001). Effects of the water column on hyperspectral reflectance of submerged coral reef features. *Bulletin of Marine Science*, Vol.69, No.2, pp. 685-699, ISSN. 0007-4977.

Jacquemoud, S., Ustin, S. L., Verdebout, J., Schmuck, G., Andreoli, G., & Hosgood, B. (1996). Estimating leaf biochemistry using the PROSPECT leaf optical properties model. *Remote Sensing of Environment*, Vol.56, No.3, pp. 194-202, ISSN. 0034-4257.

Lennon, P. J. (1989). Seagrass mapping using Landsat TM data. Avilable from *http://www.gisdevelopment.net. (accessed August 15, 2006).*

Liang, S. L. (2007). Recent developments in estimating land surface biogeophysical variables from optical remote sensing. *Progress in Physical Geography*, Vol.31, No.5, pp. 501-516, ISSN. 0309-1333.

Maritorena, S., Morel, A., & Gentili, B. (1994). Diffuse-Reflectance of Oceanic Shallow Waters - Influence of Water Depth and Bottom Albedo. *Limnology and Oceanography*, Vol.39, No.7, pp. 1689-1703, ISSN. 0024-3590.

Mobley, C. D. (1994). Light and Water: Radiative Transfer in Natural Waters. *Academic, New York*, ISBN-10: 0125027508 |.

Mobley, C. D. (1999). Estimation of the remote-sensing reflectance from above-surface measurements. *Applied Optics*, Vol.38, No.36, pp. 7442-7455, ISSN. 0003-6935.

Morel, A., & Prieur, L. (1977). Analysis of Variations in Ocean Color. *Limnology and Oceanography*, Vol.22, No.4, pp. 709-722, ISSN. 0024-3590.

Morel, A., & Gentili, B. (1991). Diffuse Reflectance of Oceanic Waters - Its Dependence on Sun Angle as Influenced by the Molecular-Scattering Contribution. *Applied Optics*, Vol.30, No.30, pp. 4427-4438, ISSN. 0003-6935.

Morel, A., & Gentili, B. (1993). Diffuse-Reflectance of Oceanic Waters .2. Bidirectional Aspects. *Applied Optics*, Vol.32, No.33, pp. 6864-6879, ISSN. 0740-3224.

Morel, A., & Gentili, B. (1996). Diffuse reflectance of oceanic waters .3. Implication of bidirectionality for the remote-sensing problem. *Applied Optics*, Vol.35, No.24, pp. 4850-4862, ISSN. 0003-6935.

Nichols, T. D., & Kyrala, G. A. (1992). Measurement of Irradiance Distribution at the Focus of a High-Irradiance Laser-Beam. *Optical Engineering*, Vol.31, No.12, pp. 2647-2656, ISSN. 0091-3286.

Orth, R. J., & Moore, K. A. (1983). Chesapeake Bay - an Unprecedented Decline in Submerged Aquatic Vegetation. *Science*, Vol.222, No.4619, pp. 51-53, ISN. 0036-8075.

Pasqualini, V., Pergent-Martini, C., Pergent, G., Agreil, M., Skoufas, G., Sourbes, L., & Tsirika, A. (2005). Use of SPOT 5 for mapping seagrasses: An application to Posidonia oceanica. *Remote Sensing of Environment*, Vol.94, No.1, pp. 39-45, ISSN. 0034-4257.

Peterson, B. J., & Fourqurean, J. W. (2001). Large-scale patterns in seagrass (Thalassia testudinum) demographics in south Florida. *Limnology and Oceanography*, Vol.46, No.5, pp. 1077-1090, ISSN. 0024-3590.

Phinn, S. R., Dekker, A. G., Brando, V. E., & Roelfsema, C. M. (2005). Mapping water quality and substrate cover in optically complex coastal and reef waters: an integrated approach. *Marine Pollution Bulletin*, Vol.51, No.1-4, pp. 459-469, ISSN. 0025-326X.

Phinn, S., Roelfsema, C., Dekker, A., Brando, V., & Anstee, J. (2008). Mapping seagrass species, cover and biomass in shallow waters: An assessment of satellite multi-spectral and airborne hyper-spectral imaging systems in Moreton Bay (Australia). *Remote Sensing of Environment*, Vol.112, No.8, pp. 3413-3425, ISSN. 0034-4257.

Sathe, P. V., & Sathyendranath, S. (1992). A Fortran-77 Program for Monte-Carlo Simulation of Upwelling Light from the Sea. *Computers & Geosciences*, Vol.18, No.5, pp. 487-507, ISSN. 0098-3004.

Yang, C. Y., Yang, D. T., Cao, W. X., Zhao, J., Wang, G. F., Sun, Z. H., Xu, Z. T., & Kumar, M. S. R. (2010). Analysis of seagrass reflectivity by using a water column correction algorithm. *International Journal of Remote Sensing*, Vol.31, No.17-18, pp. 4595-4608, ISSN. 0143-1161.

Yang, D. T. (2008). Variation of seagrass distribution in Sanya Bay impacted by land use change. *Geoinformatics 2008 and Joint Conference on GIS and Built Environment: Monitoring and Assessment of Natural Resources and Environments*, Vol.7145, pp. 714529-714529-714528

Yang, D. T., & Huang, D. J. (2011a). Impacts of Typhoons Tianying and Dawei on seagrass distribution in Xincun Bay, Hainan Province, China. *Acta Oceanologica Sinica*, Vol.30, No.1, pp. 32-39, ISSN. 0253-505X.

Yang, D. T., Y. Yang, C. Yang, J. Zhao & Z. Sun. (2011b). Detection of seagrass in optical shallow water with Quickbird in Xincun Bay, Hainan province, China. *IET Image Processing*, Vol.5, No.5,pp.363-368. ISSN.1751-9667 .

Yang, D. T., & Yang, C. Y. (2009). Detection of Seagrass Distribution Changes from 1991 to 2006 in Xincun Bay, Hainan, with Satellite Remote Sensing. *Sensors*, Vol.9, No.2, pp. 830-844, ISSN. 1424-8220.

Yang, Z. D. (1979). The geographical distribution of seagrass. *Bull. Limnol. Oceanogr*, Vol.2, pp. 41-46,

Part 2

Earth Monitoring

5

The Use of Remote Sensed Data and GIS to Produce a Digital Geomorphological Map of a Test Area in Central Italy

Laura Melelli[1], Lucilia Gregori[1] and Luisa Mancinelli[2]
[1]*Department of Earth Sciences, University of Perugia, Perugia*
[2]*Geologist, Perugia*
Italy

1. Introduction

Thematic maps in the Earth Sciences are an essential tool for the representation, analysis and visualization of geological processes. Among the large variety of thematic maps, geomorphological maps are particularly useful in understanding natural phenomena associated with human activities (Dramis & Bisci, 1998 and references within).

Geomorphological maps report the erosion and depositional relief landforms, including submarine ones, highlighting the morphographic and morphometric characters and interpreting the endogenous and exogenous morphological processes, both past or present, that produce and shape the topographic relief. In this kind of maps, the chronological sequence is also reported, distinguishing between active and inactive landforms. The geomorphological mapping, in addition to its scientific value, is the necessary starting point of different studies such applied geology and environmental protection investigations for socio-economic improvement.

A major problem with geomorphological information is that it is extremely complex to be represented due to the huge amount of data.

In particular, the reproduced information can be summarized as follows:

- Topographic, hydrographical and morphometric data;
- Lithological and structural data;
- Morphogenetic processes:
- Structural and volcanic landforms,
- Mass wasting landforms,
- Karst landforms,
- Eolic landforms,
- Glacial and nival landforms,
- Marine (emerged and submerged), lagoon and lacustrine landforms,
- Large relict and flattened areas with minor forms of complex origin associated,
- Weathering landforms,
- Anthropic landforms.

- Morpho-chronologic data;
- Morpho-evolutive data.

Often the result is an analogical map that is not easily readable, both for the large amount of information, or for the great number of symbols associated with the different landforms.

In order to adapt this kind of data to a digital file, the original map must be converted in a vector format (points, poly-lines and polygons) using a Geographical Information System or GIS software (Bocco et al., 2001; Gustavsson et al., 2006; Vitek et al., 1996).

The use of the rich symbolism available in most GIS software, improve the graphic rendering, but does not solve the problem of readability of the map.

Images acquired by remote sensing and image analysis techniques can bring a significant contribution in improving the geomorphological mapping.

The main results of this approach are:

- a static and dynamic visualization (3D visualization) of Digital Elevation Models (DEMs) derived from satellite data. These techniques allow a better view of shapes and morphogenetic processes represented in the map.
- the calculation of primary and secondary topographic attributes (slope, aspect, planar and radial curvature, roughness) closely related to the presence of some morphogenetic processes and their level of activity. The selection of meaningful ranges of attribute values enable to identify the geometry of the landforms.
- an analysis of multispectral images, with various combinations of RGB bands to highlight some specific morphogenetic processes (such as landslide prone areas).

In this paper the geomorphological map of the Subasio Mountain Regional Park (Umbria region, central Italy) is presented. The map is the result of the interaction of different datasets, both traditional and innovative in geomorphology. Aerial photos and field survey are enhanced by DEMs and satellite images to achieve a digital final product that is not only a simple thematic map, but also an interactive and upgradable Geographical Database. The geomorphological processes producing the present landscape are therefore better visible and understandable through the use of new tools: hillshade layer in transparency under different thematic maps and 3D virtual flight on the area where the map is overlaid to satellite images in a new, prospective view.

2. GIS, DEMs and remotely sensed data in computer cartography: An overview

Automatic mapping techniques are currently supported by tools with a high potential in the field of graphic representation of data such as GIS and by the use of remotely sensed data. Thematic maps produced with these methods show clear advantages, although some limits in the restitution of certain themes, in particular the geomorphological symbology, are evident.

They represent a digital geo-referenced and updatable database, i.e. a cartographic document with hyperlinks to the obtained results by the manipulation of remotely sensed data. This document can be also exportable to different platforms (handhelds PC, WebGIS) for a wide spectrum of applications.

These types of data have significant advantages over traditional methods because they: i) overlay broad areas in relatively short acquisition times; ii) have a better accuracy and precision of the measured data relative to traditional techniques; iii) are in a digital format and, therefore, are simple to elaborate for both research and application purposes; iv) can be easily updated allowing to examine the same areas at different periods and to evaluate both the possible morphological evolution and the kinetics of investigated processes.

For these reasons, research in the Earth Sciences and in geomorphology is integrating, or in some cases completely replacing, traditional techniques of acquisition of spatial information with these new tools (Schmidt & Andrew, 2005; Yongxin, 2007).

It is worth noting that the use of images and digital data, in addition to the advantages described above, opens the possibility to apply new techniques of analysis of physical variables responsible for morphogenetic processes. This being so, the spatial analysis in GIS and the most common systems of image analysis, represent a new field of Earth Sciences and not only a simple application of the theoretical traditional knowledge (Burrough & McDonnell, 1998). The huge potential offered by modern systems, allowing the simultaneous integration and analysis of a large number of spatial data by a variety of mathematical functions, investigate the spatial connections between variables and reveal new relationships and landscape evolution models (sensu Evans, 1972; Hengl & Reuter, 2009; Pike, 2000).

Two new kinds of data are particular useful for the production of geomorphological maps: DEMs and remotely sensed images.

A Digital Elevation Model (DEM) is the modelling of the Earth's surface or part of it in a digital format. Two types of DEMs exist: Triangulated Irregular Network (TIN) and Grid DEM. A TIN is a complex vector data resulting from the interpolation of a set of irregularly spaced points (Braun and Sambridge, 1997; Peucker et al., 1977; Sambridge et al., 1995; Tucker et al., 2001). A square-grid DEM is a raster data where the topography assessment is modeled in a "*gridded set of points in Cartesian space attributed with elevation values that describe the Earth's ground surface*" (Wilson, in press). Although grid DEMs show several disadvantages due to the regular spatial resolution, occasionally causing the inability to detect some topographic variations, or the impossibility of modelling particular landforms features (such stream meandering), they are used in most studies focusing on terrain analysis in geomorphological, hydrogeological (flood analysis) and environmental applications (Moore et al., 1991). Moreover, the remote sensing techniques produce new data models increasing the quality and spreading of these data. Because of these reasons grid DEMs are nowadays the most widely used in geological models requiring topographic assessment.

DEMs can be produced by different procedures (Nelson et al., 2009; Taramelli et al., 2008; Wilson, in press):

1. Vectorization of existing hard-copy topographic maps. Contour lines and spot height can be digitalized and converted in a vector format to be stored like polylines and points with location and altitude value. This procedure allows to obtain a DEM for each part of the Earth represented on a topographic map, but show several disadvantages. In particular, they are time consuming and the quality of the final product strictly depends on the original map and on the acquisition methods.

2. Ground survey methods with a set of field data (points) collected with Global Position System (GPS) or Electronic Distance Measuring (EDM). Although this method allows to save a large number of input data in areas with a strong topographic complexity, it is time consuming and sometimes very expensive. Therefore, it can be a reasonable choice only for some restricted areas.
3. Remote sensing techniques with passive and active sensors. These procedures permit to obtain data with a very high horizontal resolution and vertical accuracy for large areas. Nowadays remote sensing DEMs are the improving resource in this field research and application.

When, by transparency tools, satellite images or digital orthophotos (geological, geomorphological, land-cover e.g.) are overlaid as several thematic maps to a shaded relief, a composite visualization is achieved. In geomorphology DEMs are commonly used to calculate topographic attributes (Franklin, 1991; Moore et al., 1991; Pike, 1988; Wiebel & Heller, 1991). Among them, primary attributes are morphometric parameters deriving from DEMs, i.e. slope, aspect, plan and profile curvature. The visualization of topographic attributes and their analysis can be a very useful tool to better understand the geomorphological processes acting on a study area.

Remotely sensed imageries have a large improvement in both areal coverage and technical characteristics. Moreover, the selection of the most fitting band combination in RGB (Red, Green, and Blue) allows highlighting the required morphological characteristics and processes and facilitates landforms recognition.

3. The study area: The Subasio Mountain regional park (Umbria, central Italy)

The study area is located in the Umbria region (central Italy). This region is well-known because of its natural heritage and exceptional geological value. Twenty-seven geosites or "*any place where you can define a geological and geomorphological interest for conservation*", (Gray, 2004) are already individuated and studied. Seven regional and one national natural park are present on the territory (Figure 1).

The entire region shows a strong correlation between geological attributes and the relief energy associated with topography assessment.

The Subasio Mountain regional park covers an area of 7,200 hectares. The area has a triangular shape, is bordered to the south by the Subasio massif, a rolled and asymmetric anticline. To the west the limit follows the Tescio River. Towards NE the central part and the apex are crossed by close river networks.

In the study area outcropping lithotypes can be clustered in three main complexes and in different types of superficial deposits (Figure 2).

The first and youngest is the Fluvial Lacustrine Complex (Holocene - Pliocene) with pebbles, sand and clay sediments arranged in deposits that are heterogeneous for thickness, shape and areal extent. This complex is associated with the lowest slope values and plain areas.

The second complex is the Terrigenous one (Miocene), consisting of alternating layers of sandstone or limestone with clay or marl. According to the percentage of clay and the dip

Fig. 1. Location map of the Umbria Region (central Italy). The white circle marks the Subasio Regional Park. (1) Geosites, (2) Regional Parks.

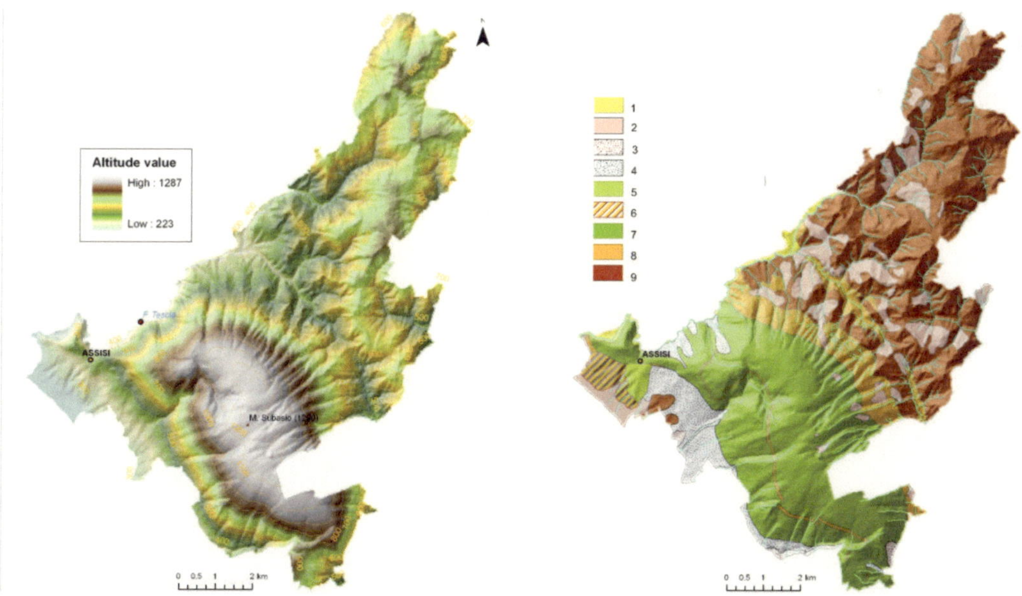

Fig. 2. Left: DEM of Subasio Mountain Regional Park with altitude values in meters a.s.l. Right: geological map. (1) Alluvial deposits, (2) Colluvial deposits; (3) Debris deposits (active); (4) Debris deposits (ancient); (5) Fluvial **L**acustrine complex; (6) Travertine; (7) Calcareous complex; (8) Terrigenous complex with prevalent clay percentage; (9) Terrigenous complex with prevalent arenaceous percentage.

direction of the layers, the energy relief shows medium values. Mass wasting processes prevail together with fluvial erosion landforms on a rolling hill landscape (Figure 3).

The oldest complex is the Calcareous one (upper Trias - Oligocene) corresponding to the mountain areas of the region and to the highest values of energy relief and altitude. The Calcareous Complex consists of a thick multilayer sequence where limestone prevails and karstic features and debris deposition at the base of the slopes are the most frequent geomorphological morphotypes (Figure 4).

Fig. 3. The Terrigenous Complex view, photographed from the top of Subasio Mountain, northwards (photo by L. Mancinelli).

Fig. 4. The Calcareous Complex on the top of the Subasio Mountain with a macro-doline in the foreground (photo by L. Mancinelli).

The geologic history of the area is tightly related with geological evolution of central Italy. From a tectonic point of view the area is the result of two different tectonic periods. In the Miocene a compressive phase originated anticlines and synclines (like the Subasio Mountain) followed, since Pliocene, by uplift with an extensional tectonic phase affecting the entire area. Because of this, a sharp increase of energy relief has forced the entrenchment of the stream network resulting in headward and stream erosion and with the simultaneous triggering of landslides along the slopes (Malinverno & Ryan, 1986, Mayer et al., 2003).

The strong heterogeneity of the substrate is responsible for the great variety of relief and geomorphological processes acting on the area. Hence, the Subasio M. Park is a perfect test-area to assess a method focusing the geomorphologic map editing.

4. The interactive geomorphologic map: A qualitative and quantitative approach in a GIS environment

The essential steps required to elaborate the final digital geomorphological map are summarized in Figure 5.

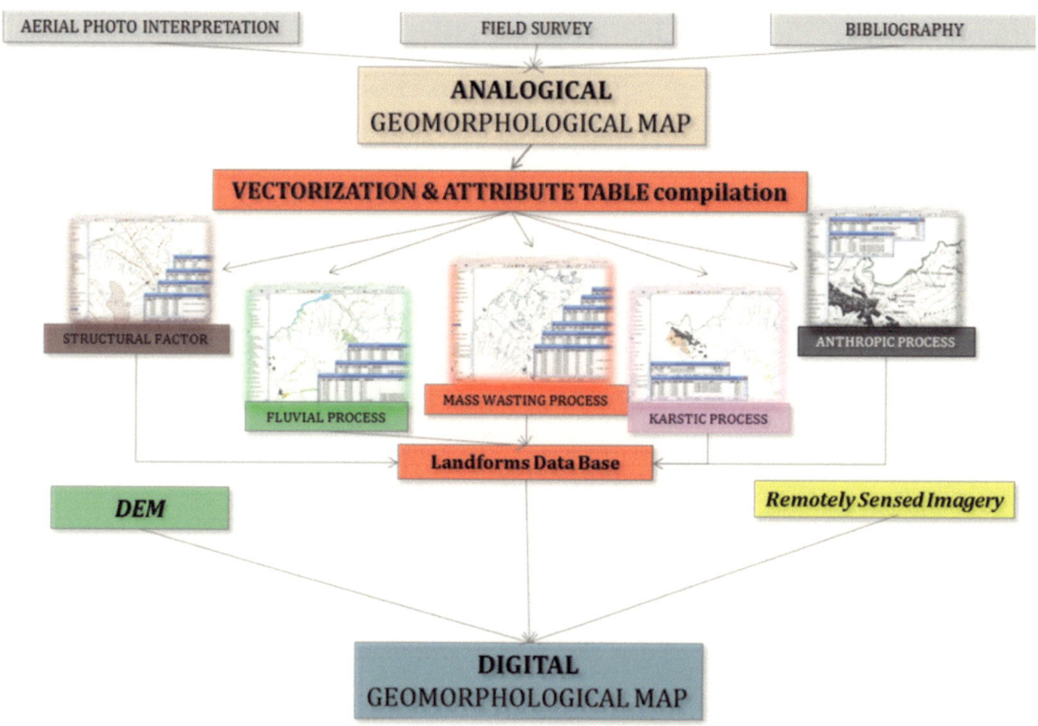

Fig. 5. Flow chart showing the steps required to produce the final digital geomorphological map starting from the analogical data.

The geomorphological map was produced to a medium scale 1:25000, with ESRI's ArcGIS 9.3 (© ESRI) with an equivalent project scale. The Spatial Reference is ED50 (European Datum) UTM (Universal Transverse Mercator) Zone 33N. The project extent is set on the mask corresponding to the polygon shape of the park boundaries. In the GIS project the background is prepared with a topographic raster image in the TIFF format (Sheet N. 123 of the Topographic Map of Italy) and the river network in a vector format (DWG). The drainage pattern is separated from the other topographic data to highlight the relationships with fluvial landforms.

The traditional working techniques are the first step. Therefore, field survey, aerial photo interpretation and collection of scientific papers focused on the study area are required. An analogical geomorphological map is the intermediate result. The map, scanned and rasterized with a high accuracy, is imported in the GIS project and then georeferenced.

The following stage, the vectorization of each single group of landforms, is particularly important. The symbology associated with a geomorphological map is complex. Thus, it is not always possible to draw symbols identical to those proposed in the traditional and official legends. The "Legend for the Geomorphogical Map of Italy" at a scale of 1:50000 is used as a reference (GLCG, 1994).

Thirty-eight vector layers are compiled. Each layer includes a variable number of landforms. Table 1 summarizes the layers and the relative information.

N.	Layer Landforms	Geomorphologic process	Shape feature	N.	Layer Landforms	Geomorphologic process	Shape feature
1	Park boundary	Topographic	Polygon	20	Fluvial lacustrine deposit	Fluvial	Polygon
2	Eluvial Colluvial deposits	Superficial deposit	Polygon	21	Fluvial scarp	Fluvial	Polyline
3	Terrigenous Complex 1	Bedrock	Polygon	22	Sheet erosion	Fluvial	Polygon
4	Terrigenous Complex 2	Bedrock	Polygon	23	Gully erosion	Fluvial	Polygon
5	Calcareous Complex	Bedrock	Polyline	24	Gully erosion	Fluvial	Polyline
6	Fault	Structural Factors	Polyline	25	Badlands	Fluvial	Polygon
7	Fractures and joints line	Structural Factors	Polyline	26	Elbow river capture	Fluvial	Polyline
8	Ridge	Structural Factors	Polyline	27	Gorge	Fluvial	Polyline
9	Peaks	Structural Factors	Point	28	Debris deposit (actual)	Mass wasting	Polygon
10	Sadde	Structural Factors	Point	29	Debris deposit (ancient)	Mass wasting	Polygon
11	Slope asymmetry	Structural Factors	Point	30	Gravitational scarp	Mass wasting	Polyline
12	Structural Scarp	Structural Factors	Polyline	31	Landslide, fall	Mass wasting	Polygon
13	Flatiron	Structural Factors	Polyline	32	Landslide, slide	Mass wasting	Polygon
14	Esplanade area	Structural Factors	Polygon	33	Landslide, slump	Mass wasting	Polygon
15	Triangular facet	Structural Factors	Polygon	34	Landslide, flow	Mass wasting	Polygon
16	River	Fluvial	Polyline	35	Travertine	Karstic	Polygon
17	Valley	Fluvial	Point	36	Doline	Karstic	Polygon
18	Alluvial deposit	Fluvial	Polygon	37	Anthropic scarp	Anthropic	Polyline
19	Alluvial fan	Fluvial	Polygon	38	Quarry	Anthropic	Point

Table 1. Layers of shapefile corresponding to geologic bedrock complexes, superficial deposits and geomorphologic features vectorized in the project.

For each landform a unique code for the graphic properties of the layer is individuated.

As an example, Table 2 reports some of the used codes.

Layer - landforms	Type	FONT	UNICODE
Layers dip direction	Character Marker Symbol	ESRI Geology AGSO 1	162
Peaks	Character Marker Symbol	ESRI Transportation & Civic	114
Slope asymmetry	Character Marker Symbol	ESRI Geology	196
Saddle	Character Marker Symbol	ESERI Cartography	164
Valley	Character Marker Symbol	ESRI Geology USGS	56
"V" shaped valley	Character Marker Symbol	Lucida Sans	86
Valley with a flat bottom	Character Marker Symbol	ESRI Geology USGS	200
Quarry (active)	Character Marker Symbol	ESRI Geometric Symbols	199
Quarry (inactive)	Character Marker Symbol	ESRI Geometric Symbols	198
Gully erosion	Character Marker Symbol	ESRI Geology AGSO 1	193
Sheet erosion	Character Marker Symbol	ESRI Geology AGSO 1	114

Table 2. Some examples of codes used for drawing the symbols in the final map, according to the features proposed in the official Italian Geomorphological Legend.

In the Attribute Tables several information are stored for each layer. In particular, for the different lithotypes and superficial deposits the following fields are included: i) a brief description of the lithology, ii) its age and iii) thickness, and iv) a link to a photo of a significant outcrop. For each landform the data included in the attribute table are: i) the main geomorphologic process responsible for landform creation, ii) the state of activity, and iii) the area and the perimeter. A link to a photo, together with a description of the most significant characteristics of the landform, are included.

5. Remotely sensed data as a support for the map creation

Several digital and analogical sources of data can be used to produce thematic maps both in the stage before the preparation of the map and in the successive stages.

Aerial photo interpretation is a well-established working tool in Earth Science research; DEMs and satellite images, on the contrary, are considered as new tools with an enormous potential, not yet fully explored. In the following paragraphs the different data are described according to their use in this work.

5.1 The aerial photo interpretation: A traditional technique for landform detection

The main goal in reading an aerial photo in the Earth Science applications is to identify and understand the physical landforms on the terrestrial surface and, in some cases, underground morphologies. Aerial photos can be in an analogical or in a digital format. Both of them are acquired by an aerial platform using a camera slipped into a mount located at the bottom of the aircraft. Analogical and digital cameras are quite similar. Analogical images, taken on a photographic film, can be in natural or black and white colours and show the topographic surface as a series of overlapping photos for a large percentage of the detected area. Digital images are taken on a strip with a linear scanner in black and white, colour (RGB format) or infrared. The most important difference is the storage device where the digital camera system uses a charge-coupled device (CCD) that can strongly vary in capacity and resolution, affecting the quality of the images. Both data sets have advantages. Digital images can have a better resolution and filter only few bands of the electromagnetic spectrum allowing the use on specific research fields. In addition they are subjected to editing and post-processing, for example to sharpen the edges of the objects represented on the image. On the contrary analogical films are more nuanced and show a better colour rendering. Moreover, in the analogical data, the images show a much more natural aspect giving the opportunity to better visualize and identify natural features on the surface.

In both cases aerial photos show a "bird's - eye" view of the Earth surface and, unlike the topographic maps that are a selective representation of reality, omitting a large number of natural features, aerial photos provide an objective idea of the arrangement of the spatial pattern.

The limits of this technique are related to the presence of clouds or haze in the atmosphere and snow on the Earth surface covering the topographic pattern. Moreover, distortion effects have to be corrected for an optimal use of the data sets.

Aerial photos are used in a wide group of applications: engineering, logistic and planning, mineral exploration, geoarchaeology, mining and resource extraction, land use and landcover analysis and so on.

In geomorphology air photos interpretation is an irreplaceable tool to detect landforms allowing to identify the type of bedrock and the main morphological processes acting in the study area and the palaeogeographic reconstruction of particular morphological situations (past river captures or the infilling of ancient lacustrine depressions). Some large landforms are more evident on the aerial photos than on the field due to the landform location or the topography arrangement.

Therefore, the aerial photo interpretation is a fundamental method in every geomorphological mapping process.

Moreover, the possibility to observe images taken in different periods of time, and with diverse scales, permits to monitor the landscape evolution (multitemporal and multi-scalar observation). Examples include the evolution of a landslide, the health status of vegetation, the rate of retreat of a cliff, the changes affecting a river drainage network.

The first elements of interpretation in geomorphology are the *size* of the objects identified and their *shape*. Also the spatial arrangement is very important, so site, situation and association are characteristics to be taken into account. *Site* is the relationship of a feature to the environment (elevation, slope, surface cover). *Situation* observes the mutual spatial relationship of the features. *Association* refers to the possibility that, when particular geomorphological processes or landforms are recorded is quite obvious to find associated features. Other important characteristics are diagnostic for geomorphogical interpretation: *tone* or colour is the brightness or the shade of gray or the colour of the detected element and depends on the amount of light that it reflects, constituting a sort of spectral signature of anthropogenic and natural objects in the area. Also, a transition between two different tones is relevant to detect a variation in some physical processes and useful to locate landform limits. *Texture* can be defined as the arrangement of tone or colour structured in a well recognizable pattern and depends strongly on the scale of the photos. When features are too small on an image to be identified, their repetition can be a clear evidence of a specific feature. So, the smoothness (uniform and homogeneous texture) or the roughness (coarse and heterogeneous texture) of an image can identify a particular vegetation cover (e.g. tree as rough, grass as smooth). *Pattern*, or the spatial arrangement of a landform, is the last characteristic used in geomorphology, particularly useful in drainage network recognition (dendritic, rectangular, parallel and so on).

In the study area aerial photo interpretation was one of the first activity carried out, joined with field survey and bibliographical research. In this project analogical photos in black and white were used at a scale of 1:33000 (year 1977) and 1:10000 (year 2004).

The use of black and white in this case is preferred because it allows to better highlight tones and textural variations on the images. At first it is useful to observe photos on a small scale (1:33000) for an overview of the area. Features due to tectonic and structural control like faults, ridge alignments, structural scarps, discontinuity along slopes are best identified in this scale. Also the river drainage pattern, any anomaly along river tracks and large landslide phenomena are well evident at this scale. In the study area these photos highlight the morphological units linked with the different bedrocks. The calcareous anticline of the Subasio Mountain shows distinctive characteristics (high slope values, low rates of drainage density), significantly different from the rest of the area, where the presence of rock types with an high clay abundance, strongly influences the morphological arrangement (i.e. high value of drainage density and medium and low slope values, high index of landslides, fluvial erosion with badlands and fluvial scarps). Photos analysed at a larger scale (1:10000) are more useful for identifying and drawing landforms. The accuracy is detailed enough for mapping the different morphological elements of a landslide (e.g. crown, main and minor scarps, the displaced material, the accumulation and so on). The choice to use two distinct years of acquisition of the images (1997 and 2004) ensure the multitemporal analysis of the area assigning a relative age to some deposits and landforms (active, inactive). The work is divided into a first phase of identification and drawing of landforms directly on aerial photo (Figure 6) and subsequent transposition of vector data in a GIS environment.

5.2 DEMs and satellite images: A new perspective to view the landscape

The resulting geomorphologic map has several advantages. The final document is upgradable and easily editable. The organization of data into layers lets the user to select, for viewing and printing operations, one or more layers simultaneously. The attribute tables associated with the themes contain alphanumeric data in unlimited quantities.

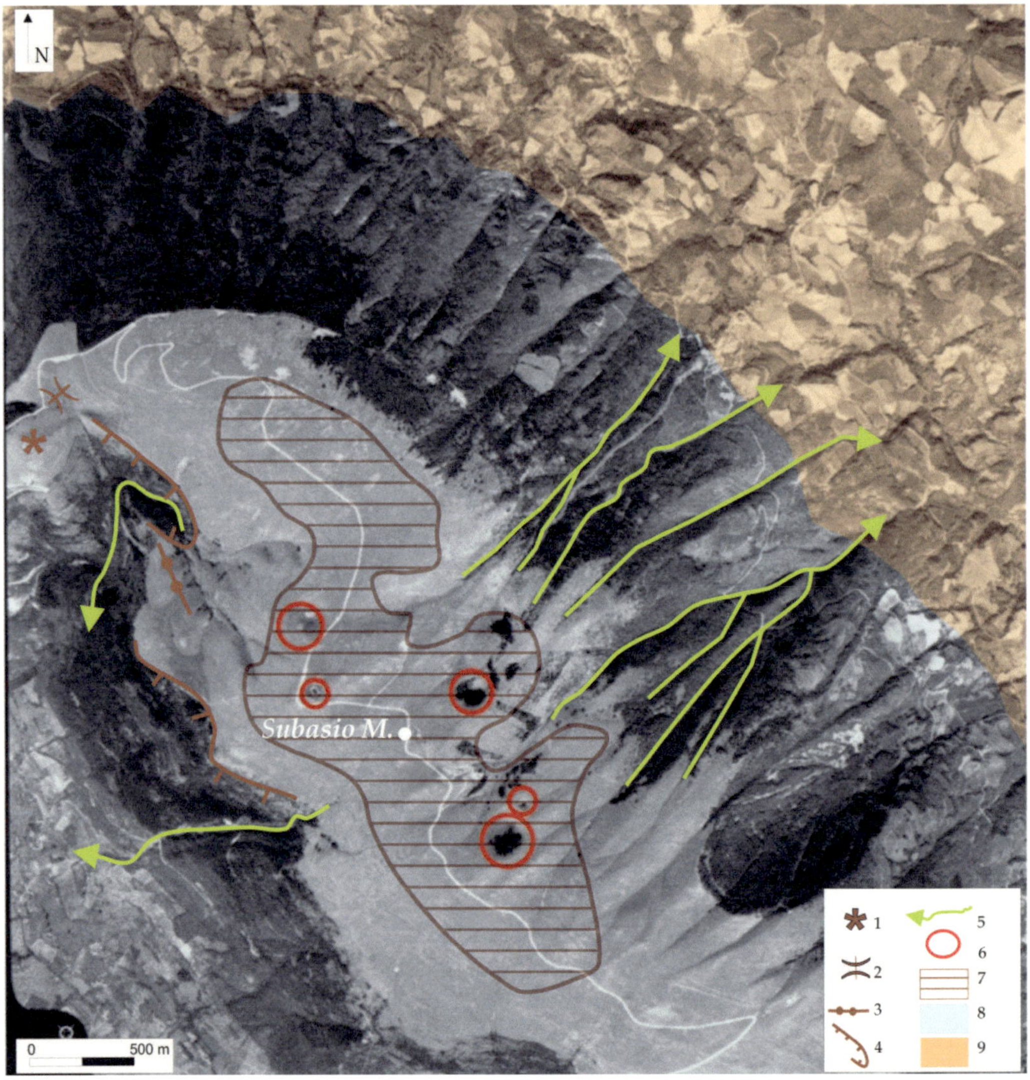

1) Peak, 2) Saddle, 3) Ridge, 4) Scarp, 5) River valley with a "V" shape, 6) Doline, 7) Structural surface, 8) Calcareous Morphological Unit, 9) Marly Morphological Unit.

Fig. 6. Aerial photo of the Subasio Mountain and the surrounding area with some examples of features identified and drawn on the photo (b/w, scale 1:33000, year 1977).

However, at this point of the project, the paper is simply a digital geomorphological map. The subsequent implementation of satellite data is an added value and offers the possibility to obtain additional useful spatial information for different types of applications.

The topographic model used in this project is the Shuttle Radar Topography Mission DEM elaborated for Italy with an horizontal resolution of about 90mx90m (Taramelli & Barbour, 2006).

Several topographic attributes including an hillshade, to better visualize the topographic surface and slope and aspect grids are derived (Figure 7).

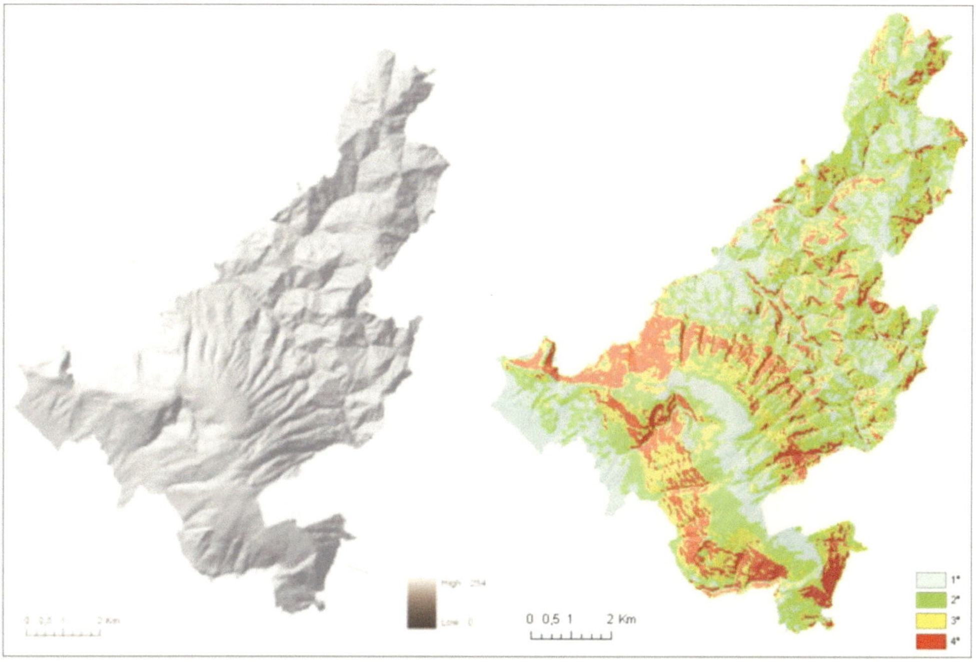

The four slope classes are: 1) 0°-13°, 2) 13°-20°, 3) 20°-27°, 4) 27°-48°.

Fig. 7. Hillshade (on the left) and slope (on the right) grids derived from SRTM DEM.

Geomorphological processes are strictly related to topographic trends and the spatial distribution of the phenomena is always significant.

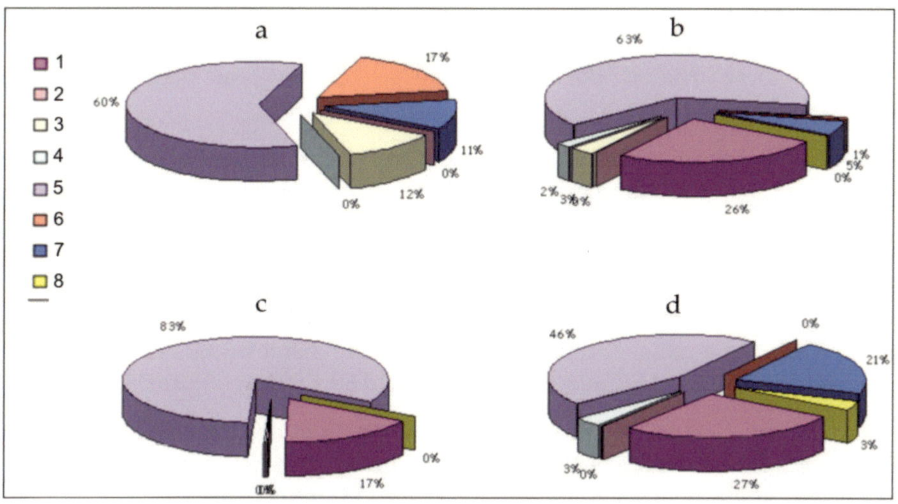

a) Falls, b) Slides, c) Flows, d) Complex landslides. 1) Eluvial and colluvial deposits, 2) Alluvial deposits, 3) Calcareous Complex, 4) Terrigenous Complex (1), 5) Terrigenous Complex (2), 6) Debris (active), 7) Debris (inactive), 8) Fluvial lacustrine deposits.

Fig. 8. Diagrams showing the spatial distribution of landslides on several lithotypes.

Spatial analysis tools can calculate the statistical distribution of the landforms, starting from the topographic grids (Melelli & Taramelli, 2010; Taramelli & Melelli, 2009). In Figure 8 a statistical distribution of the different types of landslide is shown.

To better understand to what extent the topographic parameter influences the spatial distribution of a geomorphological process a quantitative analysis is required. Therefore, the digital map, with the addition of a DEM, becomes an interactive document for further applications.

Remotely sensed data also offer further enhancements to geomorphological mapping and landscape comprehension. A different perspective view of the area, together with the overlapping of different types of data in a 3D view, is an appealing idea for a different use of geomorphological mapping, in particular for a non-specialized audience. Due to the aforementioned difficulties in interpreting the geomorphological symbolism, a backdrop layer resulting from remotely sensed images can aid in the comprehension of the landforms. The perspective view, joined with virtual flights through the area, increase even more the visualization of the landscape. The user can observe any landform in a perspective view and, with a virtual cloche, can fly near and above the feature. So it is possible to intuitively distinguish the main scarp or the convexity on a slope corresponding to the accumulation of a landslide. The transparency tool can make simultaneously visible the alignment of a fault system on the geomorphological map and the corresponding geomorphological features (scarps or triangular facets) on the underlying DEM or satellite image. In the same way a badland drawn on a map is better evident with an image overlaid, where the dense network of valleys engravings on a slope with the absence of vegetation and the grey light colours of the clay bedrock are shown.

The use of remotely sensed images can improve this kind of perception. It is well known that particular RGB arrangements can highlight different natural aspects on the ground: 432 for vegetation, 741 for the moisture content in the soil coverage and so on. So the manipulation of a remotely sensed image under the digital geomorphological map with a 3D perspective view due to the DEM addition, is the best possible analysis of a geomorphological map.

In this example, the Arcscene ESRI Tool was used to obtain a 3D view of the park and a virtual flight on the area. In order to achieve a more realistic view an ASTER image (Advanced Spaceborne Thermal Emission and Reflection Radiometer) is overlapped (Abrams, 1999; Yamaguchi et al., 1998). ASTER is an imaging instrument flying on the Terra satellite (http://asterweb.jpl.nasa.gov/index.asp). The satellite was launched in December 1999 as part of NASA's Earth Observing System (EOS). The data are in 14 bands (from the visible to the thermal infrared wavelengths) and offer high-resolution characteristics. Thanks to the swath width of the sensor, each ASTER image takes an area of 60 x 60 km (Figure 9).

All the data described above can be represented in the final double-sided printing layout of the map showing how interactive this kind of document can be (Figures 10 and 11). Figure 10 represents the first side of the map with the Figure 11 on the back.

In the final layout, the part dedicated to the geomorphogical data and the section for the grids and satellite images have the same importance. So the remote sensing information is added to make the final product in keeping with the rest of the maps, becoming a source of information for the knowledge of the territory.

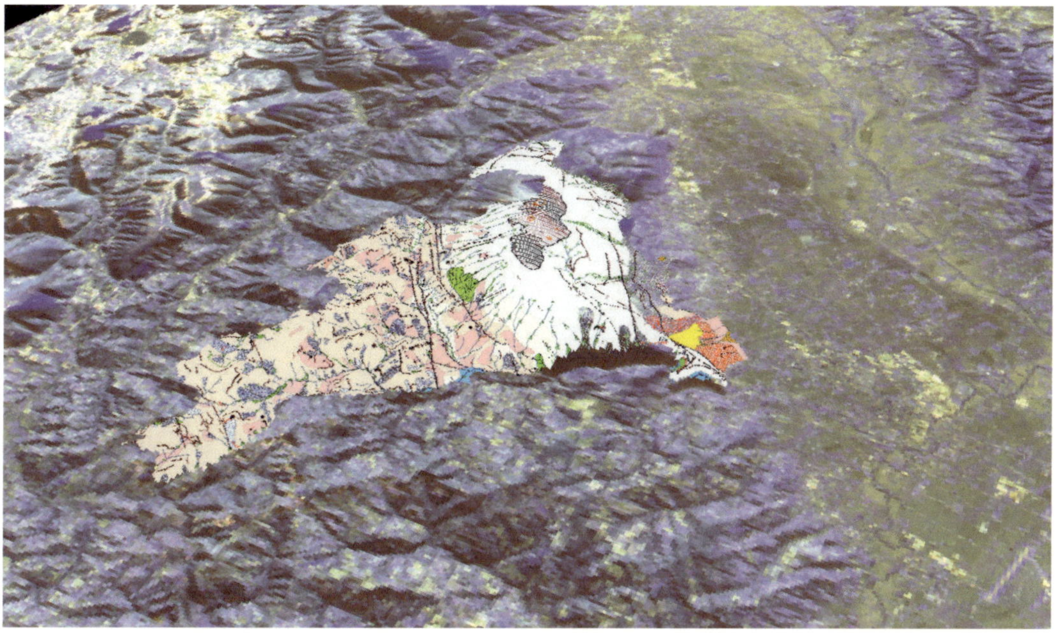

Fig. 9. A still image of the virtual flight on the park with the geomorphogical map overlapping an ASTER image (view from SW). The RGB combination is the 742. The 3D view is assured by the SRTM DEM height values.

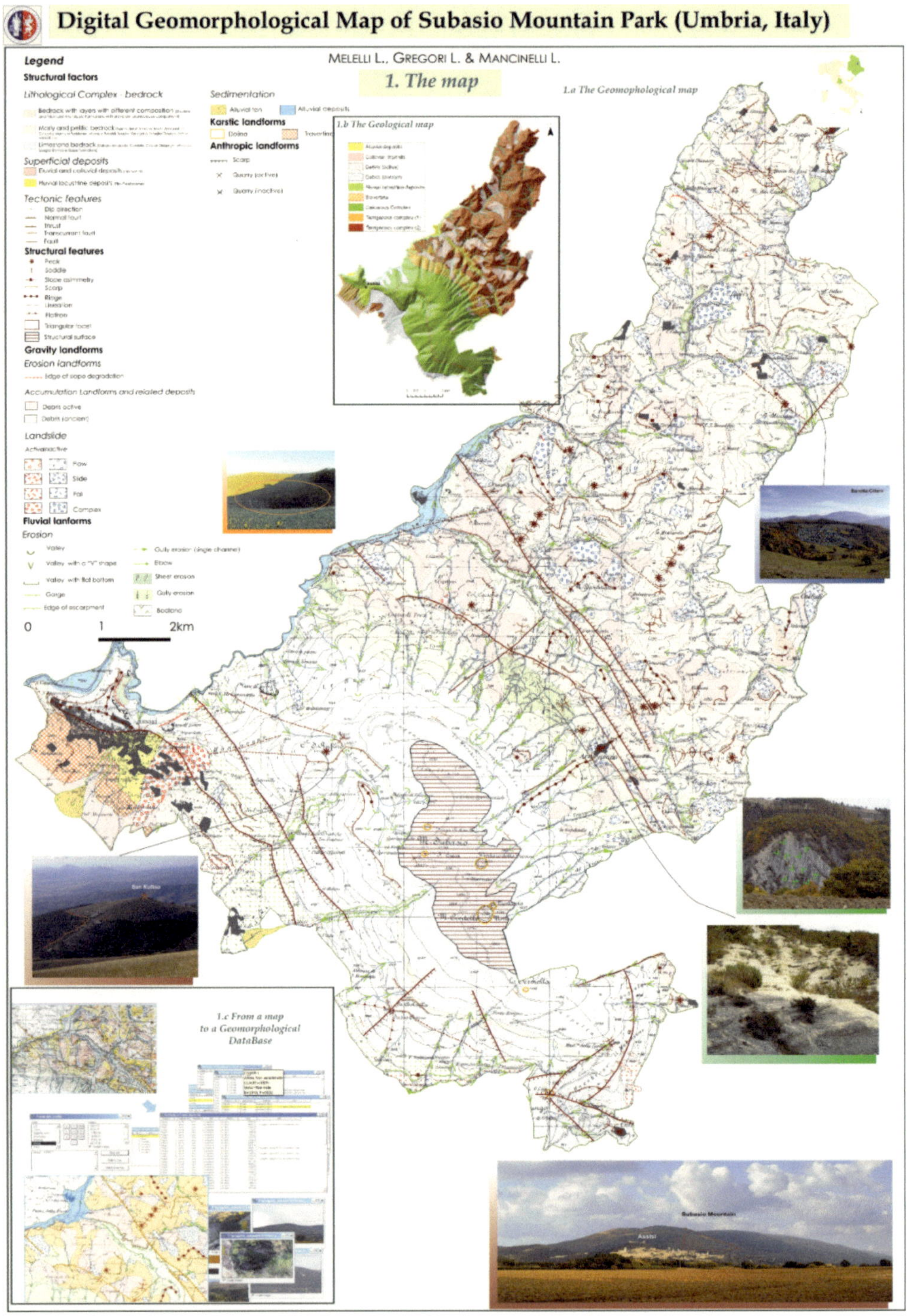

Fig. 10. The final layout of the geomorphological map (front side) with the geomorphological map, a geological sketch, some significant photos and the scheme of the transition from the analogical product to the digital one.

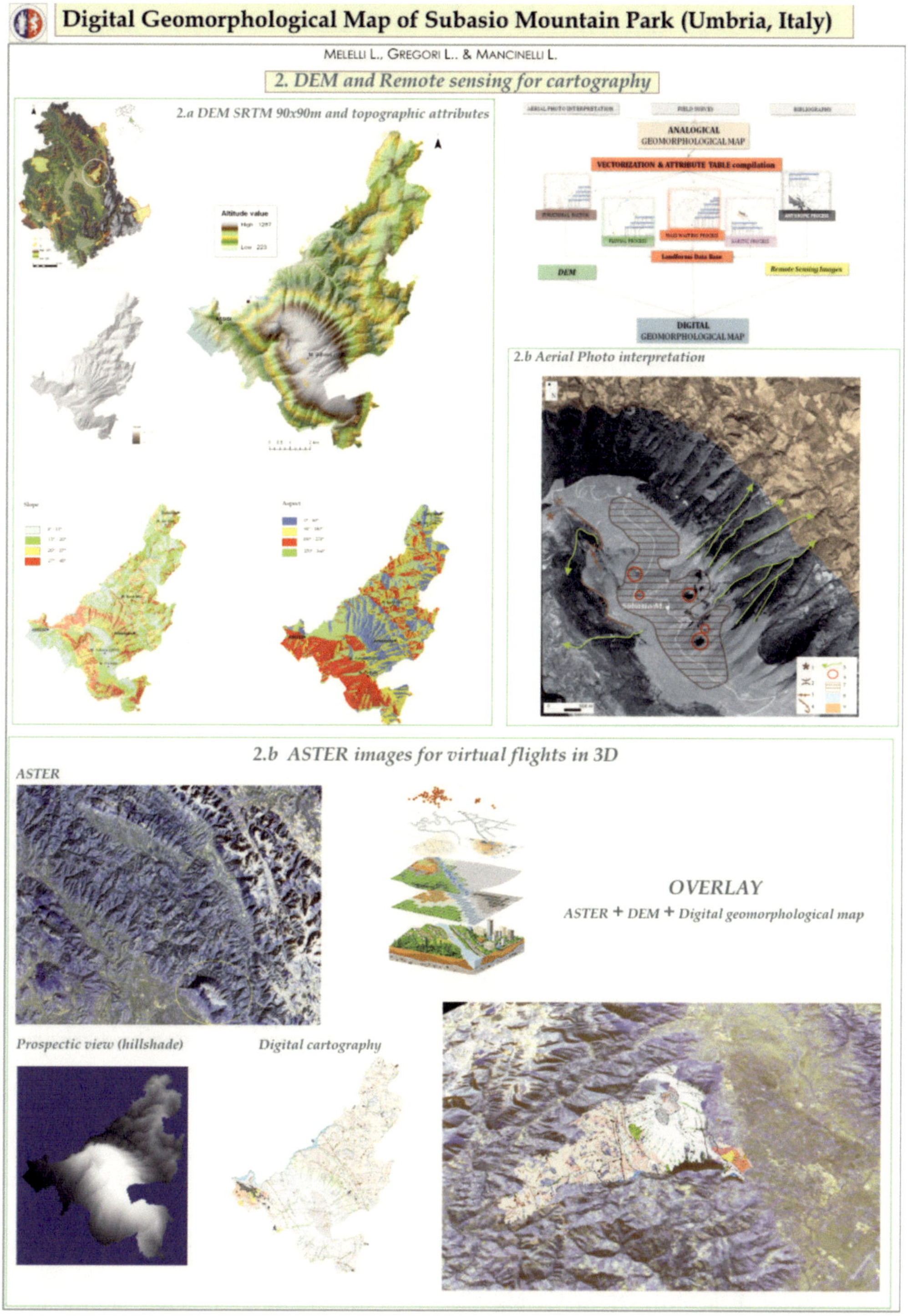

Fig. 11. The final layout of the geomorphological map (back side) including the remotely sensed data.

6. Conclusions

Cartography is experiencing an important change with the introduction of computer systems and digital images (GIS, satellite images). In particular in the Earth Sciences, geomorphological mapping begins to benefit from the digitalization of information.

From a graphical point of view, given the complexity of symbology, geomorphological maps interpretation is often difficult, especially for non-experts.

The potential offered by GIS can solve this problem. In addition, the input of satellite data allows integrating additional information to better understand the mechanisms that regulate the morphogenetic processes.

The remote spatial data acquisition techniques are also moving important steps. Therefore, the availability of data with high accuracy allows having a progressively more accurate information on the topographic attributes evaluation and for 3D observations of landforms.

Statistical distribution of landforms, morphogenetic processes and numerical calculation of quantitative indices (Melelli & Floris, 2011; Serrano & Ruiz-Flaño 2007a,b) benefit significantly from these new techniques. Today is possible to merge the information collected by traditional techniques (aerial photo-interpretation or field survey) with numerical data, obtaining final documents completely different from traditional cartography. The data can be updated, queried and displayed in various ways. They can also, with the help of statistical analysis, offer new research methods to build advanced models for morphogenetic processes of landscape evolution.

7. References

Abrams, M. (1999). The Advanced Spaceborne Thermal Emission and Reflection Radiometer (ASTER): Data Products for the High Spatial Resolution Imager on NASA's EOS-AMI Platform. *International Journal of Remote Sensing*, Sept, 1999.

Bocco, G., Mendoza, M. & Velazquez, A. (2001). Remote sensing and GIS-based regional geomorphological mapping—a tool for land use planning in developing countries. *Geomorphology*, Vol. 39, No. 3-4, pp. 211-219.

Braun, J. & Sambridge, M. (1997). Modelling landscape evolution on geological time scales: a new method based on irregular spatial discretization. *Basin Research*, Vol. 9, pp. 27–52.

Burrough, P.A. & McDonnell, R.A. (1998). *Principles of Geographical Information Systems*, 2nd Edition, Oxford University Press ISBN 0198233663, 0198233655.

Dramis, F. & Bisci, C. (1998). *Cartografia geomorfologica*. Pitagora Ed. ISBN 88-371-0809-5, Bologna.

Evans, I.E. (1972). General geomorphometry, derivatives of altitude, and descriptive statistics. In: *Spatial Analysis in geomorphology.*, Chorley, R.J., (Ed.), pp. 17-90, Methuen, London.

Franklin, S.E. (1991). Satellite remote sensing of mountain geomorphic surfaces. *Canadian Journal of Remote Sensing*, Vol. 17, No. 3, pp. 218-229.

GLCG (Gruppo di lavoro per la Cartografia Geomorfologica - Working Group for the Geomorphological Cartography) (1994). Carta geomorfologica d'Italia - 1:50.000 .

Guida al rilevamento. Servizio Geologico Nazionale. *Quaderni serie III*, Vol. 4, Istituto Poligrafico e Zecca dello Stato, Roma, 1994.

Gray, M. (2004) - *Geodiversity valuing and conserving abiotic nature*. John Wiley & Sons Ltd, Chichester, 1-434.

Gustavsson, M., Kolstrup, E. & Seijmonsbergen A.C. (2006). A new symbol-and-GIS based detailed geomorphological mapping system: Renewal of a scientific discipline for understanding landscape development. *Geomorphology*, Vol. 77, No. 1-2, pp. 90-111.

Lillesand, T.M., Kiefe, R.W. & Chipman, J.W. (2011). Front Matter. In: *Computer Processing of Remotely-Sensed Images: An Introduction*, Fourth Edition, John Wiley & Sons, Ltd, Chichester, UK. doi: 10.1002/9780470666517.fmatter.

Malinverno, A. & Ryan, W.B.F. (1986). Extension in the Tyrrhenian Sea and shortening in the Apennines as result of arc migration driven by sinking of the lithosphere. *Tectonics*, Vol. 5, No. 2, pp. 227-245.

Mayer, L., Menichetti, M., Nesci, O. & Savelli D. (2003). Morphotectonic approach to the drainage analysis in the North Marche region, central Italy. *Quaternary International*, Vol. 101–102, pp. 156–167.

Melelli, L. & Floris, M. (2003). A new Geodiversity Index as a quantitative indicator of abiotic parameters to improve landscape conservation: an Italian case study. *Geophysical research abstracts*, Vol. 13, EGU General Assembly, 2011.

Melelli, L. & Taramelli, A. (2010) – Criteria for the elaboration of susceptibility maps for DGSD phenomena in central Italy. *Geografia Fisica e Dinamica del Quaternario*, Vol. 33, pp. 179-185.

Moore, I.D., Grayson, R.B. & Ladson, A.R. (1991) Digital terrain modeling: a review of hydrological, geomorphological, and biological applications. *Hydrological Processes*, Vol. 5, pp. 3–30.

Nelson, A., Reuter, H.I. & Gessler, P. (2009). DEM Production Methods and Sources. Ch.3 inHenglandReuter, pp.65–85.

Peucker, T.K., Fowler, R.J., Little, J.J. & Mark, D.M. (1977). Digital representation of Three - Dimensional Surfaces by Triangulated Irregular networks (TIN). Tech. Rept. 10, ONR Contract N00014-75-C-0886, Dept. of Geogr., Simon Fraser U., Burnaby, B.C. Canada.

Pike, R.J. (1988). The geometric signature: Quantifying landslide-terrain types from digital elevation models. *Mathematical Geology*, Vol. 20, No. 5, pp. 491-511.

Sambridge, M., Braun, J., & McQueen, H. (1995). Geophysical parameterization and interpolation of irregular data using natural neighbors. *Geophysical Journal International*, Vol. 122, pp. 837–857.

Schmidt, J. & Andrew, R. (2005). Multi-scale landform characterization. *Area*, Vol. 37, pp. 341–350.

Serrano, E. & Ruiz-Flaño, P. (2007a). Geodiversidad: concepto, evaluación y aplicación territorial. El caso de Tiermes Caracena (Soria). *Boletín de la Asociación de Geógrafos Españoles*, Vol. 45, pp. 79–98.

Serrano, E. & Ruiz-Flaño, P. (2007b) - Geodiversity: a theoretical and applied concept. *Geographica Helvetica*, Vol. 62, pp. 140–147.

Taramelli, A. & Barbour, J. (2006). A new DEM of Italy using SRTM data. *Rivista Italiana Telerilevamento*, Vol. 36, pp. 3-15.

Taramelli, A. & Melelli, L. (2009) - Detecting alluvial fans using quantitative roughness characterization and fuzzy logic analysis using the Shuttle Radar Topography Mission data. *Int. J. of Computer Sc. and Software Techonolgy*, Vol. 2, No. 1, January-June 2009, pp. 55-67. © Int. Sc. Press, ISSN:0974-3898.

Taramelli, A., Reichenbach, P. & Ardizzone, F. (2008). Comparison of SRTM elevation data with cartographically derived DEMs in Italy. *Rev. Geogr. Acadêmica*, Vol. 2, No. 8, 41-52.

Tucker, G.E., Lancaster, S.T., Gasparini, N.M., Bras, R.L., & Rybarczyk S.M. (2001). An object-oriented framework for distributed hydrologic and geomorphic modeling using triangulated irregular networks. *Computers & Geosciences*, Vol. 27, pp. 959-973.

Vitek, J.D., Giardino, J.R. & Fitzgerald, J.W. (1996). Mapping geomorphology: A journey from paper maps, through computer mapping to GIS and Virtual Reality. *Geomorphology*, Vol. 16, No. 3, pp. 233-249.

Yamaguchi, Y.; Kahle, A.B., Tsu, H., Kawakami, T., & Pniel, M. (1998). Overview of Advanced Spaceborne Thermal Emission and Reflection Radiometer (ASTER). *Geoscience and Remote Sensing*, IEEE Transactions on. Vol. 36, No. 4, pp. 1062-1071.

Yongxin, D. (2007). New trends in digital terrain analysis: landform definition, representation, and classification. *Progress in Physical Geography*, Vol. 31, No. 4, pp. 405-419.

Weibel, R. & Heller, M. (1991). Digital Terrain Modeling. In: *Geographical Information Systems: Principles and Applications*. Maguire, D.J., Goodchild, M.F. & Rhind, D.W. (Eds.). London, Longman, pp. 269-297.

Wilson, J.P. (2011). Digital terrain modeling. in press.

6

Analysis of Land Cover Classification in Arid Environment: A Comparison Performance of Four Classifiers

M. R. Mustapha, H. S. Lim and M. Z. MatJafri
School of Physics, Universiti Sains Malaysia, USM Penang
Malaysia

1. Introduction

Arid environment is a dry landscape or region that received an extremely low amount of precipitation. Arid areas are located where vegetation cover is sparse to almost nonexistent. Almost one third of earth land surface is arid or desert. Over desert areas, a number of land cover patterns can be observed. One example is given here for the Arabian Peninsula. The located area can be found in Fig. 1. This pattern does not correlate with vegetation; the area is extremely arid with little or no vegetation. In addition, specific land cover is defined as the observed physical layer including natural and planted vegetation and human constructions, which cover the surface of the Earth. Land cover classification is a tool that fills an important informational niche for natural resource managers, decision-makers, and stakeholders. It serves to categorize natural ecosystems, managed crops, and urban areas. As a general form, land cover classifications provide the elemental information to appraise the impact of human interactions within the environment and to assess scientific foundations for sustainability, vulnerability and resilience of land systems and their use

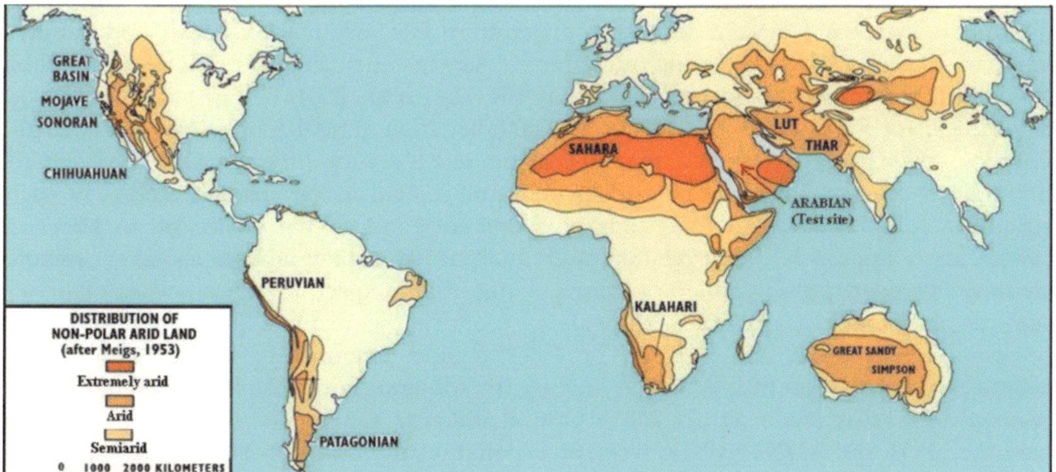

Fig. 1. Location of the arid area in the world

(Han et al., 2004). Land cover is referred to as natural vegetation, water bodies, rock/soil, artificial cover others resulting due to land transformation (Roy and Giriraj, 2008). One difficulty with land cover mapping in arid environment is the spectral similarity of their cover types. This situation leads to misclassification of land cover types.

Many classifiers have been developed, but it is difficult to identify the most appropriate approach to use for features of interest in a given study area. Different results can be attained depending on the classifiers used. In this article, four approaches—minimum-distance classifier (MD), maximum likelihood classifier (ML), artificial neural network (NN), and frequency-based contextual classifier (FBC)–were implemented to classify ALOS AVNIR-2 data in the western Saudi Arabia study area in Mecca city using identical training samples and test data sets. In the literature several studies on the classification methods comparison of multispectral remote sensing data have been reported. Some of them investigated the use of NN or contextual approaches and compared their performances with the ones of classical statistical methods. (Benediktsson et al., 1990; Gong & Howarth, 1992; Stuckens et al., 2000; Seto & Liu, 2003; Erbek et al., 2010).

The test area is composed of a variety of land-cover types, including urban, mountain, land, vegetation, ritual area and shadow. However, the major part of the Mecca province of Saudi Arabia is made up of arid environment, and only a very small portion of the area is covered by vegetation. This article is aimed at investigating the performances of statistical and advanced classification approaches using spectral and ancillary data for land-cover inventorying of a complex area in Mecca city. The different performances of the four classification approaches are evaluated in terms of overall accuracy, performance in heterogeneous area and training samples.

2. Remote sensing

Remote sensing from earth observation satellites is a powerful tool that has been used for monitoring and acquiring rapid information on land earth surfaces. Land cover mapping is one of the core areas in the remote sensing application. Remote sensing can be used to provide up to date spatial information of a wide variety of land cover assessment at multiple resolutions. In recent decades, a major effort has been made to study and monitor land cover using different satellite multispectral sensors such as SPOT, IKONOS, MODIS, QuickBird, Formosat, Landsat and ALOS AVNIR (Han et al., 2004; Wang et al., 2004; Coop et al., 2009; Avelar et al., 2009; Chen et al., 2009; Bagan et al., 2010; Mustapha et al., 2010). The land-cover mapping by using remote-sensing data is a very difficult task when complex urban areas are involved. The main difficulties are related to the characterization of such spectrally complex and heterogeneous environments and to the choice of an effective classification approach. Interpretation and analysis of urban landscapes from remote sensing, however, present unique challenges due to the spectral heterogeneity of urban surfaces and make it extremely difficult to identify the features interest in observed reflectance. Satellite remote sensing provides greater amounts of information on the geographic distribution of land cover, along with advantages of cost and time savings for regional size areas (Yuan et al., 2005). Optical imaging satellite sensor systems such as Landsat, SPOT and ALOS AVNIR, work at a spatial resolution of 10–30 m in multi-spectral bands. Ikonos and Quickbird, the latest sensor systems, provide high to very high spatial resolution data with 2–4 m resolution for the multi-spectral bands. But high or very high-resolution sensors lead to noise in generally homogeneous classes as the data contains increased information in a single pixel. For that reason the authors used the medium resolution of ALOS AVNIR data for preparing this project.

The Advanced Land Observation Satellite (ALOS) has been operating since January 24, 2006. The mission objectives of ALOS are cartography, disaster monitoring, etc. In particular, such geographical information as elevation, topography, land use, and land-cover map is necessary basic information in many practical applications and research areas. To achieve these objectives, ALOS has three mission instruments: two optical instruments, which are Panchromatic Remote-sensing Instrument for Stereo Mapping (PRISM), the Phased Array type L-band Synthetic Aperture Radar (PALSAR) and Advanced Visible and Near- Infrared Radiometer type 2 (AVNIR-2) (Tadono et al., 2009). But we only concerned with AVNIR-2 sensor for this article. AVNIR-2 has four spectral bands with about 10 m of instantaneous field of view (IFOV), 70 km (consists of 7100 pixels) of FOV, and a mechanical pointing function (by moving mirror) along the cross-track direction (+-44°) for effective global land observation (Murakami et al., 2009). One of the purposes for this sensor is to provide land cover and land-use classification maps for monitoring at regional levels. The instrument, however, does not have SWIR capabilities (Wulder et al., 2008). The information pertaining to the sensor can be found in Table 1 while ALOS satellite with their three instruments is given in Fig. 2.

ALOS AVNIR-2 Characteristics	
Orbit	Sun synchronous, descending 10:30
Repeat cycle	46 days
Altitude	691.65km
Inclination	98.16 deg
Cross rack coverage	-44~+44 deg by mirror pointing
FOV	70km
IFOV	10m
Number of band	4 (Blue, Green, Red, NIR bands)

Table 1. ALOS AVNIR-2 characteristics

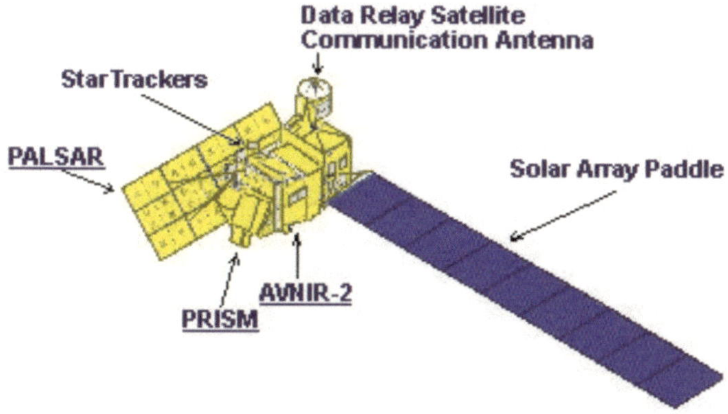

Fig. 2. ALOS satellite with three instruments (Source: Japan Aerospace Exploration Agency)

3. Classification methodology

To begin the processing of raw satellite data, remote sensing images were involved in three stages in order to complete this project. The stages are data pre-processing, image classification and data analysis as shown in Fig. 3.

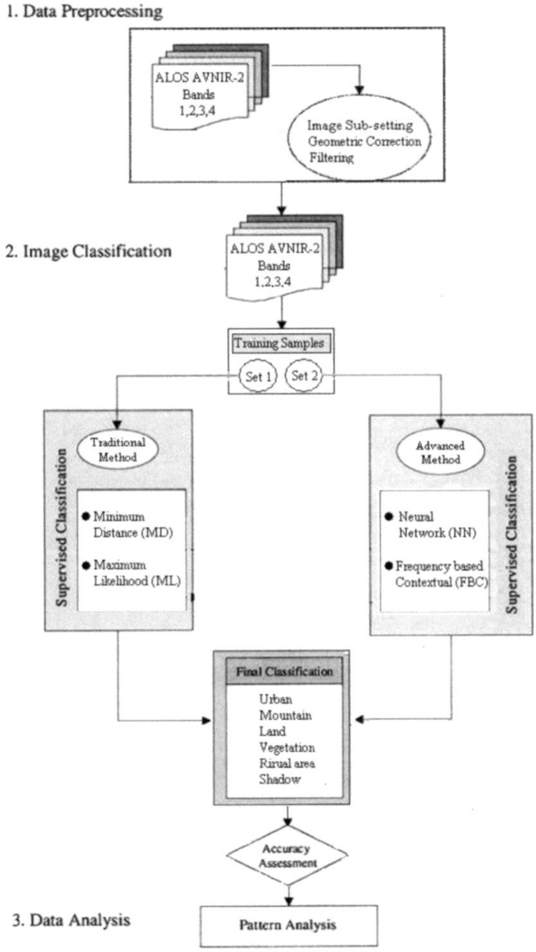

Fig. 3. Classification methodology

3.1 Image preprocessing

The application of raw remote sensing images for spatial analysis requires several pre-processing procedures. These procedures are used in order to subset the images from the original scene, to correct geometric distortion and to remove noise from the image due to error generated by the sensors. In the sub-setting process, the larger images in the original scene have been cut out to a smaller size within the desire area. Meanwhile, geometric correction was done by using second order polynomial coordinate transformation to relate the location of the reference image to the equivalent row and column positions in the ALOS AVNIR-2 images. A total of 23 ground control points were used in this process with 0.45

pixel error was obtained. On the other hand, filtering procedure was used in order to remove or reduce noisy element in the imagery. 7x7 low pass averaging filter was selected as a window to smooth the imagery. The filter applying a mathematical calculation using pixel values under selected window and replacing the central pixel with the new value.

3.2 Supervised image classification

The aim of the image classification process is to categorize all pixels in an image into their respective classes. Basically, there are two ways in order to perform the classification which are supervised and unsupervised classification methods. In a supervised classification, it requires to train a sufficient number of pixels for each class to create a representative signature. Unlike supervised classification, neither prior knowledge nor training sets are required to produce a classification map in the unsupervised or clustering methods. Therefore, the image can be automatically divided into spectrally distinct classes that still need to be interpreted in terms of land cover classes (Han et al., 2004). According (Cihlar et al., 1998), supervised classification methods are more effective in identifying complex land cover classes compared to unsupervised approaches, if detailed a priori knowledge of the study area and good training data exist. Moreover, the classification results are also influenced by a variety of factors, including availability of remotely sensed data, landscape complexity, image band selection, the classification algorithm used, analyst's knowledge about the study area, and analyst's experience with the classifiers used (Lu et al., 2004). For a given study area, selecting a suitable classifier becomes significant in improving the classification results. A comparative study of different classifiers is necessary to understand which classifier is most suitable for a specific landscape. Hence, four classifiers, ranging from simple MD to complex NN, are analyzed in this article. Different classifiers have their own advantages and disadvantages. Selecting a classifier most suitable for the characteristics of the study area can improve classification results.

The concept of image classification is often implemented based on the fact that the spectral signature of each pixel contains information on the physical characteristics of the observed materials underlying the pixel. By analyzing such information from satellite images we can infer the type of materials associated with that pixel. However, the major problem is that spectral non-homogeneity within a particular type of material or land cover makes the classification of land cover difficult (Ju et al., 2005). Taking into account physical characteristics of Mecca city, we chose to classify here the following land cover features: urban, mountain, land, vegetation, ritual area and shadow. Table 2 present the description

Class	Description
Urban	Residential, commercial, building, roadway, infrastructures, concrete and any develop areas.
Mountain	Hill, large rock, rugged terrain
Land	Bare soil, sandy soil, desert, open land
Vegetation	Trees, agriculture area, vegetated area
Ritual area	Grand mosque in Mecca and Mina tent in Mina City.
Shadow	Appearing due to the high mountain or building

Table 2. Detail description of the classes

of each class. Although ritual area can be group under urban class, but the authors decided to separate them into a new class due to the special characteristic of the class, thus, need to be appeared in the classified map. Ritual area which includes a grand mosque and a thousand of tents was a holy area for Muslims. Muslims or pilgrims need to visit to these places as part of their religious event during Hajj Season. Meanwhile, although shadow is not a pure land cover type and mostly appear in the mountainous area, the author also decided to separate them into another class due to their spectrally different against mountain. Hence, to classify the images into those six classes, the statistical minimum distance and maximum likelihood techniques representing traditional method and artificial neural network and contextual representing advanced method were applied. The details pertaining to the four classifiers will be explained in the next section.

3.3 Data analysis

In this section, the results of all classifiers will be presented. All analysis regarding the performance of four classifiers will be discussed in detail in section 8.

4. Traditional method

Of the many classifiers, MD and ML may be the most popular due to their simple theory and availability in almost any image processing or GIS software packages. Both of the classifiers also recognised as statistical method.

4.1 Minimum distance to mean (MD)

MD is a non-parametric classifier that has no assumption of data sets for features of interest. It is computationally simple and fast, only requiring the mean vectors for each band from the training data. Candidate pixels are assigned to the class that is spectrally closer to the sample mean. This method does not consider class variability; thus, large differences in the variance of the classes often lead to misclassification (Lu et al., 2004). The minimum distance algorithm allocates a pixel by its minimum Euclidean distance to the center of each class. The pixel is assigned to the closest class, or marked as unknown if it is farther than a pre-defined distance from any class mean. Though if a pixel lies on the edge of a class, it might be that the value of the pixel is closer to the mean of a neighbor class and it will be assigned to the neighbor class (Avelar et al., 2009).

4.2 Maximum likelihood (ML)

ML is a parametric classifier that assumes normal spectral distribution of data within each class. An equal prior probability among the classes is also assumed. This classifier is based on the probability that a pixel belongs to a particular class. It takes the variability of classes into account by using the covariance matrix; thus, it requires more computation per pixel compare to MD. The ML classifier considers that the geometrical shape of the set of pixels belonging to a class can be described by an ellipsoid. Pixels are grouped according to their position in the influence zone of a class ellipsoid. The probability that a pixel will be a member of each class is evaluated. The pixel is assigned to the class with the highest probability value or left as unknown if the probability value lies below a pre-defined threshold (Avelar et al., 2009).

ML requires the use of training pixels for each class and is therefore dependent on the availability of enough training pixels to produce reasonable estimates of the mean class vector (class spectral signature) and covariance matrix. For each class, training pixels were collated from all images of the same resolution, giving a pooled training sample set (Lim et al., 2009). ML requires sufficient representative spectral training sample data for each class to accurately estimate the mean vector and covariance matrix needed by the classification algorithm. When the training samples are limited, then inaccurate estimation of the mean vector and covariance matrix often results in poor classification results. Traditional pixel-based classification approaches are limited as regards the analysis of heterogeneous landscapes and lead to the reported 'salt and pepper' results (Aplin et al., 1999; Lu and Weng, 2007). Therefore, the ML classifier needs more training data to characterize the classes than the other methods (Pignatti et al., 2009).

5. Advanced method

In recent years, many advanced methods have been applied in remote sensing image classification, each of which has both strengths and limitations. We examined two classification methods, the artificial neural network with back propagation algorithm and contextual classification using frequency based approach, for each of the ALOS AVNIR-2 data sets.

5.1 Neural network (NN)

Artificial neural networks (NN) are computational systems that inspired from biological neurons, so neurons provide the information processing ability (Khan et al., 2010). NNs, like people, learn by example. NN is configured for a specific application, such as pattern recognition or data classification, through a learning process. In the last decade, NN has gained momentum in remote sensing field due to the good results obtained in many applications. NN models have two important properties: the ability to learn from input data and to generalize and predict unseen patterns based on the data source, rather than on any particular a priori model. Although there are a wide range of network types and possible applications in remote sensing, most attention has focused on the use of Multilayer Perceptron (MLP) networks trained with a back-propagation learning algorithm for supervised classification. Fig. 4 demonstrated the basic NN structure.

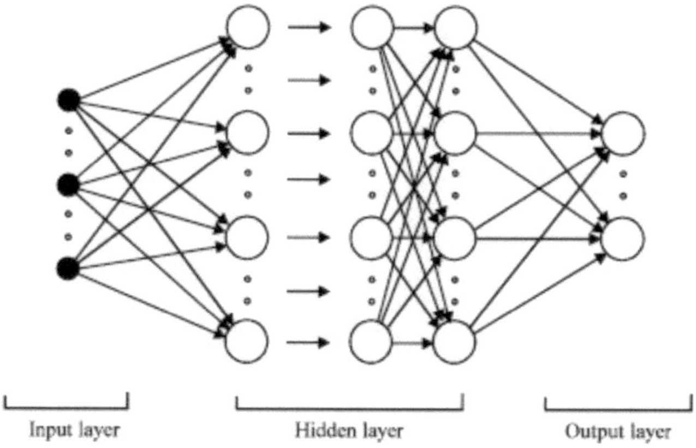

Fig. 4. Basic neural network architecture

Generally, NN require three or more layers of processing nodes: an input layer which accepts the input variables (e.g., satellite image band values) used in the classification procedure, one or more hidden layers which identify internal structure of the input data, and an output layer. The number of nodes (also called processing units or neurons) at the input layer is equal to the dimensionality of the input vector. For the purpose of land cover classification, the number of nodes at the output layer is the same as the number of the classes intended for the classification scheme. In the meantime, the size of the hidden layer can be a crucial question in network design and need to be determined carefully. Nodes between any two consecutive layers are fully connected with connection weights controlling the strength of the connections. The relationship of input - hidden layers and hidden – output layer are given by Equation 1 and 2 (Sarkheil et al., 2009):

$$[j] = f\left(\sum_{i=1}^{n} W_{ij} a_i\right) \tag{1}$$

$$[l] = f\left(\sum_{i=1}^{q} V_{ji} bj\right) \tag{2}$$

where:
a_i is the input node i of the input layer,
b_j is the output node j of the hidden layer,
W_{ij} is the weight between input and hidden layer,
V_{ji} is the weight between hidden and output layer.

The complexity of the MLP network can be changed by varying the number of layers and the number of units in each layer. Hence, the right structures of NN have to be found by experiments. It has been reported by several researchers (Lippmann, 1987; Cybenko, 1989) that a single hidden layer should usually be sufficient for most problems, especially for classification tasks. The major efforts were focused on controlling the complexity of the model in order to avoid a too complex model structure which may lead into an over fitted ANN model (Niska et al., 2010).

The non-parametric neural network classifiers have numerous advantages over the statistical methods, such as no assumption about the probabilistic models of data, the ability to generalize in noisy environments, and the ability to learn complex patterns. Other advantages of NNs are that they can classify data with a smaller training set than conventional classifiers and be more tolerant of noise present in the training patterns (Mather, 1999).

5.2 Frequency-based contextual (FBC)

Unlike three methods previously discussed, contextual technique considering both spectral and spatial information in order to perform the classification process instead of depending on spectral component alone (Mustapha et al., 2011). Classification results of spectral data can be improved by taking into account other information into the original image. The simplest way is to incorporate spatial information within the neighboring pixel. Contextual information, or so-called context for simplicity, may be defined as how the probability of presence of one object (or objects) is affected by its (their) neighbors (Tso & Olsen, 2005). There are many examples of contextual classification approach, but in this present article we only concern with FBC approach. Frequency-based contextual classification of multispectral imagery is performed by using a grey level reduced image and a set of training site bitmaps. The input layer must be 8-bit data. Any 16-bit and 32-bit data layers should be scaled to 8-bits.

There are a number of factors affecting the land cover classification accuracy of the FBC. For instance, the collection of the training area and selection of the pixel window size are very important for this approach. The training area must be representative and of a reasonable size to capture the spatial structure of any land cover type in an image. Nevertheless, pixel-window size determines the amount of spatial information that can be included in the classification. Because the optimal pixel window varies with the individual class and image resolution, it is usually difficult to determine before image classification. Therefore, an appropriate window size is usually determined empirically. Pixel window size needs to be specified specifically when performing contextual classification on each pixel. Users may have to run the contextual classifier with the same input data, but using different settings for window size until a desirable output is produced. In general, contextual classification performs better when specifying a larger window size, especially if the original input image contains complicated mixed classes (such as urban areas). If the classes are uniform and spectrally pure, then a smaller window size may sufficient. A few examples of different window sizes are shown in Fig. 5. It seems clear that the inclusion of spatial arrangement information of gray-level values in a pixel neighborhood can considerably improve the performance of the FBC, as expected by Gong and Howarth (1992). But, this classifier also has their drawback itself. Contextual classification cannot classify pixels along the edges of the image. If the output window borders the edge of the image file, then the output pixels along the edge are set to zero, to indicate unclassified or unknown pixels. Usually, the error patterns caused by the contextual classification algorithm are usually systematically located along the class boundaries. Meanwhile, the classification results demonstrate that a significant increase in overall accuracy can be achieved by combining spatial data with spectral data when comparing the results obtained from traditional method although it cannot overtake the performance of neural network algorithm.

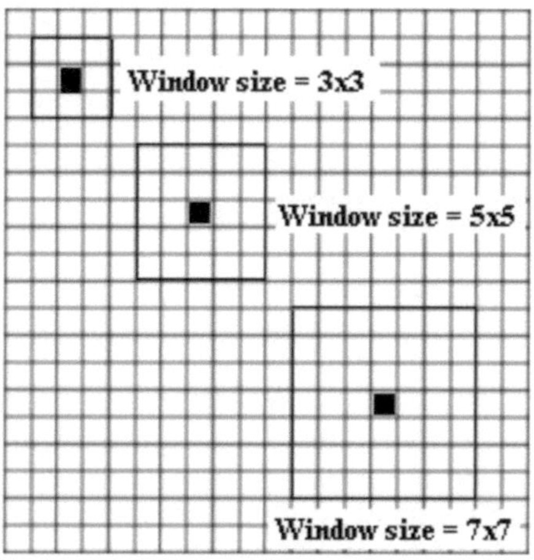

Fig. 5. Examples of window sizes used in frequency based contextual method (3x3, 5x5, 7x7). Black pixel indicates the center pixel of the specific window.

6. Training areas development

Training is the identification of a sample of pixels of known class membership obtained from reference data. These training pixels are used to derive spectral signatures for classification, and signature statistics are evaluated to ensure adequate separability. Then, the pixels of the image are allocated to the class with greatest similarity to the training data metrics (Alberti et al., 2004). The training stage of a supervised classification is designed to provide the necessary information. The training sites were used to train the supervised classification algorithm for classification process. In remote sensing, the aim of the training stage has typically been the production of descriptive statistics for each class which may then be used in the determination of class membership by the selected classifier (Foody & Mathur, 2006). Obtaining enough training data has been a tough question with land cover applications. Two sets of training data were finally prepared. The first set of data was prepared for the use of the traditional method. Meanwhile, the second set of the training data was used for the advance method. The use of the different datasets for classifying same area by using different classifier will be discussed in section 8.3.

For advanced method, knowledge of the statistical distribution is not required. Rather NNs learn it from a representative training set. In our case, the training phase of the NN was based on the back-propagation (BP) learning rule to minimize the mean square error (MSE) between the desired target vectors and the actual output vectors. Training patterns were presented to the network, and the weights of each node were adjusted so that the approximation created by the NN minimized the error between the desired output and the added output created by the network. In a network each connecting line has an associated weight. NN are trained by adjusting these input weights (connection weights), so that the calculated outputs approximate the desired. In the learning phase, input patterns from training data are fed forward through a network initiated with random synapse weights. The root-mean-square error (RMSE) is calculated between the network outputs and the desired outputs. The errors are back-propagated through the network and the synapse weights are adjusted in order to reduce the total RMSE. This process continues until a convergence criterion is satisfied (Rumelhart et *al.*, 1986). The successful generalization of the NNs used in this application is indicated by the low residual RMS errors. The training is finished when the output value is equal to the ideal output value. Mean Squares of the network Errors (MSE) is given by the Equation 3 (Moghadassi et al., 2009):

$$MSE=\frac{1}{N}\sum_{i=1}^{N}(e_i)^2=\frac{1}{N}\sum_{i=1}^{N}(\tau_i-\alpha_i)^2 \tag{3}$$

where
Target output (τ_i)
α_i is output from neuron

Meanwhile, the selection of training sets were based on field surveys, reference information from SPOT-5 images and visual inspection of the image of the particular area. Only the training samples believed to be the most useful and informative were selected for the classification. Training data acquisition can be a very costly process. Training data that are not carefully selected may introduce error. Collection of training data is the crucial step for image classification and it directly influences the classification accuracy (Wang et al., 2007). Training set size can impact greatly on classification result. However, size is only one attribute of a training set. Some of the literature suggests the use of a minimum of 10–30p cases per-class for training, where p is the number of wavebands used (Piper, 1992 & VanNiel et al., 2005). In addition, all training and test sample sites were revisited on the ground to confirm accuracy of measurement.

7. Accuracy assessment

Accuracy assessment is an important aspect of land cover mapping as a guide to map quality. The accuracy assessment sites were used to provide a statistical assessment of the accuracy produced by each of the classification mapping approaches tested for this project. The accuracy assessment sites were set aside until the map was completed and accuracy assessment was performed. This process insured that the accuracy data were completely independent of the training data (Thomas et al., 2003).

The error matrix is the standard method used to assess classification accuracy. In the error matrix, the column represents the reference data, while the rows represent the classified data (Table 3). It is typical to extract several statistics from the error matrix: overall accuracy, Kappa coefficient, producer's accuracy and user's accuracy. To conduct the accuracy assessment, a total of 500 sample plots, covering different land cover types, were randomly allocated and examined using field data, a SPOT-5 image with 5m in spatial resolution and high resolution of google earth map. Luedeling & Buerkert (2008) used the google earth map as one of their validation method. The sampling pixels used for accuracy assessment were selected using the randomly stratified sampling method. In addition, the test pixels were uniformly distributed in entire image.

Classified	Reference					
	1	2	3	4	5	Total
1	p_{11}	p_{12}	p_{13}	p_{14}	p_{15}	p_{1+}
2	p_{21}	p_{22}	p_{23}	p_{24}	p_{25}	p_{2+}
3	p_{31}	p_{32}	p_{33}	p_{34}	p_{35}	p_{3+}
4	p_{41}	p_{42}	p_{43}	p^{44}	p_{45}	p_{4+}
5	p_{51}	p_{52}	p_{53}	p^{54}	p_{55}	p_{5+}
Total	p_{+1}	p_{+2}	p_{+3}	p_{+4}	p_{+5}	

Table 3. Population error matrix with p_{ij} representing the proportion of area in the mapped land cover category i and the reference land cover category j.

Overall accuracy is the simplest and one of the most popular accuracy measures and is computed by dividing the total correct (i.e., the sum of the major diagonal) by the total number of pixels in the error matrix (Congalton, 1991). Meanwhile, Rosenfield and Fitzpatricklin (1986) identified the Kappa coefficient as a suitable accuracy measure in the thematic classification for representing class accuracy. Its strength lies in the fact that it takes all the elements (diagonal and non-diagonal) of the confusion matrix into consideration, in contrast to the overall accuracy measures which only consider the diagonal element of the matrix. In addition, Two types of thematic errors can be measured in a confusion matrix. They take into account the accuracy of individual categories. One is given by the producer's accuracy, which indicates the proportion of ground base reference samples correctly assigned. It details errors of omission, i.e., when a pixel is omitted from its correct category. The other error is given by the user's accuracy, which indicates the proportion of data from the estimation map representing that category on the ground. It is a measure of errors of commission, i.e., when a pixel is committed to an incorrect category (Avelar et al., 2009).

8. Performance evaluation

The six classes-urban, mountain, land, vegetation, ritual area and shadow were classified using four different classifiers, and classification accuracy assessments were conducted (Table 4-7). Performances of each of the classifiers that have been tested will be analyzed based on three factors. In order to make a comparison, the classifiers performance are analyse in term of their classification accuracy, training samples and performance in heterogeneous area. The area of each class estimated through various techniques was compared and evaluated with the corresponding actual area as obtained from the reference data. For lack of additional satellite data, concurrent with the periods of the field surveys, a reference dataset was generated based on the ordered SPOT-5 satellite data and expert knowledge.

8.1 Classification accuracy

From the perspective of the classification accuracy, there are four parameters could be discussed which are overall accuracy, kappa coefficient, user's and producer's accuracies (analysis per class). These parameters can be calculated from error matrix tables. A classification error matrix was computed for quantitative accuracy assessment. Table 4, 5, 6 and 7 demonstrated the error matrices table deriving from MD, ML, NN and FBC classifier. The dominant land cover types in the selected area were urban, mountain and land areas which correspond to 95% of the entire image. The remaining 5% of the image is consisted by vegetation, ritual area and shadow.

For MD algorithm which is the simplest classifier among others, the result of overall accuracy was 64.2% with 0.479 value of kappa coefficient was obtained. The user accuracy is varied between 50.7% for urban class and 100.0% for vegetation and ritual area classes. Mountain, land and shadow classes recorded 79.6%, 62.2% and 73.1% respectively. For producer accuracy, the accuracy for each class using MD approach was as follow: 67.1% for urban, 57.9% for mountain, 70.9% for land, 45.5% for vegetation (lowest), 66.7% for ritual area and 95.0% for Shadow (highest). A total of 500 random sample points were tested in order to verify the classification result with 321 points was correctly classified. Meanwhile, urban class recorded almost half of the tested pixels that correctly classified with most of the misclassified pixel go to mountain class. A total of 129 out of 162 observations had been correctly classified for mountain class and 33 points were wrongly classified with 26 points were misclassified as urban class. The high number of wrongly pixels go to urban class is due to the fact that the mountainous area in the arid environment is not cover by tree but it is filled by stones and rocks which is has a similar spectral characteristic of urban area. Nevertheless, vegetation and ritual area classes gave the perfect result by correctly classified all tested points. Both classes are easily to classify due to the significantly different on their spectral characteristics among other classes. In the other hand, 56 out of 90 observations for land class were correctly classified whereas 19 out of 26 observations for shadow class were correctly classified. Most of the incorrect pixels were classified as mountain class. This is not surprising because most of the shadow appear within mountainous area.

Meanwhile, ML algorithm was the second traditional method that has been tested in this project. It gave better result than MD classifier. The overall accuracy was 77.6% while the kappa coefficient had a value of 0.659. For each classes (user accuracy), urban recorded 69.8%, 83.4% for mountain, 82.9% for land, 92.6% for vegetation, 66.7% for ritual area (lowest) and 100.0% for shadow (highest). Although overall classification result was better than MD, but two classes (vegetation and ritual area) showing lower percentage than MD. In the meantime, producer accuracy is varied between 61.5% (shadow) and 100.0% (ritual area). Urban, mountain, land and vegetation classes had a value of 85.4%, 69.6%, 73.3% and 96.2% respectively. Further evaluation of the error matrix shows that 388 out of 500 points used from the same random samples were correctly classified. The classifier had some difficulty separating cleared land from land under construction (urban) and mountain from urban area, as exhibited by error matrix table that showed 68 points were wrongly classified to both classes (53 points for mountain, 15 points for land). This is understandable because their spectral characteristics are very similar. However, the result of urban class revealed that significant improvement (nearly 20%) was achieved compared to the MD classifier. In

	Urb	Mou	Lan	Veg	Rit	Sha	Total	UA (%)
Urb	**108**	82	22	1	0	0	213	50.7
Mou	26	**129**	1	5	0	1	162	79.6
Lan	26	6	**56**	0	2	0	90	62.2
Veg	0	0	0	**5**	0	0	5	100.0
Rit	0	0	0	0	**4**	0	4	100.0
Sha	1	6	0	0	0	**19**	26	73.1
Total	161	223	79	11	6	20	**500**	
PA (%)	67.1	57.9	70.9	45.7	66.7	95.0		
Overall accuracy = 64.2%								
Kappa coefficient = 0.479								

Table 4. Error matrix derived from Minimum Distance-to-Mean classifier

	Urb	Mou	Lan	Veg	Rit	Sha	Total	UA (%)
Urb	**164**	53	15	1	0	2	235	69.8
Mou	15	**126**	7	0	0	3	151	83.4
Lan	11	2	**63**	0	0	0	76	82.9
Veg	4	1	4	**18**	0	0	27	66.7
Rit	1	0	0	0	**2**	0	3	66.7
Sha	0	0	0	0	0	**8**	8	100.0
Total	195	182	89	19	2	13	**500**	
PA (%)	84.1	69.2	70.8	94.7	100.0	61.5		
Overall accuracy = 76.2%								
Kappa coefficient = 0.649								

Table 5. Error matrix derived from Maximum Likelihood classifier

	Urb	Mou	Lan	Veg	Rit	Sha	Total	UA (%)
Urb	**187**	48	23	0	0	0	258	72.4
Mou	2	**162**	4	1	0	1	170	95.3
Lan	0	0	**40**	0	0	0	40	100.0
Veg	0	0	0	**15**	0	0	15	100.0
Rit	0	0	0	0	**5**	0	5	100.0
Sha	0	0	0	0	0	**12**	12	100.0
Total	189	210	67	16	5	13	**500**	
PA (%)	98.9	77.1	59.7	93.8	100.0	92.3		
Overall accuracy = 84.2%								
Kappa coefficient = 0.757								

Table 6. Error matrix derived from Back-propagation Neural Network classifier

	Urb	Mou	Lan	Veg	Rit	Sha	Total	UA (%)
Urb	**163**	45	15	5	0	1	229	71.2
Mou	13	**165**	0	0	0	0	178	92.7
Lan	2	0	**58**	0	0	0	60	96.7
Veg	0	0	0	**2**	0	0	2	100.0
Rit	3	0	1	0	**6**	0	10	60.0
Sha	0	1	0	0	0	**14**	15	93.3
Unk	2	3	1	0	0	0	**6**	
Total	183	214	75	7	6	15	**500**	
PA (%)	89.1	77.1	77.3	28.6	100.0	93.3		
Overall accuracy = 81.6%								
Kappa coefficient = 0.722								

* Note: Urb = Urban, Mou = Mountain, Lan = Land, Veg = Vegetation, Rit = Ritual area, Sha = Shadow and Unk = Unknown.

Table 7. Error matrix derived from Frequency-based Contextual classifier

addition, 126 out of 151 observations were correctly classified for mountain class whereas 63 out of 76 observations were correctly classified for land class. Most of the misclassification for both classes goes to urban class. For vegetation class, although it has lower percentage over MD classifier, but the result still to be considered as a good result by obtaining over 90% with only 2 out of 27 observations were wrongly classified. Ritual area class was another category that showed their percentage lower than MD classifier. Although the class was easily to classify but they recorded only 66.7% when validation process was performed.

The lower in accuracy for the ritual area class is explained by the fact that only 3 out of 500 points were tested in that particular class meaning insufficient validation points occurred in this class. The result is expected to be higher if more validation point is added during the validation process as this class was a homogenous category. In the other hand, shadow class gave a perfect result by correctly classified all tested pixels.

However, NN approach which was one of the advanced methods tested in this project demonstrated superior result in term of overall accuracy. The NN outperformed the other classifiers for this factor. The overall accuracy was 84.2% and had a value of 0.757 for kappa coefficient. The NN method seems to do a much better job in classifying all classes than the other methods, which seems to be the primary reason for its high overall accuracy. The success of this classifier is due to the fact that four classes (land, vegetation, ritual area and shadow) have been tested correctly perfect (100% is obtained in analysis per class). In fact, the mountain class achieved a high user's accuracy (95.3 percent) using NN method with 162 out of 170 observations were correctly classified. Nevertheless, urban class recoded 72.5% in analysis per class but it still acceptable and highest among other classes. From the view point of statistical analysis, most of the pixels in urban area were confuse with mountain (48 points) and land (23 points). The reason why each of classifier always resulted urban class to the lower percentage compared to other classes will be explained in the next sub-section. Meanwhile, producer's accuracy varied between 59.7% for land class and 100% for ritual area class. Urban, mountain, vegetation and shadow classes recorded 98.9%, 77.1%, 93.8% and 92.3% respectively.

The network architecture for the NN had three layers, with twelve units in the hidden layer, four units in the input layer (one for each spectral band), and six units in the output layer (one for each class). The other parameters used in the NN algorithm are shown in Table 8. These network structures were determined through trial and error meaning the number of hidden units used in this application was determined through experimental simulations. Fig. 6 shows the variation of RMSE values at convergence as a function of the number of hidden nodes. The experiments were performed with a maximum number of iterations of 1000 and the final RMSE was between 0.009 and 0.109 with the number of hidden nodes ranging from 3 to 15 nodes with increment of 3. The minimum RMSE with the smallest number of nodes was attained adopting architecture with 12 hidden nodes. This was the architecture finally adopted for the learning and classification process.

NN architecture	4-12-6
Momentum rate	0.9
Learning rate	0.1
Iteration	1000 epochs

Table 8. Parameters used in the neural network algorithm

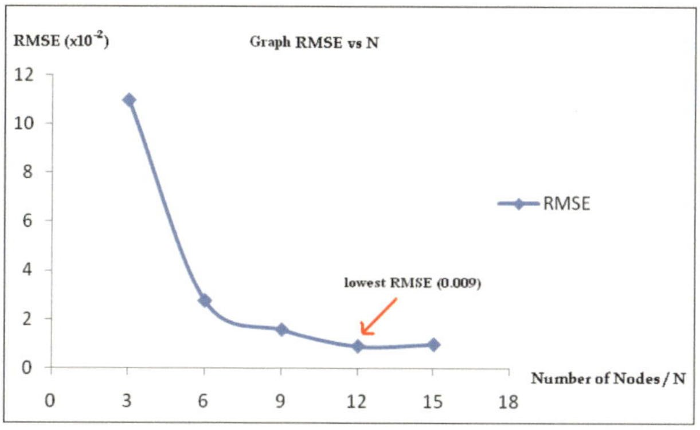

Fig. 6. Graph of RMSE versus Number of Nodes

For FBC classifier, it is required to determine window sizes before begin the classification process. This window size was very important because it will impact on how much the spatial information will be included during decision making process. But the exactly size is not easy to determine. Hence, it was determined by experiment. From the experimental simulations, 9x9 window size was determined as an optimum window. But we decided to choose window sizes of 7x7 instead of 9x9 as the images used for other three classifiers were filter out using 7x7 averaging filter to reduce the noisy effect from the original image. Moreover, there are no significant differences on overall accuracy between these two windows as well as for the remaining of window sizes beyond 9x9 as shown in Fig 7. The FBC techniques used in this project achieved higher overall accuracy and kappa coefficient (81.6 percent and 0.722) rather than traditional method. Even though it cannot overtake the performance of NN but their result is still good and acceptable. Further evaluation of the error matrix shows that the additional of contextual information increases map accuracy. The high quality of the spatial information had a large impact on the success of this method. However, there are some extremely difficult types of confusion to map. Desert landscaping often consist of gravel, and certain types of gravel can be spectrally indistinguishable from urban. To make the confusion even more complex, some part of mountainous area were located within urban area. This situation would lead the misclassification to be occurred since their spectral characteristic is similar. In the meantime, the class specific producer accuracy varied between 28.6% for vegetation class and 100.0% for ritual area class. User accuracy reached the highest value of 100.0% for vegetation class. Lowest values were obtained for the class ritual area with 60.0%. The integration of contextual information showed its benefits in the sharp improvement in accuracy for the mountain and land classes compared to traditional method. Other behavior of FBC method that it can be seen from the classification result that pixels at the edge of different land cover type are mostly misclassified. At the center of each land cover type, most classes are correctly classified. By evaluating error matrix table, it revealed that urban and mountain classes were confused each other. Shadow, land and vegetation classes were easily classified with all observation points were classified correctly for vegetation class. Nevertheless, the unexpected result was achieved by ritual area class where it gave lower result (60 percent) although this class was considered as homogenous area with has uniformly in their spectral characteristic. The sharp decrease in accuracy for that class is explained by the fact that it was not suitable to use the current window sizes due to the homogenous behavior of the class. For this situation, smaller window size is more suitable and could be expected to increase the class accuracy.

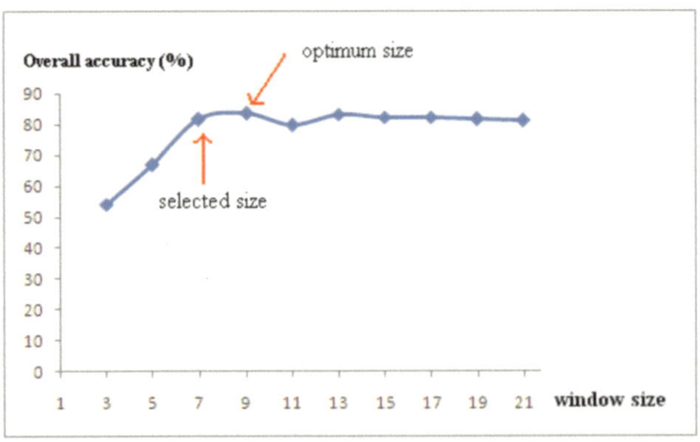

Fig. 7. Graph showing experimental result of FBC using different window sizes

In general, the NN approach generally provided the highest accuracies for all classes. Considering the overall accuracy, NN provided the best classification results with 84.2% and MD provided the poorest results with overall accuracy of 64.2%. Fig. 8 provides a comparison of kappa and overall accuracy results among the different classifiers. It indicates that NN and FBC have a significantly better accuracy than do MD and ML classifiers. MD produced lower classification accuracy because it only used the mean vector and ignored the covariance between the classes. ML produced a relatively higher accuracy than did MD because it takes the covariance into account in its algorithm. However, ML assumes a normal distribution for the histograms of the classes, which is not always true. Both MD and ML only consider per-pixel information, ignoring texture or contextual information. Comparing the two approaches (traditional and advanced methods), the proposed NN classifier proved to be more effective, with a 6.6% and 20.0% increase in accuracy compared to ML and MD classifier whereas FBC could increased their accuracy up to 4.0% and 17.4% compared to the same classifiers.

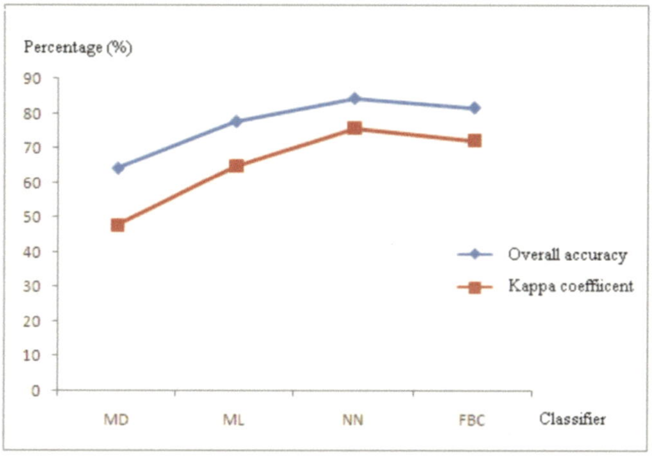

Fig. 8. Comparison of overall accuracy and kappa coefficient using different classifier

8.2 Performances in heterogeneous area

The heterogeneity environment in an image is a major problem in classification where a pixel contains more than one land cover class. In our case, urban class is considered as a heterogeneous area instead of homogenous area for remaining of the five other classes. In this section, we will explain the reason why the percentage of urban class in this work is always lower compared to other classes. This is due to the urban factor itself. The spectral characteristics of urban surfaces are known to be complex. This is due to the fact that much information could be extracted from the urban class. Urban areas are characterized by a large variety of built-up environments and natural vegetation covers which not only determine the surface features of a city, such as land use patterns, but also influence ecological, climatic and energetic conditions of land surface processes (Chen et al, 2009). For instance, sites under construction possess a more varied high reflectance resulting from building construction foundations and construction materials. Cleared land exhibits high uniform spectral reflectance which is characteristic of bare soil, while some vegetated area also located in urban environment. These numbers of information in a single class would create the high possibility of mixed pixel to be occurred. Mixed pixel problem increases the difficulty in classification process and lead to has misclassification pixels and reduce the classification accuracy. As stated by Small (2005), the highly heterogeneous nature of urban surface materials is problematic at multiple spatial scales, resulting in a high percentage of mixed pixels in moderate resolution imagery and even limiting the utility of high spatial resolution imagery. Furthermore, Alberti et al., (2004) in their article mentioned that interpretation and analysis of urban landscapes from remote sensing, however, present unique challenges due to the characteristics of urban land cover which amplify the spectral heterogeneity of urban surfaces and make it extremely difficult to identify the source of observed in observed reflectance.

The greatest challenge for each of the classifiers is to accurately determine various materials that make up urban surface reflectance. Hence, the classification result could be increased extremely if any classifiers can performed well in these urban mixing surfaces as it represented almost one third of the entire image. Fig. 9 shows a comparison of different classifiers for each class. It is evident that all four classifiers have produced noisy results, although the results generated by the NN and FBC are slightly less noisy compared to that of the ML and MD methods. The increasing of accuracy in urban class of advanced methods compared to traditional methods is mainly attributed to better identification between urban features and mountain, leading to significant increases in overall classification accuracy is achieved. The results indicate that the NN with back-propagation algorithm network is able to adjust the values of the network connections so that the activations of the output neurons match more closely the desired output values. Meanwhile, the key to successful mapping from the FBC method is it able to utilize the advantages of neighborhood information to enhance classification performance. We may conclude that the inclusion of contextual information can considerably improve the remotely sensed imagery classification performance and visual interpretation if the model is well defined and the relating parameter is carefully chosen. From the visual inspection, result of ML and MD classification performance yet preserves the potential difficulty in interpreting the classified images in a meaningful way because the different class pixels are still mixing and resulting in a noisy image view as shown in Fig 10. Significant confusion occurred between urban and

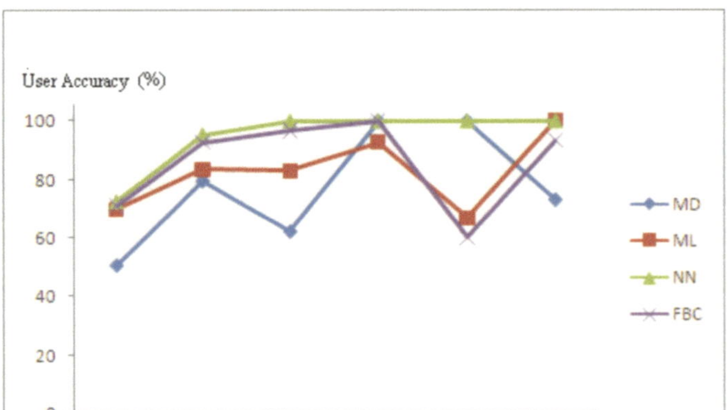

Fig. 9. Comparison of four classifiers for each class

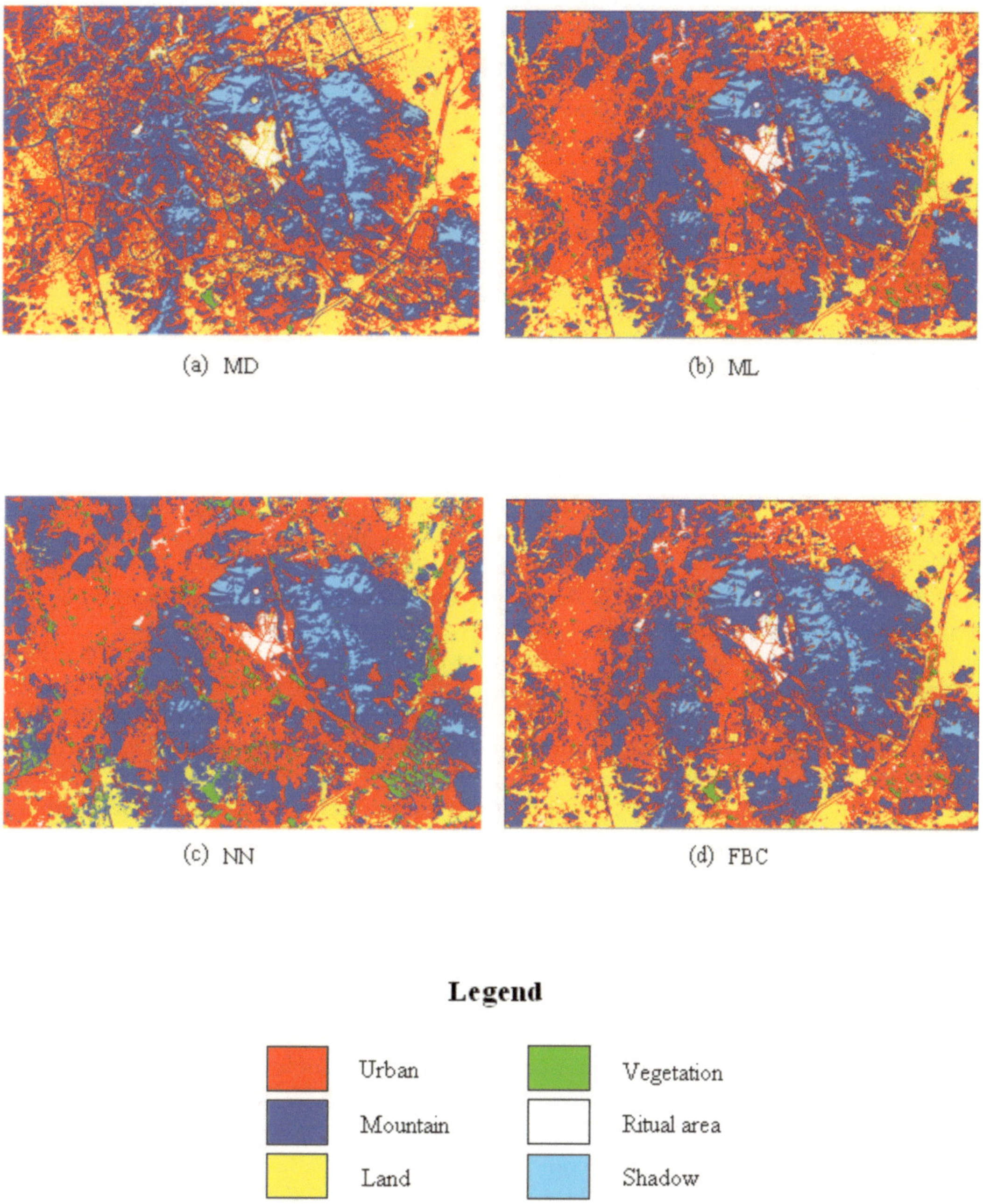

Fig. 10. Result of land cover classification using (a) MD, (b) ML, (c) NN and (d) FBC

mountain classes as their spectral characteristic are very similar. So that traditional per pixel classifiers such as ML and MD are not recommended to be used when the image contain large portion of heterogeneity area surfaces. The MD too broadly classified class by often overlapping another class because the classifier lacks sophisticated spectral discrimination between very complex features. The ML is more sophisticated, but being a per-pixel classifier, created a "salt and pepper" pattern classification, which showed misclassification has been occurred.

8.3 Training sample

In supervised classification approach, training stage become a major part in the decision making process as it will affected the outcome of the classification result. In order to analyse the performance of the four classifiers in term of training sample, two sets of training data were prepared. Training samples were chosen across the study area and the number of samples for each land cover type was listed in Table 9. However, the classification results greatly depended on the quality of training datasets and required abundant and accurate field measurements from all classes of interest. One difficulty encountered in particularly heterogeneous areas, such as the urban class, is related to the difficulty of identifying a sufficient number of pure pixels for classifier training and validation. Unlike the other classes, particularly on the vegetation, ritual area and shadow classes were easy to identify due to the spectrally different among each other. The use of different training data sets for the classification of the same images is due to the differences of the classifier characteristic behavior in the decision making process. For example, traditional method needs more training data as this type of method was a statistical approach. With a large number of the training data, it can generate the statistical information for the classification process. Meanwhile, advanced method do not required a large number of training data as it not a statistical approach. They have their own way to handle the training stage. For instance, the training of a network by back-propagation involves three stages: the feed forward of the input training pattern, the calculation and back-propagation of the associated error, and the adjustment of the weights (Rezapour et al., 2010). In fact, the weights are usually randomized at the beginning of the training.

Evaluation on table 9 demonstrated that traditional method needs almost double size of pixels in order to perform classification compared to advanced method. We also conducted experiment for traditional method by using the dataset that prepared for advance method (data set 2). The experimental results revealed that both classifiers cannot perform well with this training dataset as their overall accuracy were decreased from 77.6% to 68.0% and 64.2% to 57.0% for ML and MD classifiers. The amounts of seven to nine percent reduction were obtained. This indicates that the small number of training samples is not sufficient for both of classifiers. The experiment shows the strong evidence that the traditional classifier needs a large number of training samples in order to perform the classification.

In addition, the training samples of ML and MD were selected in their raster layer. Any repeatable on experiments are without difficulty. The training process is not take long time to complete although they have a large number of training data. Unlike NN and FBC, their training samples were collected in bitmap layer. The number of bitmap layer is corresponding to the number of intended classes. The training process is time consuming

Class	Traditional method (ML and MD) Set 1	Advanced method (NN and FBC) Set 2
Urban	6230	3032
Mountain	7128	3751
Land	5579	2767
Vegetation	3157	1186
Ritual area	1898	541
Shadow	1875	763
Total samples	25867	12040

Table 9. Number of training samples for each class

especially for NN classifier. This is due to the fact that the repeatable on experiments required all the parameter settings and also the first set of random weights. If the structure has more than one hidden layer, hence, more time is needed to finish the training process. Lippman (1987) suggested that NN with more than one hidden layer are harder to use because they add the problem of hidden structures and lengthen training time. For FBC, it also takes longer time in training stage but not too longer as NN. Thus, NN was found the least friendly in training and the most expensive in terms of time requirement although they have less number in training sample.

9. Conclusion

In this article, four different approaches to the classification of complex areas by use multispectral data have been described. The main purpose of our investigation was to quantitatively assess, also from the viewpoint of statistical significance, the capabilities of the four approaches to exploit ALOS AVNIR-2 satellite data in an effective way. Some interesting conclusions can be drawn from the obtained results. Different classifiers have their own advantages and disadvantages. For a given research topic, deciding which classifier is more appropriate depends on a variety of factors. Even though some classifiers provide more accurate results than others, all four used in this research are useful in extracting land-cover information. However, of the four classifiers tested, NN and FBC are the two most recommended approaches when classifying the image that surrounding with desert environment especially for urban class. Experimental results confirm the significant superiority of the advanced method in the context of multispectral data classification over the conventional classification methodologies. Sophisticated algorithms are needed to successfully discriminate distinct features in complex environments. In this case, classification problems will be either related to spatial/spectral aspects or to spectral mixtures at a given resolution. Our results show that NN and FBC had the best performance to address the land cover heterogeneity of the study area. These two classification approaches have proved to be suited for classification of complex areas. NN method was preferred because they are capable of handling large amounts of data and do not require simplifying hypotheses on the statistical distribution. The NN approach provided better overall accuracy than did the FBC, ML and MD approaches. On the other hand, the NN approach requires a complex and expensive design phase (e.g., concerning the correct size of the hidden layers and parameter settings) and a much longer training time. For FBC, the contribution of the spatial information (neighboring pixels) to the digital satellite imagery for land cover mapping was very valuable instead of depending on multispectral data alone. Although there are several limitations, the results of the classification procedures performed highlight the accuracy improvement compared to traditional method. The traditional classification methods, in this case ML and MD, reach their limitations in urban systems due to the high spectral heterogeneity of urban features. The misclassification of some urban features came therefore as no surprise, since high-quality buildings, streets and their surroundings are very heterogeneous.

In conclusion, remote sensing has been shown to be a useful tool for evaluating the performances of different classifiers in arid environment. Remote sensing classifications should be considered the technique of choice for land cover study and monitoring. In many instances remotely sensed data are used to derive information on a specific land cover class of interest. Although a conventional classifier may be used to derive this information but it cannot handle the complex mixture environment and always produced noisy image in that particular environment such as in the urban class. Urban environments represent one of the most challenging areas for remote sensing analysis due to high spatial and spectral diversity of surface materials. Finally, future study are planned that will compare the results of this study to those that can be obtained using object based approaches. Additionally, research will be conducted on the use of high-resolution image and applying it to more extensive remote sensing data such as hyperspectral images.

10. Acknowledgment

The authors would like to acknowledge the Universiti Sains Malaysia (USM) for funding this project. We would also like to thank JAXA for providing the satellite images. The authors would like to thank the anonymous referees for their helpful comments and suggestions.

11. References

Aplin, P.; Atkinson, P. M. & Curran, P. J. (1999). Fine Spatial Resolution Simulated Satellite Sensor Imagery for Land Cover Mapping in the United Kingdom. *Remote Sensing of Environment,* Vol.68, No.3, pp. 206-216, ISSN 0034-4257

Avelar, S.; Zah, R. & Tavares-Correa, C. (2009). Linking Socioeconomic Classes and Land Cover Data in Lima, Peru: Assessment through the Application of Remote Sensing and GIS. *International Journal of Applied Earth Observation and Geoinformation,* Vol.11, No.1, pp. 27-37, ISSN 0303-2434

Bagan, H.; Takeuchi, W.; Kinoshita, T.; Bao, Y. & Yamagata, Y. (2010). Land Cover Classification and Change Analysis in the Horqin Sandy Land from 1975 to 2007.

IEEE Journal of Selected Topics in Applied Earth Observations and Remote Sensing, Vol.3, No.2, pp. 168-177, ISSN 1939-1404

Benediktsson, J. A.; Swain, P. H. & Ersoy, O. K. (1990). Neural Network Approaches Versus Statistical Methods in Classification of Multisource Remote Sensing Data. *IEEE Transactions on Geoscience and Remote Sensing,* Vol.28, No.4, pp. 540-552, ISSN 0196-2892

Cihlar, J.; Xia, Q. H.; Chen, J.; Beaubien, J; Fung, K. & Latifovic, R. (1998). Classification by Progressive Generalization: A New Automated Methodology for Remote Sensing Multichannel Data. *International Journal of Remote Sensing,* Vol.19, No.14, pp. 2685-2704, ISSN 0143-1161

Chen, H.; Chang, N.; Yu, R. & Huang, Y. (2009). Urban Land Use and Land Cover Classification Using the Neural-fuzzy Inference Approach with Formosat-2 Data. *Journal of Applied Remote Sensing,* Vol.3, pp. 033558, ISSN 1931-3195

Congalton, R. G. (1991). A Review of Assessing the Accuracy of Classification of Remotely-sensed Data. *Remote Sensing of Environment,* Vol.37, No.1, pp. 35-46, ISSN 0034-4257

Coops, N. C.; Wulder, M. A. & Iwanicka, D. (2009). Exploring the Relative Importance of Satellite-derived Descriptors of Production, Topography and Land Cover for Predicting Breeding Bird Species Richness over Ontario, Canada. *Remote Sensing of Environment,* Vol.113, No.3, pp. 668-679, ISSN 0034-4257

Cybenko, G. (1989). Approximation by Superposotions of a Sigmoidal Function. *Mathematics of Control, Signals and Systems,* Vol.2, No.4, pp. 303-314, ISSN 0932-4194

Erbek, F. S.; Ozkan, C. & Taberner, M. (2010). Comparison of Maximum Likelihood Classification Method with Supervised Artificial Neural Network Algorithms for Land Use Activities. *International Journal of Remote Sensing,* Vol.25, No.9, pp. 1733-1748, ISSN 0143-1161

Foody, G. M. & Mathur, A. (2006). The Use of Small Training Sets Containing Mixed Pixels for Accurate Hard Image Classification: Training on Mixed Spectral Responses for Classification by a SVM. *Remote Sensing of Environment,* Vol.103, No.2, pp. 179-189, ISSN 0034-4257

Gong, P. & Howarth, P. J. (1992). Frequency-based Contextual Classification and Gray Level Vector Reduction for Land Use Identification. *Photogrammetric Engineering & Remote Sensing,* Vol.58, No.4, pp. 423-437, ISSN 0099-1112

Han, K. S.; Champeaux, J. S. & Roujean, J. L. (2004). A Land Cover Classification Product over France at 1 km Resolution using SPOT4/VEGETATION Data. *Remote Sensing of Environment,* Vol.92, No.1, pp. 52-66, ISSN 0034-4257

Ju, J.; Gopal, S. & Kolaczyk, E. D. (2005). On the Choice of Spatial and Categorical Scale in Remote Sensing Land Cover Classification. *Remote Sensing of Environment,* Vol.96, No.1, pp. 62-77, ISSN 0034-4257

Khan, L.; Javed, K. & Mumtaz, S. (2010). ANN Based Short Term Load Forecasting Paradigms for WAPDA Pakistan. *Australian Journal of Basic and Applied Sciences,* Vol. 4, No.5, pp. 932-947, ISSN 1991-8178

Lim, A.; Hedley, J. D.; LeDrew, E.; Mumby, P. J. & Roelfsema, C. (2009). The Effects of Ecologically Determined Spatial Complexity on the Classification Accuracy of

Simulated Coral Reef Images. *Remote Sensing of Environment,* Vol.113, No.5, pp. 965-978, ISSN 0034-4257

Lippmann, R.P. (1987). An Introduction to Computing with Neural Nets, *IEEE ASSP Magazine,* Vol.4, No.2, pp. 4 – 22, ISSN: 0740-7467

Lu, D. & Weng, Q. (2007). A Survey of Image Classification Methods and Techniques for Improving Classification Performance. *International Journal of Remote Sensing,* Vol.28, No.5, pp. 823-870, ISSN 0143-1161

Lu, D.; Mausel, P.; Batistella, M. & Moran, E. (2004). Comparison of Land-Cover Classification Methods in the Brazilian Amazon Basin. *Photogrammetric Engineering & Remote Sensing,* Vol.70, No.6, pp. 723-731, ISSN 0099-1112

Luedeling, E. & Buerkert, A. (2008). Typology of Oases in Northern Oman Based on Landsat and SRTM Imagery and Geological Survey Data. *Remote Sensing of Environment,* Vol.112, No.3, pp. 1181-1195, ISSN 0034-4257

Mather, P. (1999). *Computer Processing of Remotely-Sensed Images an Introduction,* John Wiley & Sons, New York

Moghadassi, A.; Parvizian, F. & Hosseini, S. (2009). A New Approach Based on Artificial Neural Networks for Prediction of High Pressure Vapor-liquid Equilibrium. *Australian Journal of Basic and Applied Sciences,* Vol.3, No.3, pp. 1851-1862, ISSN 1991-8178

Murakami, H.; Tadono, T.; Imai, H.; Nieke, J. & Shimada, M. (2009). Improvement of AVNIR-2 Radiometric Calibration by Comparison of Cross-Calibration and Onboard Lamp Calibration. *IEEE Transactions on Geoscience and Remote Sensing,* Vol.47, No.12, pp. 4051-4059, ISSN 0196-2892

Mustapha, M. R.; Lim, H. S. & MatJafri, M. Z. (2010). Comparison of Neural Network and Maximum Likelihood Approaches in Image Classification. *Journal of Applied Sciences,* Vol.10, No.22, pp. 2847-2854, ISSN 1812-5654

Mustapha, M. R.; Lim, H. S.; MatJafri, M. Z. & Syahreza, S. (2011). Comparison of Frequency-based Contextual and Maximum Likelihood Methods for Land Cover Classification in Arid Environment. *Journal of Applied Sciences,* Vol.11, No.17, pp. 3177-3184, ISSN 1812-5654

Niska, H.; Skon, J.; Packalen, P.; Tokola, T.; Maltamo, M. & Kolehmainen, M. (2010). Neural Networks for the Prediction of Species-Specific Plot Volumes Using Airborne Laser Scanning and Aerial Photographs. *IEEE Transactions on Geoscience and Remote Sensing,* Vol.48, No.3, pp. 1076-1085, ISSN 0196-2892

Pignatti, S.; Cavalli, R. S.; Cuomo, V.; Fusilli, L.; Pascucci, S.; Poscolieri, M. & Santini, F. (2009). Evaluating Hyperion Capability for Land Cover Mapping in a Fragmented Ecosystem: Pollino National Park, Italy. *Remote Sensing of Environment,* Vol.113, No.3, pp. 622-634, ISSN 0034-4257

Piper, J. (1992). Variability and Bias in Experimentally Measured Classifier Error Rates. *Pattern Recognition Letters,* Vol.13, No. 10, pp. 685-692, ISSN 0167-8655

Rezapour, O. M.; Shui, L. T. & Ahmad, D. (2010). Review of Artificial Neural Network Model for Suspended Sediment Estimation. *Australian Journal of Basic and Applied Sciences,* Vol.4, No.8, pp. 3347-3353, ISSN 1991-8178

Rosenfield, G. H. & Fitzpatrick-Lins, K. (1986). A Coefficient of Agreement as a Measure of Thematic Classification Accuracy. *Photogrammetric Engineering & Remote Sensing,* Vol.52, No.2, pp. 223-227, ISSN 0099-1112

Roy, P. S. & Giriraj, A. (2008). Land Use and Land Cover Analysis in Indian Context. *Journal of Applied Sciences,* Vol.8, No.8, pp. 1346-1353, ISSN 1812-5654

Rumelhart, D.E., Hinton, G. E & Williams, R. J. (1986). Learning Internal Representation by Error Propagation, in: Parallel Distributed Processing: Explorations in the Microstructures of Cognition, Rumelhart, D. E & McClelland, J. L (Eds.), 318-362MIT Press, Cambridge, Massachusetts

Sarkheil, H.; Hassani, H. & Alinia, F. (2009). The Fracture Network Modeling in Naturally Fractured Reservoirs Using Artificial Neural Network Based on Image Loges and Core Measurements. *Australian Journal of Basic and Applied Sciences,* Vol.3, No.4, pp. 3297-3306, ISSN 1991-8178

Seto, K. C. & Liu, W. (2003). Comparing ARTMAP Neural Network with the Maximum-Likelihood Classifier for Detecting Urban Change. *Photogrammetric Engineering & Remote Sensing,* Vol.69, No.9, pp. 981-990, ISSN 0099-1112

Small, C.; (2005). A Global Analysis of Urban Reflectance. *International Journal of Remote Sensing,* Vol.26, No.4, pp. 661-681, ISSN 0143-1161

Stuckens, J.; Coppin, P. R. & Bauer, M. E. (2000). Integrating Contextual Information with per-Pixel Classification for Improved Land Cover Classification. *Remote Sensing of Environment,* Vol.71, No.3, pp. 282-296, ISSN 0034-4257

Tadono, T.; Shimada, M.; Murakami, H. & Takaku, J. (2009). Calibration of PRISM and AVNIR-2 Onboard ALOS "Daichi" . *IEEE Transactions on Geoscience and Remote Sensing,* Vol.47, No.12, pp. 4042-4050, ISSN 0196-2892

Thomas, N.; Hendrix, C. & Congalton, R. G. (2003). A Comparison of Urban Mapping Methods Using High Resolution Digital Imagery. *Photogrammetric Engineering & Remote Sensing,* Vol.69, No.9, pp. 963-16, ISSN 0099-1112

Tso, B. & Olsen, R. C. (2005). A Contextual Classification Scheme Based on MRF Model with Improved Parameter Estimation and Multiscale Fuzzy Line Process. *Remote Sensing of Environment,* Vol.97, No.1, pp. 127-136, ISSN 0034-4257

VanNiel, T. G.; McVicar, T. R & Datt, B. (2005). On the Relationship between Training Sample Size and Data Dimensionality of Broadband Multi-temporal Classification. *Remote Sensing of Environment,* Vol.98, No.4, pp. 468-480, ISSN 0034-4257

Wang, C.; Menenti, M., Stoll, M.;Belluco, E. & Marani, M. (2007). Mapping Mixed Vegetation Communities in salt Marshes using Airborne Spectral Data. *Remote Sensing of Environment,* Vol.107, No.4, pp. 559-570, ISSN 0034-4257

Wang, L.; Sousa, W. P.; Gong, P. & Biging, G. S. (2004). Comparison of IKONOS and QuickBird Images for Mapping Mangrove Species on the Caribbean Coast of Panama. *Remote Sensing of Environment,* Vol.91, No.3-4, pp. 432-440, ISSN 0034-4257

Wulder, M. A.; White, J. C.; Goward, S. N.; Masek, J. G.; Irons, J. R.; Herold, M.; Cohen, W. B.; Loveland, T. R. & Woodcock, C. E. (2008). Landsat Continuity: Issues and Opportunities for Land Cover Monitoring. *Remote Sensing of Environment,* Vol.112, No.3, pp. 955-969, ISSN 0034-4257

Yuan, F.; Sawaya, K. E.; Loeffelholz, B. C & Bauer, M. E. (2005). Land Cover Classification and Change Analysis of the Twin Cities (Minnesota) Metropolitan Area by Multitemporal Landsat Remote Sensing. *Remote Sensing of Environment,* Vol.98, No. 2-3, pp. 317-328, ISSN 0034-4257

7

Application of Remote Sensing for Tsunami Disaster

Anawat Suppasri[1], Shunichi Koshimura[1], Masashi Matsuoka[2], Hideomi Gokon[1] and Daroonwan Kamthonkiat[3]
[1]*Tsunami Engineering Laboratory, Disaster Control Research Centre, Graduate School of Engineering, Tohoku University*
[2]*National Institute of Advanced Industrial Science and Technology*
[3]*Department of Geography, Faculty of Liberal Arts, Thammasat University*
[1,2]*Japan*
[3]*Thailand*

1. Introduction

This chapter aims to introduce an application of remote sensing to recent tsunami disasters. In the past, acquiring tsunami damage information was limited to only field surveys and/or using aerial photographs. In the last decade, remote sensing was applied in many tsunami researches, such as tsunami damage detection. Satellite remote sensing can help us survey tsunami damage in many ways. In general, the application of remote sensing for tsunami disasters can be classified into three stages depending on time and disaster-related information. In the first stage, general damage information, such as tsunami inundation limits, can be obtained promptly using an analysis combined with ground truth information in GIS. The tsunami inundation area is one of the most important types of information in the immediate aftermath of a tsunami because it helps estimate the scale of the tsunami's impact. Travel to a tsunami-affected area for field surveys takes a lot of time, given the presence of damaged roads and bridges, with much debris as obstacles. In the second stage, detailed damage interpretation can be analysed; i.e., classification of the building damage level. Recently, the quality of commercial satellite images has improved. These images help us clarify, i.e., whether a house was washed away or survived; they can even classify more damage levels. The third stage combines the damage and hazard information obtained from a numerical simulation, such as the tsunami inundation depth. The damage data are compiled with the tsunami hazard data via GIS. Finally, a tsunami vulnerability function can be developed. This function is a necessary tool for assessing future tsunami risk.

The contents of this chapter are arranged in three sections:

- Satellite image analysis for detecting tsunami-affected areas
- Tsunami damage level classification by visual interpretation and image analysis
- Development of a tsunami vulnerability function by applying a numerical model

2. General satellite image analysis for tsunami-affected areas

2.1 NDVI analysis using optical high-resolution satellite imagery

Tsunami inundation limit

Recent advances in remote sensing technologies have expanded the capabilities of detecting the spatial extent of tsunami-affected areas and damage to structures. The highest spatial resolution of optical imageries from commercial satellites is up to 60–70 centimetres (QuickBird owned by DigitalGlobe, Inc.) or 1 metre (IKONOS operated by GeoEye). Since the 2004 Sumatra-Andaman earthquake tsunami, these satellites have captured images of tsunami-affected areas, and the images have been used for disaster management activities, including emergency response and recovery. To detect the extent of a tsunami inundation zone, NDVI (Normalised Difference Vegetation Index) is the most common index obtained from the post-event imagery, focusing on the vegetation change due to the tsunami penetration on land. The NDVI is calculated from these individual measurements as follows:

$$NDVI = \frac{NIR - R}{NIR + R} \tag{1}$$

where R and NIR stand for the spectral reflectance or radiance in the visible (red) and near-infrared bands, respectively. Focusing on the existence of tsunami debris, 100 points were sampled to identify the NDVI threshold to classify the tsunami inundation zone. As shown in Fig. 1, the NDVI values are calculated within a range 0.34 ± 0.05. As a result, the extent of the tsunami inundation zone is determined by the supervised classification based on the NDVI threshold. As shown in Figure 2 (a), the QuickBird imagery clearly detects the vegetation change between pre- and post-tsunami. Tsunami debris can be seen along the edge of the tsunami inundation zone. Figure 2 (b) shows the result of the detection of the tsunami inundation zone by applying the threshold value of NDVI, and the result is consistent with the field survey.

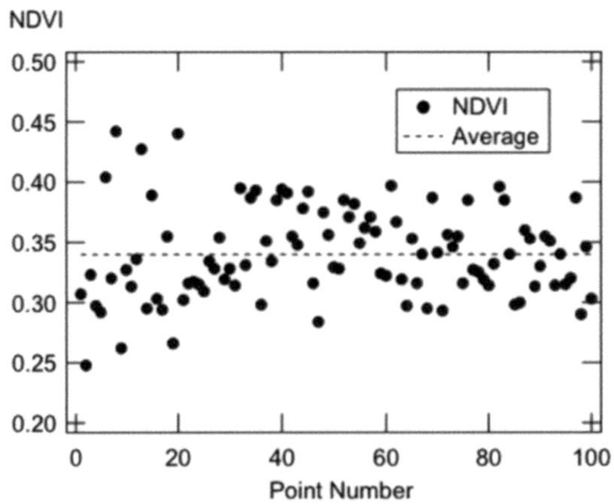

Fig. 1. Threshold value of NDVI within the tsunami inundation zone obtained from the analysis of the post-tsunami satellite imagery.

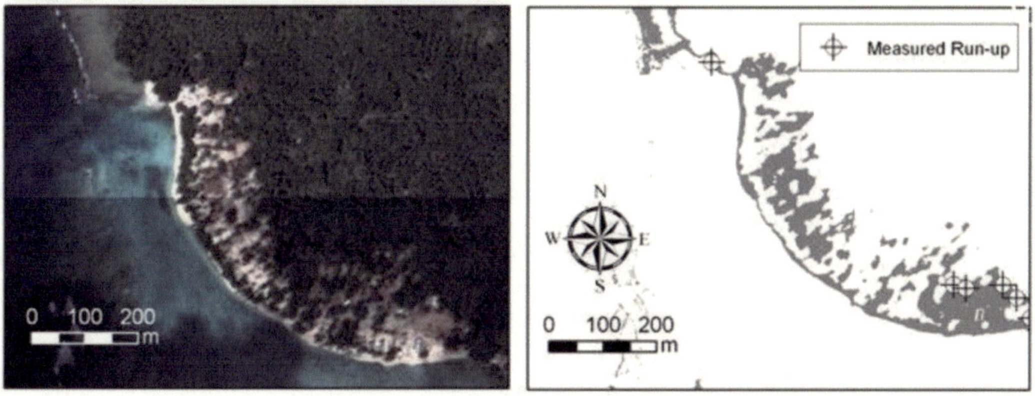

Fig. 2. (a) Vegetation change found from pre- and (b) post-event imageries and estimated extent of tsunami inundation zone by the supervised classification of NDVI.

Damage and recovery monitoring of mangrove

Because mitigation and protection against the 2004 Indian Ocean Tsunami was one of the important services that mangrove ecosystems provided in the affected areas, a six-year program to conserve and rehabilitate mangrove forests in the tsunami-impacted areas was implemented by the Thai Government after the tsunami. However, information on mangrove restoration and reforestation is limited to field surveys. Monitoring proposals were applied for a damaged mangrove area. Kamthonkiat et al. (2011) used ASTER images acquired in 2003, 2005 (two months after the 2004 Indian Ocean tsunami), 2006 and 2010 and the analysis using NDVI to monitor the mangrove recovery in tsunami-impacted areas in the southern part of Thailand. Figure 3 depicts the area of mangroves in 2003 in red and the area impacted by the tsunami in 2005 in dark blue and white for the same location. After the mangrove trees were uniformly or homogeneously replanted in the same location in the last quarter of 2005 in Takuapa District, the areas marked in red increased in 2006 and increased still further in 2010, as shown in Fig. 3 (*Note*: red represents vegetation or mangroves, white represents bare soil/sand, and blue/dark blue represent water). The recovery process can be detected, as some parts in light blue became red in 2006, and most became red in 2010 meaning the mangroves recovered to nearly the normal condition before the tsunami attack. These results show the abilities of geoinformatic technologies, especially regarding the time series analysis.

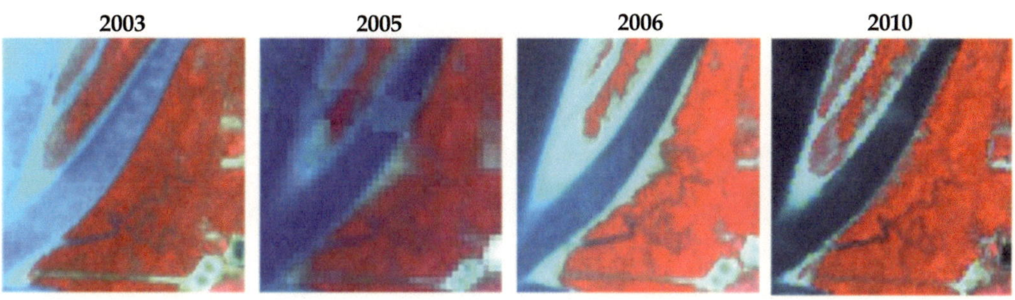

Fig. 3. Damage and recovery process due to the 2004 tsunami in Takuapa, Thailand

2.2 TerraSAR-X image analysis

Among the various sensors, SAR (Synthetic Aperture Radar) is remarkable for its ability to record the physical value of the Earth's surface (Henderson and Lewis, 1998). Unlike passive optical sensors, SAR enables the observation of surface conditions day or night, even through clouds. SAR interferometric analyses using phase information have successfully provided quantification of relative ground displacement levels due to natural disasters (Massonnet et al., 1993). More importantly, intensity information obtained from SAR represents a physical value (backscattering coefficient) that is strongly dependent on the roughness of the ground surface and the dielectric constant. Based on this idea, models for satellite C- and L-band SAR data were developed to detect building damage areas due to earthquakes by clarifying the relationship between the change in the backscattering coefficient from pre- and post-event SAR images (Matsuoka & Yamazaki, 2004; Matsuoka & Nojima, 2009) and then applying the models to tsunami-induced damage areas (Koshimura & Matsuoka, 2010). TerraSAR-X, which is the first German radar satellite with high-resolution X-band, was successfully launched on June 15, 2007, and has been in operation for data acquisition since early 2008. The day after the event, TerraSAR-X observed the coastal area in the affected regions by the StripMap-mode, which captures the Earth's surface with an approximately 3-metre resolution. Typically, man-made structures show comparatively high reflection due to the cardinal effect of structures and the ground. Open spaces or damaged buildings have comparatively low reflectance because they scatter the microwaves in different directions. Buildings may be reduced to debris by earthquake ground motion, and in some cases, the debris of buildings may be removed, leaving the ground exposed. Thus, the backscattering coefficient determined after building collapse is likely to be lower than that obtained prior to the event (Matsuoka & Yamazaki, 2004; Nojima et al., 2006). Inundated areas also show a lower backscattering coefficient because of the smooth surface and the dielectric constant of water bodies (Fig. 4 centre). By examining the backscattering characteristics of tsunami damage in typical areas, however, the reverse case occurred in some damaged areas in farmlands and controlled forests. To explain these anomalies in the post-tsunami TerraSAR-X image, several factors need to be considered, such as changes of the Earth's surface and its materials. Scattered debris from collapsed buildings, visible in the farmlands and bare ground in the post-tsunami image, show brighter reflections than in the pre-tsunami image (Fig. 4 centre).

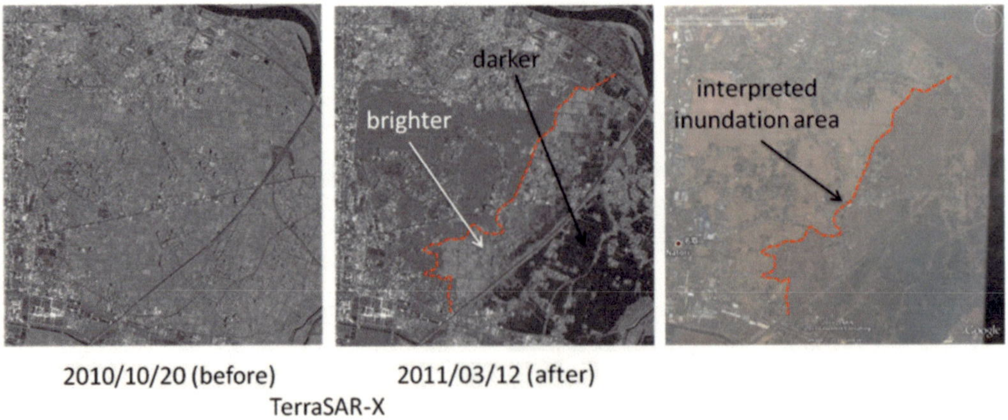

Fig. 4. Comparison between TerraSAR-X image and IKONOS (GeoEye) image

For the areas where some trees in the forest were washed away, the significant backscattering characteristics changed from volume scattering to surface scattering with significant roughness. These kinds of characteristics affecting the backscattering echo were identified in the tsunami-affected areas in the TerraSAR-X image. Following Nojima et al. (2006), the regression discriminant function for building damage was calculated from two characteristic values, the correlation coefficient and the difference in backscattering coefficient for pre- and post-event SAR images. First, following the accurate positioning of the two SAR images, a speckle noise filter with a 21×21 pixel window (Lee, 1980) was applied to each image. The difference value, d, is calculated by subtracting the average value of the backscattering coefficient within a 13×13 pixel window in the pre-event image from the post-event image (after – before). The correlation coefficient, r, is also calculated from the same 13×13 pixel window (Matsuoka & Yamazaki, 2004). The result of applying regression discriminant analysis, using the d and r, is shown in Equation (2).

$$Z_{R1} = -A \cdot d - B \cdot r \quad (2)$$

Here, Z_{R1} represents the discriminant score from the SAR images where the values of parameter A and B are 1.21 and 4.36, respectively. The pixels whose Z_{R1} value is positive (red) are interpreted as suffering severe damage (Fig. 5 left). Because both coefficients are negative, higher and negative d or smaller r produce larger Z_{R1} values. A preliminary formula for the C-band dataset was used because that for the X-band was unavailable. For this reason, the backscattered echoes were stronger in the post-tsunami image. To detect such damaged areas using image analysis, cases where the reverse occurs need to be considered. Therefore, the following Equation (3) was also calculated based on a positive value for the difference in backscattering coefficient d.

$$Z_{R2} = A \cdot d - B \cdot r \quad (3)$$

Here, Z_{R2} represents another discriminant score where the values of parameters A and B are 1.21 and 4.36, respectively. Using this formula, the pixels whose Z_{R2} value is positive (red) might be assigned as damaged areas (Fig. 5 centre).

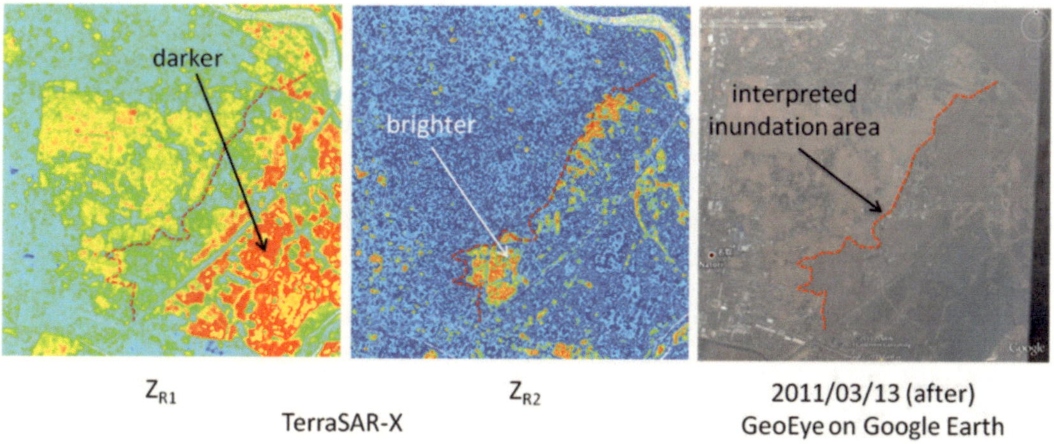

Fig. 5. Computed Z_{R1} and Z_{R2} from TerraSAR-X image to determine inundation area

Two discriminant scores, Z_{R1} and Z_{R2}, were calculated for the TerraSAR-X image pair using the described procedure. The threshold values for Z_{R1} and Z_{R2} were determined to be 6 and 0, respectively. The extracted areas where the Z_{R1} is larger than 6 or the Z_{R2} is larger than 0 are shown in red in Fig. 6.

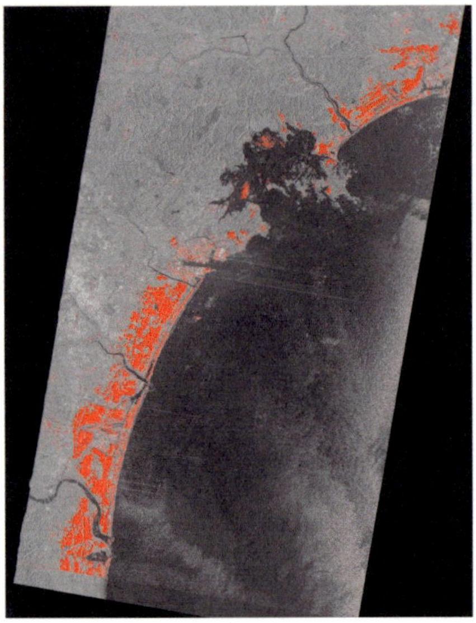

Fig. 6. Threshold values for Z_{R1} and Z_{R2} in the case of the 2011 Tohoku tsunami in Miyagi prefecture. The tsunami inundation area was extracted when the Z_{R1} was larger than 6 or the Z_{R2} was larger than 0.

3. Tsunami damage detection and classification by remote sensing

This section mainly focuses on how remote sensing is used for further research on the detailed classification of tsunami damage areas using structural damage as an example. By taking advantage of satellite remote sensing, the spatial distribution of structural damage by a tsunami can be identified. SAR images are widely used to determine tsunami-affected or inundated areas using the reflection property or backscattering coefficient as mentioned in the previous section. However, through inspecting a set of pre- and post-tsunami satellite images visually or manually, the presence of building roofs can be interpreted. The highest spatial resolution of commercial optical satellite imaging is up to 60-70 cm (QuickBird) or 1 m (IKONOS). The advantage of using high-resolution optical satellite images for damage interpretation is the capability of understanding structural damage visually. These images also enable us to comprehend the spatial extent of damage at the regional scale, where post-tsunami surveys hardly penetrate because of limited of survey time and resources. However, note that no structural types were identified by the interpretation of the satellite images. Additionally, the damage feature that can be identified from the satellite images is only structural destruction or major structural failure, which reveals the change of a roof's shape, namely "collapsed" and "major or severe damage." Accordingly, the interpretation "Destroyed" means "Collapsed" or "Major or severe damage," and "Survived" is classified as "Moderate," "Minor," "Slight" and "No" damage. An example of building damage classification is shown in Fig. 7.

Damage class	Pre and Post satellite images	Criteria for classification
Survived	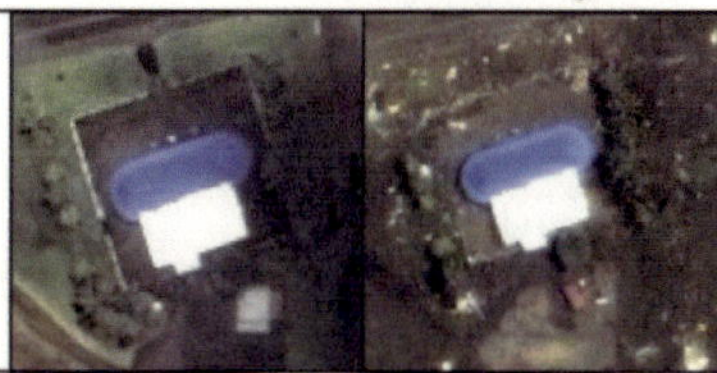	Change of the roof between pre and post buildings can not be found.
Major		Change of the roof between pre and post buildings can be found clearly but in a small scale.
Collapsed		Change of the roof between pre and post buildings can be found clearly but in a large scale.
Washed away		All of the buildings are washed away, and only the foundation of the buildings can be found.

Fig. 7. Example of building damage classification criteria for the 2009 Samoa tsunami

3.1 The 1993 Hokkaido Nansei-Oki tsunami

In 1993, a tsunami accompanied by a M7.8 earthquake off the south-west coast of Hokkaido, Japan, struck Okushiri Island, which is 30 kilometers west of Hokkaido, within 5 minutes after the quake, causing more than 200 casualties. In particular, the Aonae district in the southernmost area of Okushiri Island suffered devastating damage due to an approximately 11-m tsunami that struck from the west coast of the island as well as fire caused during and after the tsunami attack (Murosaki, 1994). Visual damage inspection was conducted using pre- and post-tsunami aerial photographs acquired on 29 October 1990 and 14 July 1993 (one day after the event occurred), as shown in Fig. 8. Because the Aonae district suffered from extensive fire during and after the tsunami attack, it is not possible to discriminate between tsunami and fire damage by the aerial photographs alone. Thus, focusing on the existence of house roofs, the structural damage was categorised into five classes according to the damage area, whether flooded or burned, reported by Shuto (2007). The number of inspected houses and structures was 769, and the result of the structural damage interpretation in Aonae district is shown in Table 1 (Koshimura et al., 2009a). The method to detect the damaged area using SAR image analysis was applied to the tsunami-affected area in Okushri Island. Using a set of pre- and post-tsunami SAR images acquired by JERS (Japanese Earth Resources Satellite), Matsuoka & Yamazaki (2002) calculated the correlation and difference in the backscattering coefficient to represent the changes in the tsunami-affected area. To detect the impacted area, the discriminant score, Equation (2), was

Damage category	Cause	Number of houses
Destroyed or Major damage	Flooded by tsunami	417
Destroyed, Burned or Major damage	Flooded by tsunami and burned by fire	123
Burned or Major damage	Burned by fire	75
Destroyed	Unknown	11
Survived (Moderate, slight or no damage)	–	143

Table 1. Results of structural damage interpretation in Aonae district, Okushiri Island

incorporated, and the values of parameters A and B were modified to 1.277 and 2.729, respectively. Fig. 8 shows a comparison among the results of the visual damage interpretation of the aerial photographs, the post-tsunami JERS/SAR image (Fig. 8(b)) and the discriminant score Z_{R1} (or Z_{Rj}). It is found that Z_{R1} represents relatively larger values in severely impacted areas and that Z_{R1} (Fig. 8(c)) is likely to be fairly consistent with the results of the visual interpretation (Fig. 8(a)). To increase the capability of the SAR image analysis to detect the tsunami impacted area, further discussion is required to explore the relationships between Z_{R1} and the structural damage probability by correlating both with regard to the JERS/SAR resolution.

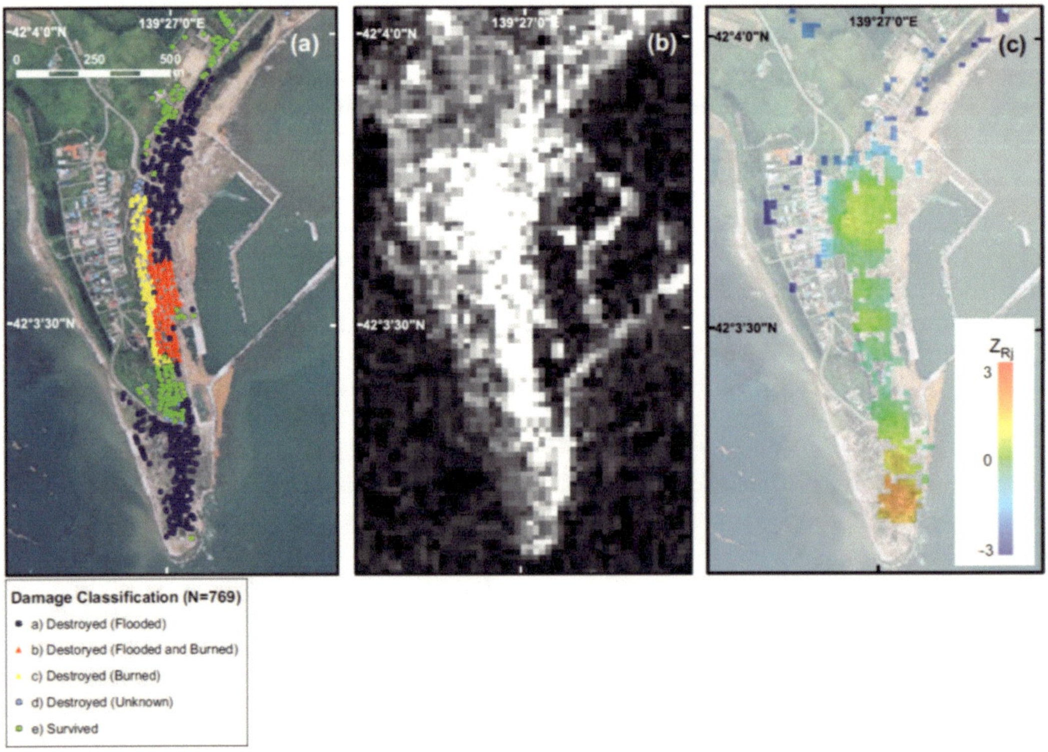

Fig. 8. Comparison among (a) the result of the visual damage interpretation of the aerial photographs, (b) the post-tsunami JERS/SAR image and (c) the discriminant score Z_{R1}

3.2 The 2004 Indian Ocean tsunami

The 2004 Indian Ocean megathrust earthquake occurred on 26 December 2004, creating a gigantic tsunami striking coastal communities over a large area. The earthquake, with a magnitude of 9.3 was the second largest ever recorded and caused the deadliest tsunami disaster in history. The tsunami devastated 11 Asian and African countries, and at least 282,517 people lost their lives. There were two locations to which satellite images were applied for tsunami damage detection, Indonesia and Thailand.

3.2.1 Banda Aceh, Indonesia

Banda Aceh, a city in northern Sumatra, Indonesia, suffered more than 70,000 casualties and 12,000 house damage incidents during the 2004 event. We acquired the post-tsunami survey data from JICA (2005), which was based on a visual interpretation of the pre- and post-tsunami satellite imageries (IKONOS) with some random field checks, focusing on the existence of the individual structures' roofs. Figure 16 indicates the post-tsunami survey result in terms of structural damage in the city by JICA (2005). As shown in the right panels of the figure, the use of high-resolution optical satellite images has the capability to detect individual damages and be utilised as a promising technology for post-disaster damage investigation. Throughout the visual inspection of the two satellite images, the remaining roofs were interpreted as "Survived" and the roofs that disappeared as "Destroyed". The total number of inspected buildings in the tsunami-inundated area was 48,910, of which 16,474 were interpreted as destroyed and 32,436 as survived, as shown in Fig. 9 (Koshimura et al., 2009c).

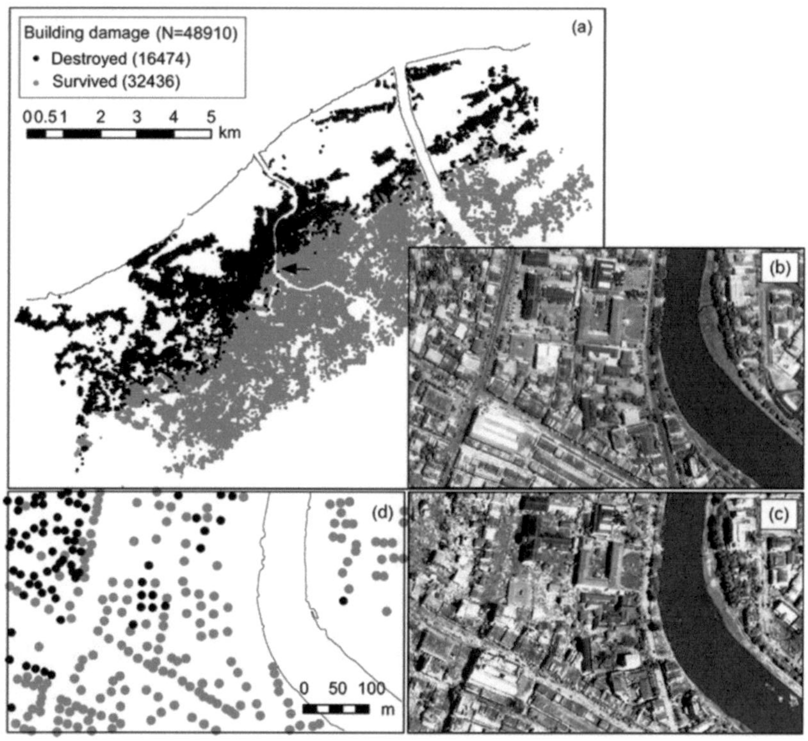

Fig. 9. Visual damage inspection results in Banda Aceh, Indonesia

3.2.2 Phang Nga and Phuket, Thailand

Phang Nga and Phuket were two of six southern Andaman coast provinces that were damaged by the tsunami. They provinces are famous for sightseeing areas such as Khao Lak and Patong. Therefore, reinforced concrete (RC) buildings are common in this area. Regarding structural damage, 4,806 houses were affected by the tsunami, of which 3,302 houses were destroyed completely, and as many as 1,504 were partly damaged. The maximum water level of approximately 15 m reported at Khao Lak in the Phang Nga province and of 7 m at Kamala and Patong Beach in Phuket gave these areas their respective distinction as the worst and second-worst areas, with structural damage to 2,508 and 1,033 houses, respectively. High-resolution satellite images (IKONOS) taken before and after the tsunami event were used for visual damage interpretation. The pre-event images were acquired on 13 January 2003 and 24 January 2004 for Phang Nga and Phuket; the post-event images were both acquired on 15 January 2005. In a recent study (Gokon et al., 2011), four damage levels were classified "Not collapsed" (moderate, slight or no damage), "Major damage", "Collapsed" and "Washed away," using a QuickBird satellite image with a 0.6×0.6 m^2 resolution. However, the 1.0×1.0 m^2 resolution of the IKONOS satellite image is not fine enough for a visual interpretation to differentiate the damage levels of buildings. Therefore, the classification of the building damage in this study was limited to "Not destroyed" and "Destroyed" (Koshimura et al., 2009c). The remaining roof buildings were interpreted as "Not destroyed" and those that had disappeared were classified as "Destroyed". Note that the buildings classified as "Not destroyed" may have had some sort of Damage that could be identified by the satellite images. The results of the building damage inspection in residential areas are presented in Fig. 10 (Suppasri et al., 2011a), which shows damaged buildings in residential areas in Khao Lak, Phang Nga province (1,722 destroyed and 1,285 not destroyed) and the populated residential areas in Kamala and Patong, Phuket province (233 destroyed and 1,356 not destroyed). The visual interpretation data resulted in an accuracy of more than 90 per cent after being checked with the investigation data.

Fig. 10. Visual damage inspection results in Phang Nga and Phuket, Thailand

3.3 The 2007 Solomon Islands tsunami

The 2007 Solomon Islands earthquake took place on 1 April 2007 near the provincial capital of Ghizo on Ghizo Island, in the Solomon Islands. The magnitude of this earthquake was calculated by the United States Geological Survey (USGS) as 8.1 on the moment magnitude scale. The tsunami that followed the earthquake killed 52 people. The structural/house damage was focused on Ghizo Island and was caused by the tsunami. First, the QuickBird pan-sharpened composite images of Ghizo Island were acquired pre- and post-tsunami (23 September 2003 and 5 April 2007) to build house inventories for visual damage inspection, as shown in Fig. 11 (Koshimura et al., 2010). The extent of the tsunami inundation zone is determined by the supervised classification based on the NDVI of the post-tsunami satellite imagery (Fig. 12), as already shown in section 2.1.

Fig. 11. The structural damage interpretation is divided into four classes: slight/no damage, substantial damage, collapsed and washed away

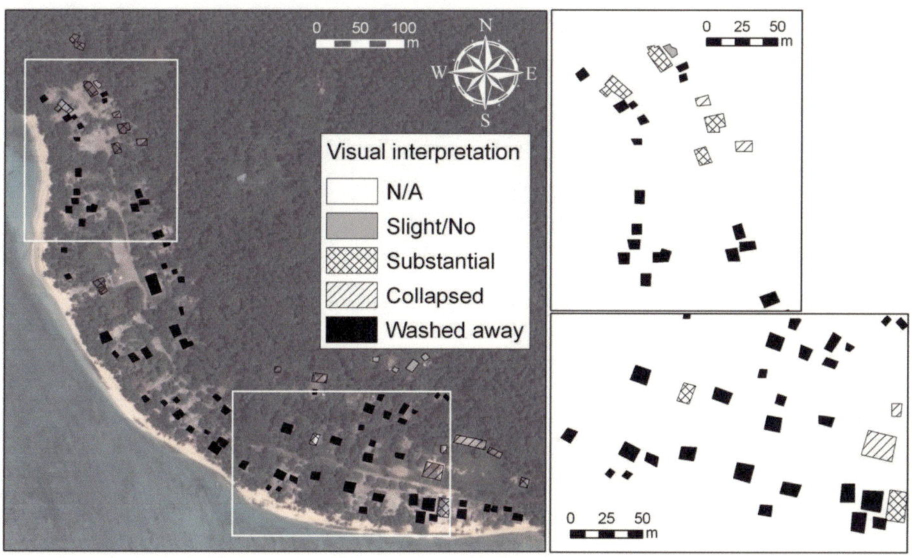

Fig. 12. Visual damage inspection results for Ghizo Island, Solomon Islands

3.4 The 2009 Samoa Islands tsunami

In 2009, a tsunami accompanied by a M8.1 earthquake off the southwest coast of Tutuila Island, American Samoa, struck the Samoa and Tonga islands and caused a total of 184 deaths and 7 missing. A visual damage inspection was conducted using pre- and post-tsunami QuickBird images acquired on 15 April 2007, 24 September 2009, 29 September 2009, 02 October 2009 and November 2009. The damaged structures were classified into four categories: washed-away, collapsed, major damage and survived (as previously mentioned in Fig. 7). The number of inspected houses and structures in the four study areas, namely, Pago Pago, Amanave, Poloa and Leone, totalled 451, and the results are summarised in Fig. 13 and Table 2 (Gokon et al., 2011).

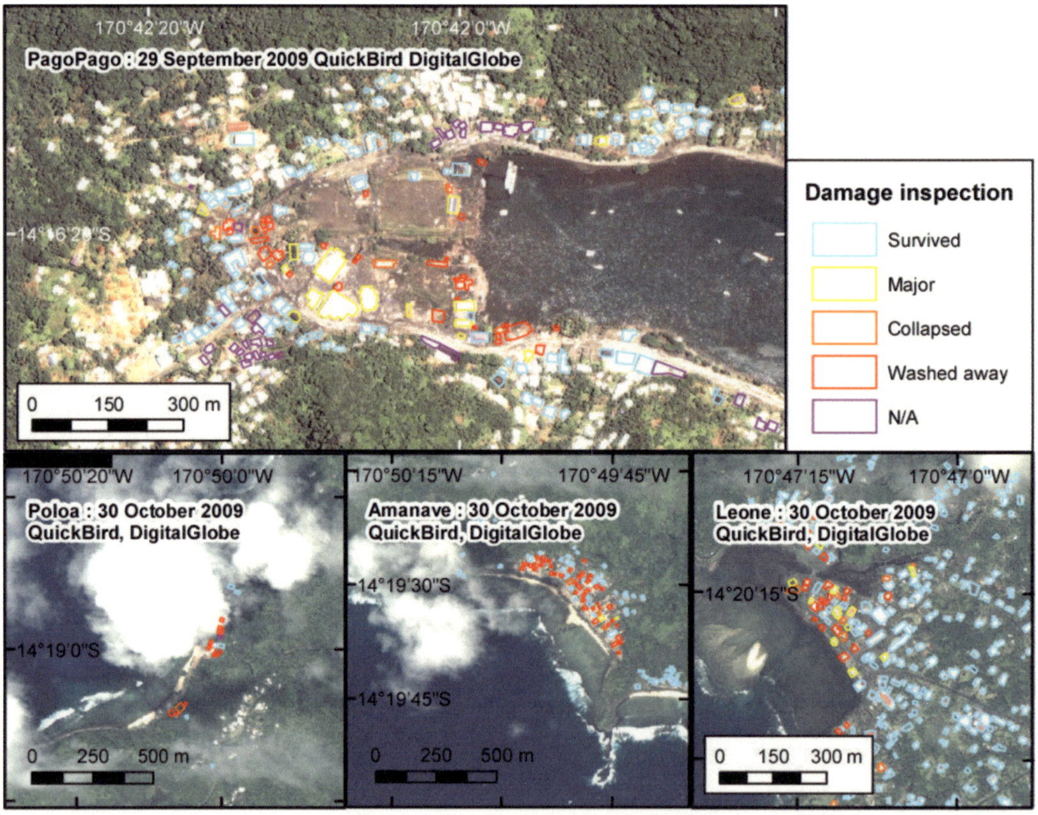

Fig. 13. Visual damage inspection results in Tutuila Island, American Samoa

Damage category	Number of houses (Pago Pago/Amanave/Poloa/Leone/Total
Washed-away	34/42/13/28/117
Collapsed	7/3/1/7/18
Major damage	142/0/12/28
Survived	5434/4/196/288

Table 2. Results of structural damage interpretation in Tutuila Island, American Samoa

3.5 The 2010 Chile tsunami

A moment magnitude 8.8 earthquake struck the central region of Chile on February 27, 2010. The earthquake produced a tsunami that caused major damage in locations spanning over 500 km of coastline, from Tirúa to Pichilemu. The coastal locations were affected by both ground shaking and the tsunami. As of May 2010, 521 people had died and 56 persons were still missing. The earthquake and tsunami destroyed over 81,000 houses, and another 109,000 were severely damaged. Following Matsuoka & Nojima (2009), the regression discriminant function for building damage was calculated from two characteristic values: the correlation coefficient and the difference in the backscattering coefficient for pre- and post-event SAR images (Matsuoka & Koshimura, 2010). First, following the accurate positioning of the two SAR images, a speckle noise filter with a 21×21 pixel window was applied to each image. The difference value, d, is calculated by subtracting the average value of the backscattering coefficient within a 13×13 pixel window in the pre-event image from the post-event image (after - before). The correlation coefficient, r, is also calculated from the same 13×13 pixel window. The result of applying regression discriminant analysis, using the d and r from the building damage dataset of the 1995 Kobe earthquake, is shown in Equation (2), where the values of parameters A and B are modified to 1.277 and 2.729, respectively. Here, Z_{R1} in Equation (2) represents the discriminant score from the SAR images. The pixels whose Z_{R1} value is positive are interpreted as suffering severe damage. Because both coefficients are negative, higher and negative d or smaller r produce larger Z_{R1} values. However, in the tsunami damage areas in the PALSAR images in the abovementioned examination, the backscattered echoes were stronger in the post-tsunami image. To detect such damaged areas using image analysis, cases where the reverse occurs need to be considered. Therefore, the absolute value of the difference in the backscattering coefficient, $|d|$, was calculated, which changed the coefficient of the difference to positive values, as shown in Equation (3), where the values of parameters A and B are modified to 1.277 and 2.729, respectively. Here, Z_{R2} represents the modified discriminant score. Using this formula, the pixels whose Z_{R2} value is positive might be assigned as areas damaged not only by earthquakes but also by tsunamis. Using the procedure described above and the PALSAR images of the 2010 Chile earthquake tsunami, discriminant scores Z_{R2} were calculated in the areas shown to be vulnerable on the inundation susceptibility maps, and the tsunami damage distribution was estimated.

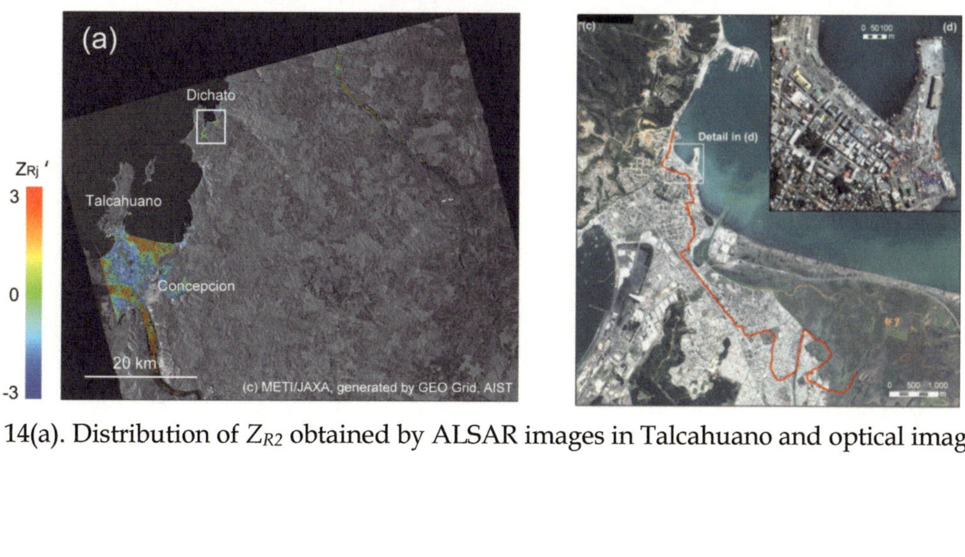

Fig. 14(a). Distribution of Z_{R2} obtained by ALSAR images in Talcahuano and optical images

The results are shown in Fig. 14 (a) and (b). The sections on the sea are masked, but the areas where the river could not be masked have large Z_{R2} values because of the surface changes caused by the flow of water. The wetlands near Talcahuano and Llico, where the Z_{R2} values are large, seem to be affected by the tsunami. Figure 14(b) shows a close-up Z_{R2} image of the Dichato area, with a comparison pre- and post-tsunami from an optical image (Koshimura et al., 2011).

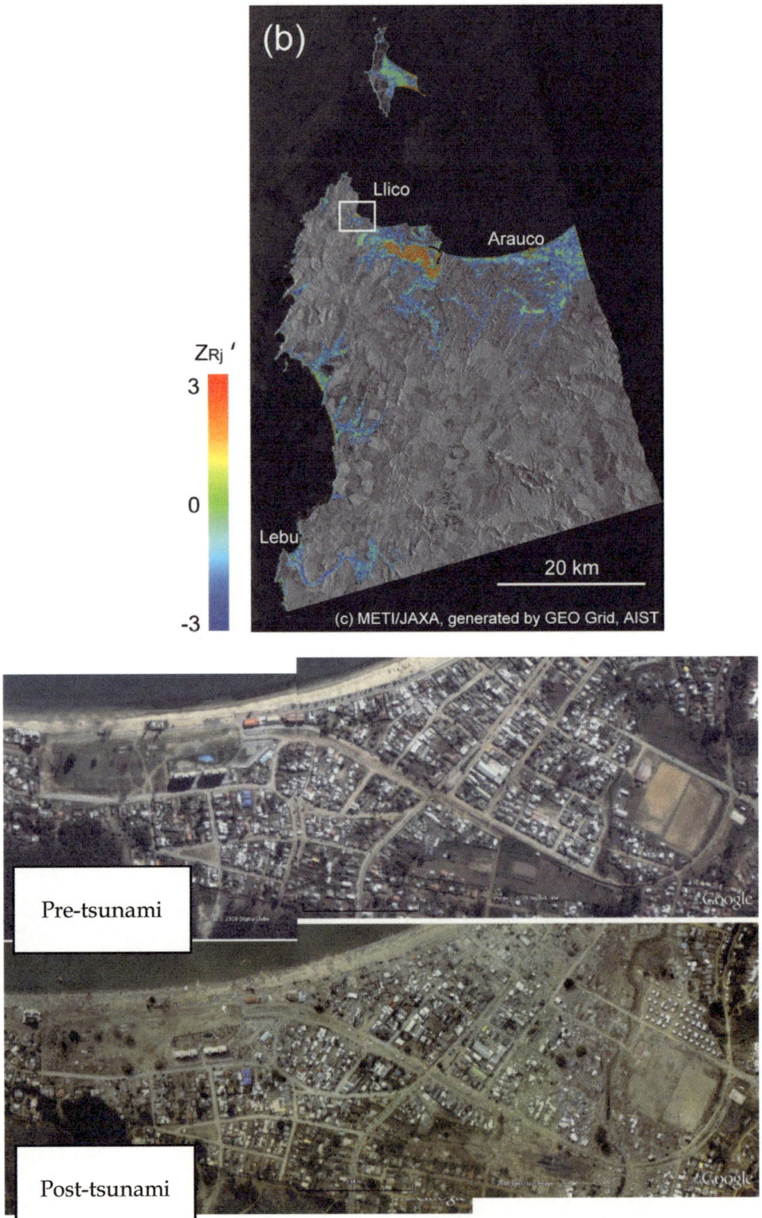

Fig. 14(b). Distribution of Z_{R2} in a close-up of the Dichato area and comparison of optical images pre- and post-tsunami

3.6 The 2011 Tohoku tsunami

On March 11, 2011, a giant earthquake of M9.0, whose epicentre was located off the eastern part of Miyagi prefecture, Japan, caused catastrophic damage to the coastal area facing the Pacific Ocean of the Tohoku district. This earthquake caused an enormous tsunami with a run-up height that reached 40 m and destroyed approximately 270,000 houses. Aerial photos that were captured on March 12, 13 and 19 and April 1 and 5 in 2011 by GSI were used to classify the electronic building map into 2 classes: washed-away or surviving (Gokon & Koshimura, 2011). First, the panels of ortho photos, with a resolution of 80 cm/pixel, are combined with mosaic image processing. Then the electronic map and the aerial photos were integrated into the same coordinate system in ArcGIS. Finally, a visual inspection was performed for the building damage one by one (washed-away or surviving) for all the buildings in the inundation area in Miyagi prefecture, Japan. Housing damage characteristics can be explained by bathymetry conditions as follows: the Ria coast, i.e., the towns of Minami-Sanriku in Fig. 15 (upper-left), has the potential to amplify the tsunami height. As a result, the probabilities of the washed away houses in the inundation area are estimated to be over 70%. In Ishinomaki city, the number of washed away houses is small in an area located behind the breakwaters and control forests. The effect of the breakwaters and control forests in reducing tsunami damage is shown in Fig. 15 (lower-left). Most of the buildings in Matsushima town and Shiogama city, located in a bay with a small opening and almost 270 small Islands acted as natural barrier, survived the tsunami, as shown in Fig. 15 (right).

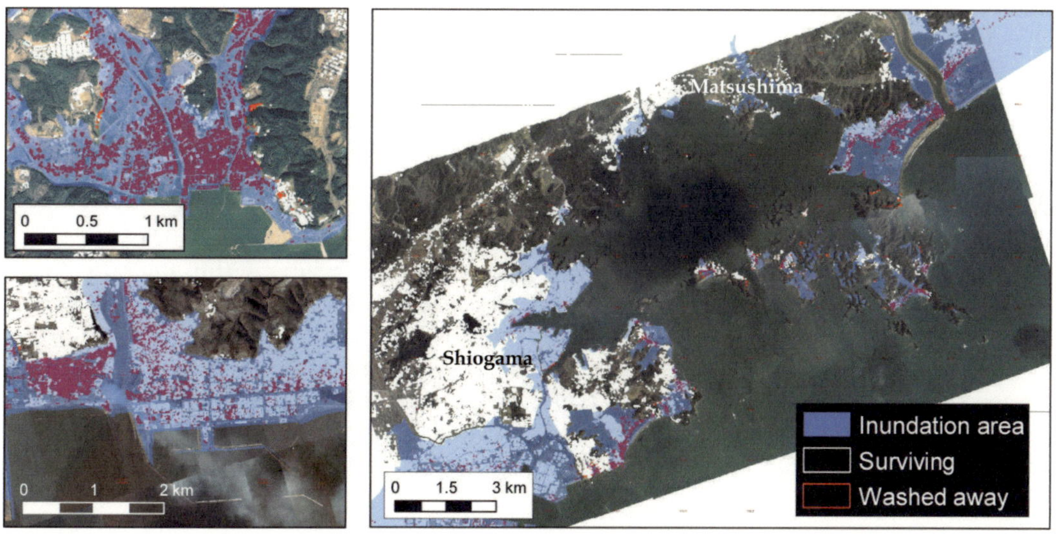

Fig. 15. Visual damage inspection results in Minami-Sanriku town (upper-left), Ishinomaki city (lower-left), and Matsushima town and Shiogama city (right). The red rectangles show washed away houses and the blue areas indicate tsunami inundation areas.

4. Developing a tsunami vulnerability function by applying a numerical model

The next step is to apply the previous damage inspection data with the tsunami numerical model. One method is to develop a fragility curve (Koshimura et al., 2009b). The tsunami fragility curve is a function used to estimate the structural fragility against tsunami hazards. Visual inspections of satellite images taken before and after tsunami events are to be used to classify whether the buildings were destroyed or not based on the remaining roofs. Then a tsunami inundation model is created to reconstruct the tsunami features, such as inundation depth, current velocity, and hydrodynamic force of the event. For the tsunami inundation model, a set of nonlinear shallow water equations are discretised using the Staggered Leap-frog finite difference scheme (Imamura, 1995), with the bottom friction in the form of Manning's formula according to a land use condition. In general, two methods exist for modelling flow resistance depending on the relation between the scale of an obstacle and the grid size: the topography model and the equivalent roughness model. The topography model is used when the grid size is finer than the obstacle. The tsunami in the model simulation will not pass into a grid space that is occupied by an obstacle. Then the flow around an obstacle and the contracting flow between obstacles can be simulated. However, in a larger grid size, such as that of this study, the obstacle is smaller than the grid size. The equivalent roughness model is then appropriate for this problem. In a non-residential area, the roughness coefficient is inferred from land use, and it is used to quantify the Manning's roughness coefficient ($s \cdot m^{-1/3}$). The lowest Manning's roughness coefficient is 0.02 for smooth ground, followed by 0.025 for shallow water or natural beach and by 0.03 for vegetated area. However, Manning's roughness coefficient in a densely populated area is highly affected by the number of buildings in each computational grid. In a densely populated town, in which the building occupation ratio is high, the resistance law with the composite equivalent roughness coefficient according to land use and building conditions was first studied by Aburaya & Imamura (2002), as shown in Equation (4).

$$n = \sqrt{n_0^2 + \frac{C_D}{2gw} \times \frac{\theta}{100 - \theta} \times D^{4/3}} \tag{4}$$

In this equation, n_0 signifies the Manning's roughness coefficient (n_0=0.025, $s \cdot m^{-1/3}$), θ denotes the building/house occupation ratio in percentage varying within the range from 0 to 100 in the finest computational grid of 52 m and obtained by calculating the building area over the grid area using GIS data. C_D represents the drag coefficient (C_D=1.5), w stands for the horizontal scale of houses, and D is the modelled flow depth. Fragility curves can be developed for various types, such as building material (wood, block or reinforced concrete), number of floors and country. Developed tsunami fragility curves are crucial for future tsunami risk assessment when tsunami hazards and exposure data are given.

4.1 Method and procedure for developing tsunami fragility curves

To develop tsunami fragility curves, a statistical approach is used with a synergistic use of the numerical model results and damage data by the procedure itemised below.

Damage data acquisition

The damage data was obtained from pre- and post-tsunami aerial photographs (e.g., number of destroyed or surviving structures).

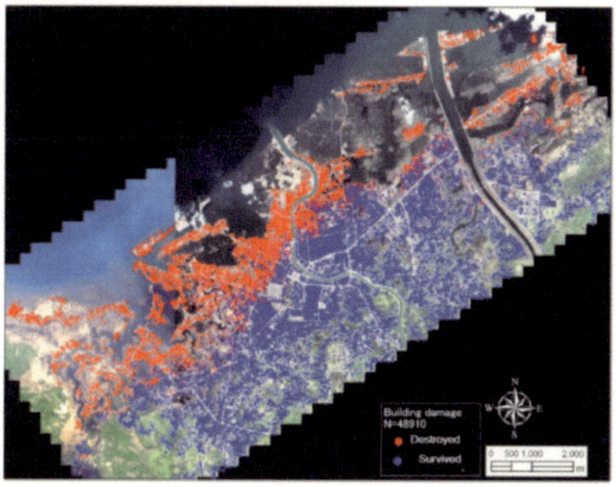

Fig. 16. Tsunami damage detected by the visual interpretation of IKONOS pre- and post-tsunami imageries. The red dots indicate totally damaged houses and the blue dots not-damaged.

Tsunami hazard estimation

Speculate the hydrodynamic feature of tsunamis by numerical modelling.

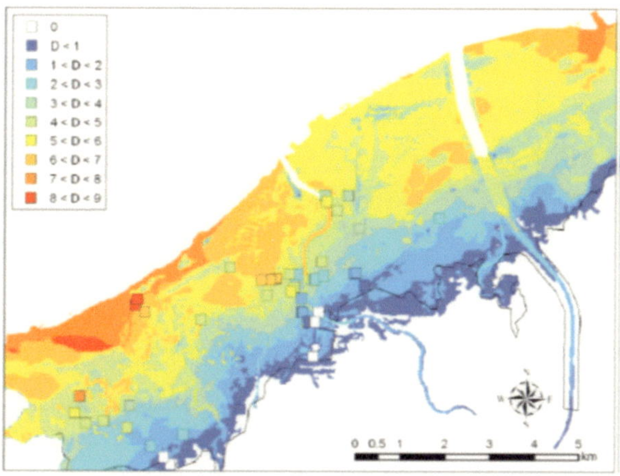

Fig. 17. Modelled tsunami inundation in the city of Banda Aceh. The result is validated by the measured flow depth shown with the squares in the figure.

Data assimilation between the damage data and tsunami hazard information

Correlate the damage data and the hydrodynamic features of tsunami inundation through the GIS analysis.

Sample determination

Sample sorting by the level of hydrodynamic features to explore an arbitrary range of these features such that each range includes the determined number of samples; check the data distribution.

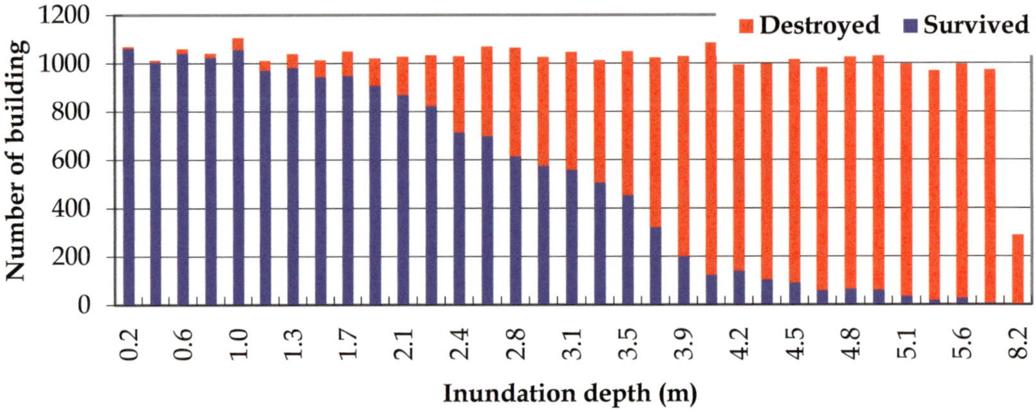

Fig. 18. Histogram of damaged and not-damaged houses to calculate the damage probability

Calculating damage probability

Calculate the structural damage probabilities by counting the number of destroyed or surviving structures within each range of the tsunami hydrodynamic features described above.

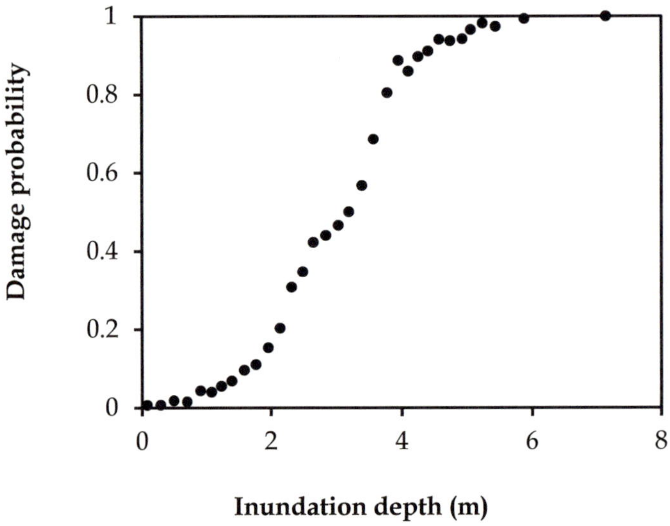

Fig. 19. The plot of damage probabilities and the median values of inundation depths that were compiled from sample data

Regression analysis

Determine the fragility curves by the regression analysis of the discrete set of the structural damage probabilities and hydrodynamic features of a tsunami. The damage probabilities of buildings and a discrete set were calculated and shown against a median value within a range. Linear regression analysis was performed to develop the fragility function.

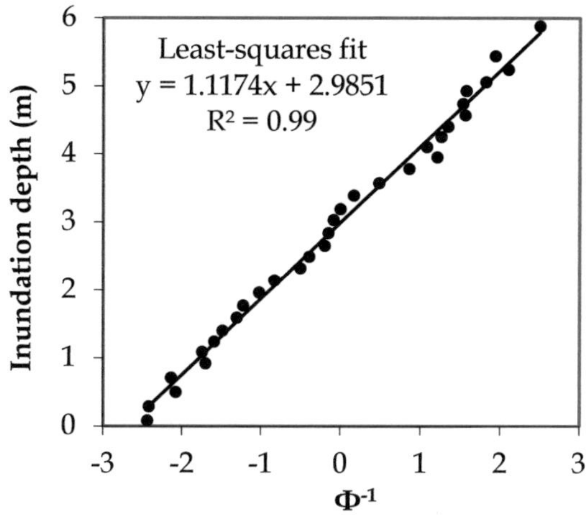

Fig. 20. An example of the plot on normal probability paper

Taking as an analogy earthquake engineering studies, the cumulative probability *PD* of damage occurrence is assumed to be given with two statistical parameters, (μ, σ) or (μ', σ'). The cumulative probability *P* of the occurrence of damage is given either by Equation (5) or (6):

$$P(x) = \Phi\left[\frac{x-\mu}{\sigma}\right] \tag{5}$$

$$P(x) = \Phi\left[\frac{\ln x-\mu'}{\sigma'}\right] \tag{6}$$

In these equations, Φ represents the standardised normal (log-normal) distribution function, x stands for the hydrodynamic feature of the tsunami (e.g., inundation depth, current velocity and hydrodynamic force), and μ and σ (μ' and σ'), respectively, signify the mean and standard deviation of x ($\ln x$). Two statistical parameters of the fragility function, μ and σ (μ' and σ'), are obtained by plotting x ($\ln x$) against the inverse of Φ on normal or log-normal probability papers and performing a least-squares fitting of this plot. Consequently, two parameters are obtained by taking the intercept (= μ or μ') and the angular coefficient (= σ or σ') in Equations (7) or (8):

$$x = \sigma\Phi^{-1} + \mu \tag{7}$$

$$\ln x = \sigma'\Phi^{-1} + \mu' \tag{8}$$

Throughout the regression analysis, the parameters are determined as shown in Table 8 to obtain the best fit of fragility curves with respect to the inundation depth, the maximum current velocity and the hydrodynamic force on structures per unit width. Here, the hydrodynamic force acting on a structure is defined as its drag force per unit width as

$$F = \frac{1}{2}C_D\rho u^2 D \tag{9}$$

where C_D denotes the drag coefficient (C_D = 1.0 for simplicity), ρ is the density of water (= 1,000 kg/m^3), u stands for the current velocity (m/s), and D is the inundation depth (m). From this result, all the fragility functions with respect to the inundation depth, current velocity and hydrodynamic force are given by the standardised lognormal distribution functions with μ' and σ'. It should be noted that because the damage interpretation using the pre- and post-tsunami satellite images focused on whether the houses' roofs remained, we supposed that the structural damage was caused by the tsunami inundation. Additionally, note that the tsunami damage to structures was caused by both hydrodynamic force/impact and the impact of floating debris, i.e., these facts are reflected in the damage probabilities but not in the numerical model results (the estimated hydrodynamic features). In that sense, the present fragility functions might indicate overestimation in terms of the damage probabilities to the hydrodynamic features of the tsunami inundation flow.

4.2 Tsunami fragility curves for Okushiri Island, Japan

The task of discriminating between the damage caused by tsunami inundation or by fire was quite speculative. Thus, fragility curves were developed using 523 houses within the inundation zone estimated by the numerical model. A relationship between the damage probability and the tsunami's hydrodynamic features were obtained as a discrete set of structural damage probabilities using a range of approximately 50 buildings and the tsunami hazard. The relationship was explored with the form of a fragility curve by performing the regression analysis. Structural damage is severe when the inundation depth is greater than 3 m, the current velocity is greater than 4 m/s and the hydrodynamic force is greater than 25 kN/m (Fig. 21).

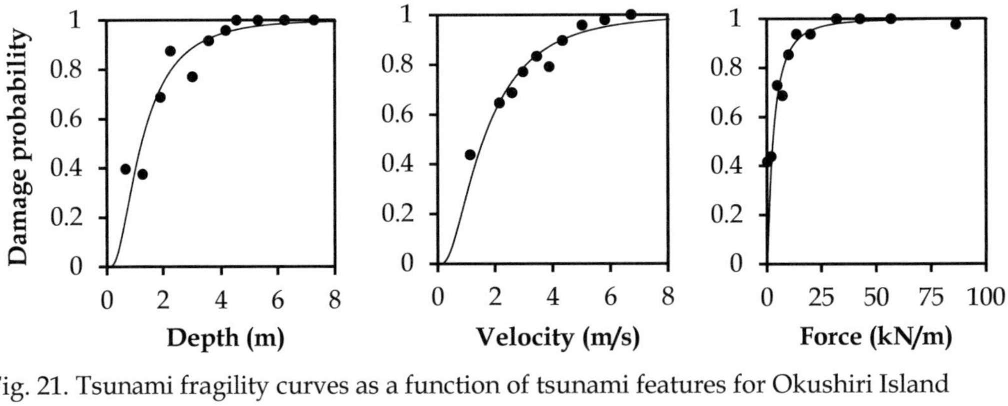

Fig. 21. Tsunami fragility curves as a function of tsunami features for Okushiri Island

4.3 Tsunami fragility curves for Banda Aceh, Indonesia

The number of destroyed buildings in Banda Aceh is 16,474, and the number of surviving building is 32,436 based on the remaining roofs. The damage probabilities of buildings and a discrete set were calculated and shown against a median value within a range of approximately 1,000 buildings. Linear regression analysis was performed to develop the fragility function. As a result, the fragility curves are obtained as Figure 20, indicating the damage probabilities according to the hydrodynamic features of the tsunami inundation flow in Banda Aceh. For instance, the structures were significantly vulnerable when the local inundation depth exceeds 2 or 3 m, the current velocity exceeds 2.5 m/s or the hydrodynamic load on a structure exceeds 5 kN/m (Fig. 22).

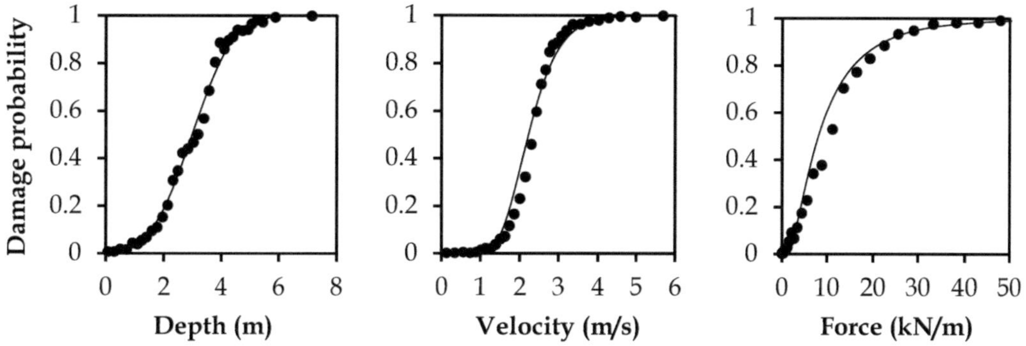

Fig. 22. Tsunami fragility curves as a function of tsunami features for Banda Aceh

4.4 Tsunami fragility curves for Phang Nga and Phuket, Thailand

From the visual inspection of damaged buildings based on the remaining roof structures, a histogram of tsunami features (inundation depth, current velocity, and hydrodynamic force) and the number of buildings, including those not destroyed and those destroyed, was plotted. The damage probabilities of buildings and a discrete set were calculated and shown against a median value within a range of approximately 100 buildings in Phang Nga and 50 buildings in Phuket. Linear regression analysis was performed to develop the fragility function. The differences in damage characteristics of the buildings in Phang Nga and Phuket due to the construction materials are represented by the developed fragility curves in this study (Fig. 23 (a) and (b)). Although the inundation depth of 6 m engenders 100% damage probability in both locations, a lower inundation depth of 2 m is more fragile in Phang Nga: the damage probability would be 25% in Phuket but would be as high as 35% in Phang Nga.

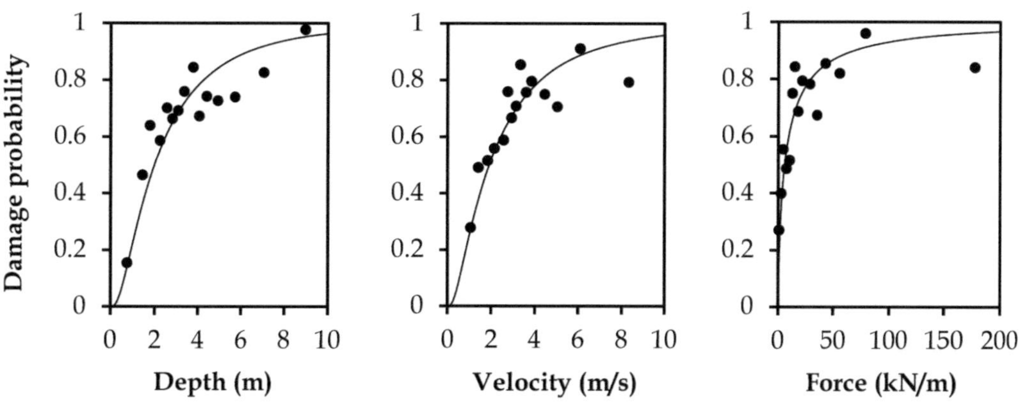

Fig. 23(a). Tsunami fragility curves as a function of tsunami features for Phang Nga

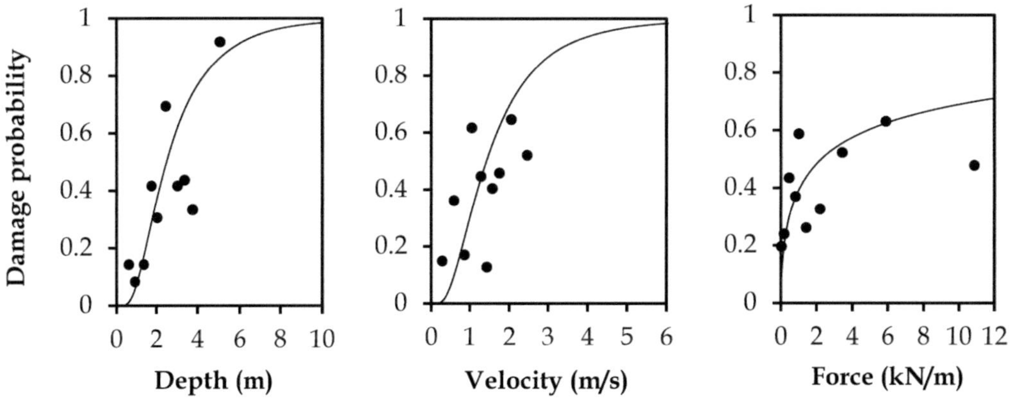

Fig. 23(b). Tsunami fragility curves as a function of tsunami features for Phuket

4.5 Tsunami fragility curves for American Samoa, USA

A visual inspection shows that there were 134 damaged and 210 surviving houses. The damage probabilities were calculated using a range of 20 buildings, and a linear regression analysis was performed. From Fig. 24, 80% of the buildings were damaged when the inundation depth exceeds 6 m. More than half of the buildings were damaged if the current velocity exceeds 2 m/s. The damage due to the hydrodynamic force increased rapidly up to 10 kN/m.

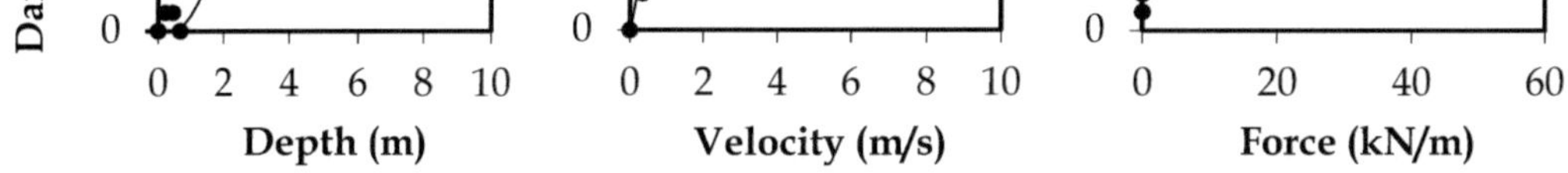

Fig. 24. Tsunami fragility curves as a function of tsunami features for American Samoa

4.6 Summary of statistical parameters for developed fragility curves

Tsunami fragility curves were developed using a numerical model and a visual inspection of satellite images in several countries (Japan, Indonesia, Thailand and American Samoa) with different building materials (wood or reinforced concrete). The necessary statistical parameters for plotting the fragility curves with inundation depth, current velocity and hydrodynamic force are summarised in Table 3.

Event (Year)	Location	Tsunami feature	Building type	μ	σ	μ'	σ'	R2
Nansei Hokkaido (1993)	Okushiri Island	Inundation depth	Mainly wood			0.216	0.736	0.82
		Current velocity				0.475	0.776	0.89
		Hydrodynamic force				1.033	1.186	0.92
Indian Ocean (2004)	Banda Aceh	Inundation depth	Mainly wood & brick	2.985	1.117			0.99
		Current velocity				0.799	0.278	0.97
		Hydrodynamic force				2.090	0.791	0.99
Indian Ocean (2004)	Phang Nga	Inundation depth	Some RC			0.689	0.903	0.80
		Current velocity				0.649	0.952	0.72
		Hydrodynamic force				1.748	1.937	0.75
	Phuket	Inundation depth				0.917	0.642	0.62
		Current velocity				0.352	0.675	0.32
		Hydrodynamic force				0.821	3.000	0.50
Samoa (2009)	American Samoa	Inundation depth	Some RC			1.170	0.691	0.89
		Current velocity				0.541	1.650	0.73
		Hydrodynamic force				1.070	3.160	0.72

Table 3. Summary of statistical parameters for developed fragility curves

5. Conclusion

This chapter introduced how remote sensing can be applied for tsunami research fields. In general, remote sensing is used for rapid and large-scale damage detection to understand the scale of a tsunami, especially when accessibility to disaster-affected areas is limited in the immediate aftermath. Some of the general applications shown in this chapter are related to the tsunami inundation limit, damaged buildings/debris and mangrove recovery monitoring. SAR images are used to determine tsunami-affected areas using the reflection property or backscattering coefficient as mentioned in the previous section. The next step focused on damage classification in a tsunami affected area, i.e., structural damage of housing or buildings. The benefit of high-resolution images from the sky helps tsunami researchers interpret the tsunami damage level based on roofs. A one-metre resolution, such as that of IKONOS, could help classify buildings as destroyed or not destroyed. In addition, a very high-resolution satellite image such as QuickBird (0.6 m resolution) was used to classify a number of levels, i.e., washed-away, collapsed, major damage or survived. Some recent research on tsunami events was introduced, namely, the 1993 Hokkaido Nansei-oki tsunami, the 2004 Indian Ocean tsunami, the 2007 Solomon tsunami, the 2009 Samoa tsunami, the 2010 Chile tsunami and the most recent 2011 Tohoku tsunami. However, information from the sky has some limitations because it is impossible to make a detailed damage inspection of a structural member, and it might have some errors compared with an actual field survey. Finally, classified structural damage data from a visual interpretation of high-resolution satellite images were used in combination with the tsunami numerical simulation to develop tsunami vulnerability curves called tsunami fragility curves. Tsunami features during inundation, such as inundation depth, current velocity and hydrodynamic force can be simulated by the numerical model. The tsunami fragility function can be constructed by combining the inspected damage data and simulated tsunami features using a statistical approach. The developed tsunami fragility curves for each location could be important tools for tsunami risk assessment against potential future tsunamis. However, applying tsunami fragility for future risk evaluation should be performed with care. The structural characteristics and behaviour of housing and buildings differ by country (Fig. 24). For example, an RC-frame building with brick walls is common in Southeast Asian countries. However, wooden walls are commonly used in Japan because of their light weight for reducing damage from earthquakes. These differences cause the tsunami damage characteristics to be different (Suppasri et al., 2011b).

Fig. 25. Examples of building damage in the case of the 2004 Indian Ocean tsunami in Thailand and the 2011 Tohoku tsunami in Japan

6. Acknowledgment

QuickBird images are owned by DigitalGlobe, Inc. and IKONOS images are operated by GeoEye. ASTER and PALSAR images are owned by METI/NASA and METI/JAXA, respectively, and both are processed by GEO Grid, AIST. JERS-1/SAR image is also owned by METI/JAXA. TerraSAR-X image is the property of Infoterra GmbH and distributed by PASCO Corporation. We express our deep appreciation to the Industrial Technology Research Grant Program in 2008 (Project ID: 08E52010a) from the New Energy and Industrial Technology Development Organization (NEDO), the Willis Research Network (WRN) under the Pan-Asian/Oceanian tsunami risk modelling and mapping project and the Ministry of Education, Culture, Sports, Science and Technology (MEXT) for the financial support for this study.

7. References

Aburaya, T. & Imamura, F. (2002). The proposal of a tsunami run-up simulation using combined equivalent roughness, *Proceedings of the Coastal Engineering Conference (JSCE)*, 49, 276–280 (in Japanese)

Gokon, H. & Koshimura, S. (2011). Mapping of buildings damage of the 2011 Tohoku earthquake tsunami, *Proceedings of the 8th International Workshop on Remote Sensing for Post Disaster Response*, Standford University, California, United States, 15-16 September 2011

Gokon, H.; Koshimura, S.; Matsuoka, M. & Namegaya, Y. (2011). Developing tsunami fragility curves due to the 2009 tsunami disaster in American Samoain, *Proceedings of the Coastal Engineering Conference (JSCE)*, Morioka, 9-11 November 2011 (in Japanese)

Henderson, F. M. & Lewis, A. J. (1998). Principles and applications of imaging radar, Manual of Remote Sensing, 2, John Wiley & Sons, Inc., New York

Imamura, F. (1995). Review of tsunami simulation with a finite difference method, Long-Wave Runup Models, *World Scientific*, 25–42

Kamthonkiat, D.; Rodfai, C.; Saiwanrungkul. A.; Koshimura, S. & Matsuoka, M. (2011). Geoinformatics in mangrove monitoring: damage and recovery after the 2004 Indian Ocean tsunami in Phang Nga, Thailand, *Natural Hazards and Earth System Sciences*, 11, 1851–1862

Koshimura, S.; Matsuoka, M. & Kayaba, S. (2009a). Tsunami hazard and structural damage inferred from the numerical model, aerial photos and SAR imageries, *Proceedings of the 7th International Workshop on Remote Sensing for Post Disaster Response*, University of Texas, Texas, United States, 22–23 October 2009

Koshimura, S.; Namegaya, Y. & Yanagisawa, H. (2009b). Tsunami fragility - A new measure to assess tsunami damage, *Journal of Disaster Research*, 4, 479–488

Koshimura, S.; Oie, T.; Yanagisawa, H. & Imamura, F. (2009c). Developing fragility curves for tsunami damage estimation using numerical model and post-tsunami data from Banda Aceh, Indonesia, *Coastal Engineering Journal*, 51, 243–273

Koshimura, S.; Kayaba, S. and Matsuoka, M. (2010). Integrated approach to assess the impact of tsunami disaster, *Safety, Reliability and Risk of Structures, Infrastructures and Engineering Systems,,* Taylor & Francis Group, London, pp. 2302 - 2307, ISBN 978-0-415-47557-0

Koshimura, S. & Matsuoka, M. (2010). Detecting tsunami affected area using satellite SAR imagery, *Journal of Japan Society of Civil Engineers*, Ser. B2 (Coastal Engineering), 66(1), 1426–1430 (in Japanese with English abstract)

Koshimura, S.; Matsuoka, M.; Matsuyama, M.; Yoshii, T.; Mas, E.; Jimenez, C; & Yamazaki, F. (2011). Field survey of the 2010 tsunami in Chile, *Proceedings of the 8th International Conference on Urban Earthquake Engineering,* Tokyo Institute of Technology, Japan, 7–8 March 2011

Lee, J. S. (1980). Digital image enhancement and noise filtering by use of local statistics, IEEE Trans. Pattern Analysis and Machine Intelligence, 2, 165-168

Massonnet, D.; Rossi, M.; Carmona, C.; Adragna, F.; Peltzer, G.; Fiegl, K.; & Rabaute, T. (1993). The displacement field of the Landars earthquake mapped by radar interferometry, *Nature*, 364, pp.138-142

Matsuoka, M. & Nojima, N. (2009). Estimation of building damage ratio due to earthquakes using satellite L-band SAR imagery, *Proceedings of the 7th International Workshop on Remote Sensing and Disaster Response*, University of Texas, Texas, United States, 22–23 October 2009

Matsuoka, M. & Yamazaki, F. (2002). Application of a methodology for detecting building damage area to recent earthquakes using satellite SAR intensity imageries and its validation, *Journal of Structural and Construction Engineering*, Vol. 558, pp.139–147 (in Japanese)

Matsuoka, M. & Yamazaki, F. (2004). Use of satellite SAR intensity imagery for detecting building areas damaged due to earthquakes, *Earthquake Spectra*, Vol. 20, No. 3, pp.975–994

Matsuoka, M. & Koshimura, S. (2010). Tsunami damage area estimation for the 2010 Maule, Chile earthquake using ASTER DEM and PALSAR images with the GEO grid system, *Proceedings of the 8th International Workshop on Remote Sensing for Post Disaster Response*, Tokyo Institute of Technology, Tokyo, Japan, 31 October - 1 November 2010

Murosaki, Y. (1994). Great fire in Okushiri in case of the 1993 Hokkaido Nansei–Oki earthquake, in Survey and Research on the 1993 Hokkaido Nansei–Oki Earthquake, Tsunami and Damages, Report No. 05306012, pp.161–170 (in Japanese)

Nojima, N.; Matsuoka, M.; Sugito, M. & Esaki, K. (2006). Quantitative estimation of building damage based on data integratoin of seismic intensities and satellite SAR imagery, *Journal of Structural Mechanics and Earthquake Engineering*, Japan Society of Civil Engineers, 62(4), 808–821 (in Japanese with English abstract)

Shuto, N. (2007). Damage and Reconstruction at Okushiri Town Caused by the 1993 Hokkaido Nansei–oki Earthquake Tsunami, *Journal of Disaster Research*, Vol. 2, No.1, pp.44–49

Suppasri, A.; Koshimura, S. & Imamura, F. (2011a). Developing tsunami fragility curves based on the satellite remote sensing and the numerical modeling of the 2004 Indian Ocean tsunami in Thailand, *Natural Hazards and Earth System Sciences*, 11, 173–189

Suppasri, A.; Koshimura, S. & Imamura, F. (2011b). Tsunami risk assessment for building using numerical model and fragility curves, *Proceedings of the Coastal Engineering Conference (JSCE)*, Morioka, International session, 9-11 November 2011

Part 3

Sensors and Systems

8

GNSS Signals: A Powerful Source for Atmosphere and Earth's Surface Monitoring

Riccardo Notarpietro[1], Manuela Cucca[1] and Stefania Bonafoni[2]
[1]*Politecnico di Torino, Electronics Dept.*
[2]*Università di Perugia, Electronics and Information Engineering Dept.*
Italy

1. Introduction

It is well known that Global Navigation Satellite Systems signals (which include for example the U.S. GPS and its modernization, the Russian GLONASS, the future European Galileo, the Chinese COMPASS), commonly processed for navigation purposes, can also be used to characterize media where they propagate in. In the last decade, GNSS atmospheric and Earth's surface remote sensing become more and more important, thanks to technical improvements applied to the processing of such "free-of-charge", everywhere available and weather insensitive signals.

For example, remote sensing of wet part of troposphere is possible "extracting" the atmospheric delays from GNSS observations. These delays are associated to water vapour and are accumulated by the signal along its propagation path. In the double difference phase observation adjustment (a standard GNSS signal pre-processing) it is possible and quite easy to estimate the wet contribution to atmospheric total delay mapped into the zenith direction, the so-called Zenith Wet Delay. From one side the estimate of propagation delays is essential to improve the accuracy of the height determination in the geodetic positioning framework (Kleijer, 2004). From the remote sensing point of view, Zenith Wet Delay may be then transformed into the so-called Integrated Precipitable Water Vapour (IPWV). Therefore, the knowledge of the temporal behaviour of IPWV above a GPS receiver network allows meteorologists to know the evolution of total water vapour content in atmosphere, which is one of the variable operatively used in Numerical Weather Prediction Models. These aspects are described in section 2.

A second important application allows to add vertical variability information to the atmospheric parameter distribution with respect to the previous one, which represents an "integrated" quantity. The amplitude and phase variations experienced by GNSS signal crossing the atmospheric "limb" and received on-board a Low Earth Orbit satellite, can be used to infer temperature and water vapor profiles, thanks to the GNSS Radio Occultation technique (Melbourne et al., 1994; Ware et al., 1996; Kursinski et al., 1997; Hajj, 2002). Even if aspects related to such very important Remote Sensing technique are not treated in the present chapter (a comprehensive tutorial can be found in Liou et al (2010), while review of results obtainable can be found in Anthes et al. (2008) and Luntama et al. (2008)) a mention is due. When signals cross in this way the atmosphere, they are delayed and their path is bent: therefore, the signal can be received also below the terrestrial limb, when the satellites are not yet in view. GNSS Radio Occultation is based on the inversion of the excess-phase (carrier phase in excess with respect the one experienced considering vacuum propagation) and amplitude evolution measured on the received signal when it is "occulted" with respect to the transmitter. Applying Geometric Optics algorithms or Wave Optics algorithm and Fourier operators to such observables, time evolutions of two important parameters identifying each trajectory followed by the signal can be derived: its total bending and its impact parameter, which is the distance of the trajectory asymptotes from the Earth's mass centre. Such quantities are in turn related to the integral of the atmospheric refraction index vertical profile, in a mathematical formulation that is invertible in a closed form. Result of the inversion is a very-accurate and high-resolved (up to about 100 m) atmospheric refractivity vertical profile, from which the corresponding temperature and humidity profiles can be inferred.

The second technique described in this chapter adds a further spatial variability characterization possibility with respect to that given by IPWV and Radio Occultation. It deals with the three-dimensional reconstruction of atmospheric refractivity and, thus, water vapour density, applying tomographic techniques to phase delays measurements collected by small (but dense) networks of GPS receivers. Because of volume dimensions, inhomogeneity spatial distribution and geometric constraints, all the weak points of tomography emerge in characterizing neutral atmospheric parameter distributions using GNSS signals. Results and comments are given in section 3.

The last application we will describe (section 4) is the most recent and maybe the most challenging one. It foresees the use of GNSS signals reflected off from lands and oceans for characterizing the Earth's surface at L-band frequencies. The signal is received under bistatic geometry since the received signal power is that which is forward scattered from the Earth's surface towards the GNSS-R (GNSS-Reflectometry) receiver. The reflected signal contains many differences with respect to the direct one, in terms of delay, Doppler shift, power strength and polarization. Once the reflected signal is received, it is processed using hardware or software correlators. The reflecting surface features are dipped inside the shape, the magnitude and the maxima location (which is related to the propagation delay) of the obtained correlation function. Among the possible remote sensing applications we list: ocean altimetry (from delay); wind speed and ocean scatterometry (from shape and spreading), ice topography and monitoring (from delay and magnitude); soil moisture (from magnitude).

2. Integrated precipitable water vapour

Water vapour is one of the main constituents of the atmosphere and its accurate and frequent sampling is obviously of great use for climatological research as well as operational weather forecasting. Moreover, water vapour is one of the most variable atmospheric constituents, fundamental in the transfer of energy in atmosphere: improving knowledge of its distribution is fundamental to set good initial conditions in numerical weather forecast. In addition, water vapor fluctuations are a major error source in ranging measurements through the Earth's atmosphere, and therefore the principal limiting factor in space geodesy applications such as GNSS, very long baseline interferometry, satellite altimetry, and

Interferometric Synthetic Aperture Radar (InSAR). Several techniques are well established to derive the vertically Integrated Precipitable Water Vapor (IPWV)[1], in particular using ground-based and spaced-based radiometers, radiosonde observations and GNSS receivers.

Radiosonde observations produce an accurate measurement of the water vapour profile, but the temporal and spatial resolution is rather poor. Radiosondes are typically launched every 6 to 12 hours, which may cause significant variations in water vapour to go undetected.

Ground-based microwave radiometers show problems during periods of rain fall and space-based radiometer observations can be degraded in the presence of clouds. This prevents reliable measurements during periods where changes in water vapour could be quite great. Besides these limitations, all systems involve considerable costs.

The technique to estimate IPWV by means of GNSS receivers is based on measurements of the tropospheric delay time of navigation signals. Therefore the delay, regarded as a nuisance parameter by geodesists, can be directly related to the amount of water vapour in the atmosphere, and hence is a product of considerable value for meteorologists. Furthermore, water vapour estimation with ground-based GNSS receivers is not affected by rain fall and clouds, and can therefore be considered an all-weather system.

So, GNSS is a valuable complement to radiosondes and radiometers, taking into account that GNSS IPWV estimates come from an existing GNSS infrastructure and frequently from quite dense receiver networks.

2.1 Description of observables, theoretical basis and retrieval technique

The use of GNSS receivers to estimate IPWV is based on measurements of the delay affecting the navigation signals during their propagation in troposphere (neutral atmosphere) from the GNSS satellites to the receivers on ground. The dispersive ionospheric effect can be removed with a good level of accuracy by a linear combination of dual frequency data.

Such a technique is founded on the non-dispersive refractive characteristics of the neutral atmosphere, governed by its composition. The water vapour molecules in atmosphere are polar in nature possessing a permanent dipole moment. All the other gases are non-polar molecules and a dipole moment is induced among these gases when microwave propagates through atmosphere. These molecules reorient themselves according to the polarity of propagating wave. In the retrieval technique to be described the atmosphere is considered as the sum of a dry component (mainly due to O_2) and a wet component.

Consequently, the neutral delay due to the troposphere can be decomposed into the hydrostatic delay associated with the induced dipole moment of the atmosphere constituents and the wet delay associated with the permanent dipole moment of water vapour (Askne & Nordius, 1987; Brunner & Welsch, 1993; Treuhaft & Lanyi, 1987). The zenith hydrostatic delay (ZHD) has a typical magnitude of about 2.4 m at sea level, and it grows with increasing zenith angle reaching about 9.3 m for elevation angle of 15°. With

[1] Consider the total amount of atmospheric water vapour contained in a vertical column of unit cross section: if this water vapour were to condensate and precipitate, the equivalent height of the liquid water within the column is the Integrated Precipitable Water Vapour, usually measured in cm or in g/cm^2

simple models and accurate surface pressure measurements, it is usually possible to predict accurately the ZHD. The zenith wet delay (ZWD) can vary from a few millimeters in very arid condition to more than 350 mm in very humid condition, and it is not reliable to predict the wet delay with an useful degree of accuracy from surface measurements of pressure, temperature and humidity.

Therefore, from GNSS radio signals the total tropospheric delay is provided and, measuring the ZHD, it is possible to retrieve the remaining ZWD, incorporating mapping functions which describe the dependence on path orientation. The ZWD time series are then directly transformed into an estimate of IPWV: GNSS receivers can estimate IPWV with a temporal resolution of 30 min or better and with an accuracy better than 0.15 cm.

2.1.1 Retrieval algorithm

In this section the retrieval algorithm used for the estimation of IPWV from GNSS observations is presented.

Using GNSS methods of path delay correction, developed for geodetic applications, it is possible to estimate time-varying atmospheric zenith neutral delay ZTD (excess path length due to signal travel in the troposphere at zenith) defined as:

$$ZTD = ZHD + ZWD = 10^{-6} \int_{H}^{\infty} N(s)ds \tag{1}$$

where ds has units of length in the zenith, H is the surface height and N(s), usually expressed in parts per million (ppm), is the refractivity of air given by (Thayer, 1974):

$$N = k_1\left(\frac{P_d}{T}\right)Z_d^{-1} + k_2\left(\frac{e}{T}\right)Z_w^{-1} + k_3\left(\frac{e}{T^2}\right)Z_w^{-1} \tag{2}$$

where P_d is the dry air pressure (hPa), T is the air temperature (K), e is the partial pressure of water vapour (hPa), Z_d and Z_w are the dry air and water vapour compressibility factors, that consider the departure of air from an ideal gas. Values for inverse dry and wet compressibility factors differ from unity of about one part per thousand, and are given by:

$$\begin{aligned} Z_d^{-1} &= 1 + P_d\left[57.90 \cdot 10^{-8}\left(1 + \frac{0.52}{T}\right) - 9.4611 \cdot 10^{-4}\frac{T_c}{T^2}\right] \\ Z_w^{-1} &= 1 + 1650 \cdot \left(\frac{e}{T^3}\right)\left(1 - 0.01317 \cdot T_c + 1.75 \cdot 10^{-4} \cdot T_c^2 + 1.44 \cdot 10^{-6} \cdot T_c^3\right) \end{aligned} \tag{3}$$

where T_c is temperature in Celsius.

Several authors have given values for the empiric constants k_1, k_2 and k_3 of eq. 2: a typical choice is k_1=77.604 ($K \cdot hPa^{-1}$), k_2=64.79 ($K \cdot hPa^{-1}$) and k_3=$3.776 \cdot 10^5$($K^2 \cdot hPa^{-1}$) (Thayer, 1974).

In eq. 2, the first two terms of N are due to the induced dipole effect of the neutral atmospheric molecules (dry gases and water vapour), and the third term is caused by the permanent dipole moment of the water vapour molecule. Therefore, the hydrostatic part is described by:

$$ZHD = 10^{-6} \cdot k_1 \int \frac{P_d}{T} Z_d^{-1} ds \tag{4}$$

and the wet part is:

$$ZWD = 10^{-6} \cdot k_2 \int \frac{e}{T} Z_w^{-1} ds + 10^{-6} \cdot k_3 \int \frac{e}{T^2} Z_w^{-1} ds \tag{5}$$

In the estimation algorithm of IPWV we can identify four principal steps:

1. We start with the estimation of the neutral zenith path delay from GNSS observations (Bevis et al., 1992), which are elaborated using a specific GNSS software (e.g. Bernese GPS software or others). The neutral radio path delay has to be estimated using precise orbit ephemerides, choosing a proper cut-off angle (e.g. 15 degrees), resolution time (e.g. 30 minutes), and a suitable law as dry and wet mapping functions. Different kinds of mapping functions exist and they are different in number of meteorological parameters involved (Herring, 1992; Ifadis, 1986; Niell, 1996).
2. Computation of the ZHD component of the atmosphere, that is the greater component in magnitude of ZTD but it is less variable with respect to ZWD.

If atmospheric profiles of temperature and dry pressure are available near the GPS station, the ZHD can be computed using eq. 4. Since such an availability is difficult in time and space, alternative and more simple procedures can be adopted for a reliable estimation of the hydrostatic delay.

If surface pressure is known with an accuracy of 0.3 hPa or better, *ZHD* can be estimated through simple models to better than 1 mm (Elgered et al., 1991), e.g. using the Saastamoinen model (Saastamoinen, 1972):

$$ZHD = {0.22768\ P_s} / {f(\lambda, H)} \tag{6}$$

$$f(\lambda, H) = (1 - 0.00266 \cdot \cos(2\lambda) - 0.00028\ H) \tag{7}$$

where ZHD depends on actual surface pressure P_s (hPa), on latitude λ (rad) and on the surface height H (km). The error introduced by the assumption of hydrostatic equilibrium in the model formulation is typically of the order of 0.01%, corresponding to 0.2 mm in the zenith delay.

3. Then ZWD is computed by subtracting ZHD from ZTD.
4. Finally, it is possible to retrieve IPWV using the relationship:

$$IPWV = \Pi \times ZWD \tag{8}$$

Typical values for the parameter Π are approximately 0.16, so 6 mm of ZWD is equivalent to about 1 mm of IPWV.

The parameter Π is a function of various physical constants and of the weighted mean temperature T_m of the atmosphere (Askne & Nordius, 1987; Davis et al., 1985):

$$T_m = \frac{\int (e/T)dz}{\int (e/T^2)dz} \tag{9}$$

$$\Pi = 10^6 \Big/ \rho R_v[(k_3 / T_m) + (k_2 - mk_1)] \tag{10}$$

where ρ is the density of liquid water, R_v is the specific gas constant for water vapour, m is the ratio of molar masses of water vapour and dry air, and k_1, k_2, k_3 are the constants defined previously.

The transformation described in eq. 8 assumes that the wet path delay is entirely due to water vapour and that liquid water and ice do not contribute significantly to the wet delay (Duan et al., 1996).

2.2 State of the art

The '90s witnessed the fast increasing of the use of the tropospheric delay time of GNSS signals to estimate the Integrated Precipitable Water Vapour (Bevis et al., 1992; Bevis et al., 1994; Businger et al., 1996; Coster et al., 1997; Davies & Watson, 1998; Duan et al., 1996; Emardson et al., 1998; Kursinski, 1994; Rocken et al., 1993; Ware et al., 1997; Yuan et al., 1993).

Although the IPWV retrieval algorithm from ZTD measurements is well-established, different strategies were adopted for the time-varying parameter Π. Anyway, Π can be estimated with such an accuracy that very little uncertainty is introduced during the computation of eq. 8.

Bevis et al. (1994) provided an error budget for Π and showed that in most practical conditions the uncertainty for this parameter is essentially due to the uncertainty for T_m (usually predicted from the surface temperature T_s on the basis of regressions), leading to a relative error in Π of the order of 2%. In fact, exact calculations of T_m require profiles of atmospheric temperature and water vapor, as from radiosoundings or analysis from Numerical Weather Prediction Models (e.g the global European model, ECMWF). Since those data are not easily available, T_m is commonly estimated using station data of surface air temperature with empirical linear or more complicated relationship (the so-called T_m-T_s relationship) that can be site-dependent and may vary seasonally and diurnally (Bevis et al., 1994).

A simple and alternative approach can be considered for Π estimation: the use of a linear regression (ZWD and IPWV as predictors and predictands, respectively) from historical data base of radiosoundings or ECMWF available near the site of interest for the water vapour estimation, leading again to a relative error in Π just above 2%. Considering monthly averages of Π the uncertainty is around 1.5% (Basili et al., 2001). This approach does not need measurements of surface temperature for each computation of Π.

Estimation of water vapour features by GNSS is valuable from the point of view of climate monitoring, atmospheric research, and other applications such as ground-based and satellite-based sensor calibration and validation. GNSS tropospheric delays are also useful for operational weather prediction models (Gutman & Benjamin, 2001; Macpherson et al., 2008; Smith et al., 2000).

The IPWV retrieval by means of a GNSS ground-based receiver can be used to monitor *in situ* water vapour time series, or to compare the IPWV values estimated by co-located ground-based sensors (e.g. microwave radiometer, photometer). Networks of GNSS receivers can be used to monitor the water vapour field, mapping its horizontal distribution.

The possibility of mapping IPWV measured by GNSS networks has been explored (de Haan et al., 2009; Morland & Matzler, 2007), also combining IPWV data retrieved from GNSS receivers and from satellite-based radiometers to produce IPWV maps over extended areas (Basili et al., 2004; Lindenbergh et al, 2008).

2.3 Results

The degree of accuracy in IPWV estimation by GNSS receivers exploiting the tropospheric propagation delay at L-band is usually around 0.10-0.20 cm. The horizontal resolution of zenith columnar water vapour associated to a single receiver using standard methods (azimuthally symmetric weighting functions) is in the order of tens of kilometers, roughly corresponding to the aperture of the cone which includes all the lines of sight of the various GNSS satellites observed at different elevation angles.

Besides GNSS, several techniques are well established to derive the vertically IPWV, such as ground-based microwave radiometers (MWR), radiosonde observations (RAOBs), analysis data from Numerical Weather Prediction Models (e.g ECMWF). Some examples of IPWV comparisons among different techniques during experimental campaigns are reported in this sub-section.

For instance, during an experimental campaign in Rome, Italy (20 September - 3 October, 2008), different instruments managed by the Sapienza University of Rome were operative at the same site: a GPS receiver (included in the Euref Permanent Network) a MWR (a dual-channel type, 23.8 and 31.4 GHz, model WVR-1100, Radiometrics) and six RAOBs (Pierdicca et al., 2009). Also, analysis data from ECMWF nearest the site were considered. The *IPWV* time series for the entire campaign are plotted in Fig. 1.

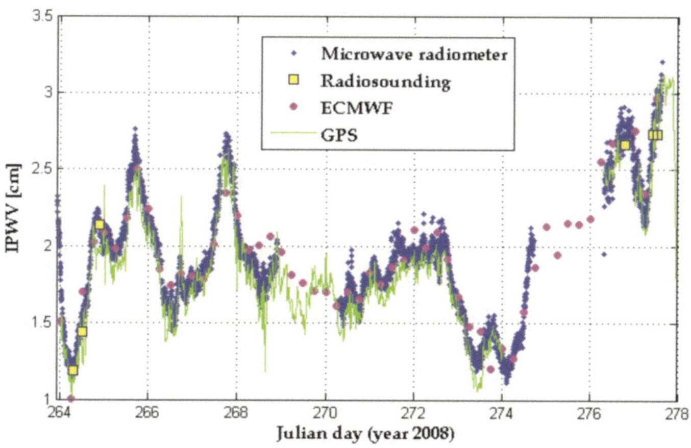

Fig. 1. Rome, Sapienza University of Rome (41.89 N and 12.49 E, 72 m a.s.l.), 20 September - 3 October, 2008. Time series of IPWV from MWR (blue dots), GPS (green), RAOBs (yellow squares) and ECMWF (magenta circles).

The IPWV root mean square (rms) difference of GPS compared with MWR is 0.10 cm, with RAOBs and ECMWF is around 0.15 cm.

With reference to an Italian ground-based network of GPS receivers, managed by the Italian Space Agency (ASI), another experimental campaign was conducted in Cagliari (Italy), during the whole 1999 (Basili et al., 2001). The experimental site was selected at the Cagliari GPS station where a ground-based dual-channel microwave radiometer (WVR-1100) was operated for the whole campaign of measurements. Also, data from RAOBs released at Cagliari every six hours were available. Results of the experiment for the whole 1999 are shown in Fig. 2, gathered in non-precipitating conditions to avoid problems with the radiometer measurements. The comparison is performed considering a sampling time of 6 hours, in coincidence with RAOB releases.

This long-term comparison has shown a fairly good agreement among the two remote sensors and the RAOBs, with an error standard deviation similar to other experiments reported in literature.

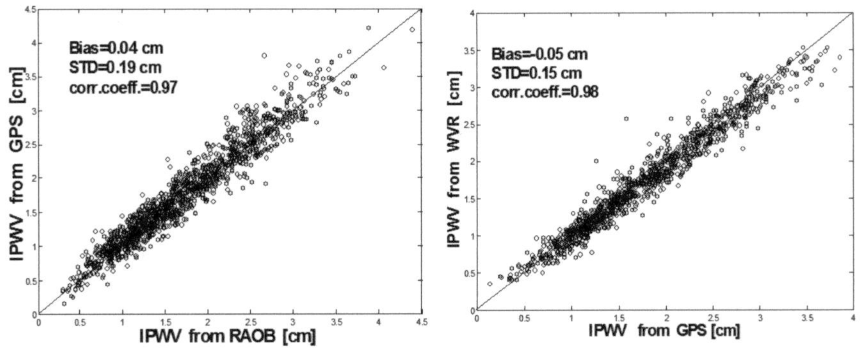

Fig. 2. Scatterplots of IPWV computed at Cagliari, 1999, by three different instruments (RAOB, GPS and WVR). Left: IPWV-GPS vs. IPWV-RAOB; right: IPWV-WVR vs. IPWV-GPS (Basili et al. 2001). Bias and STD refer to the mean difference and to the standard deviation of the difference.

3. Wet atmospheric refractivity maps through tomography

As it has already been shown in Section 2, the remote sensing of "wet" troposphere is possible by estimating the wet contribution to atmospheric total delay mapped into the zenith direction, the ZWD, in the general adjustment of double difference phase observations. Following a step ahead, it is also possible to try to extract some information on the three dimensional distribution of atmospheric parameters, from total delay observations taken by different line of sights. Tomography deals with the inversion of integral measurements collected from a great variety of directions, for the extraction of non-homogeneous signatures inside the analyzed volume. Requirements necessary to make tomographic inversion procedures effective are well known. The geometry of the signal paths is crucial for the stability of the inversion procedure. All the voxels (volume pixels) have to be crossed by a lot of rays coming from different directions. Horizontal resolution can be improved only considering quite dense GNSS networks. Vertical resolution can be improved if receivers are deployed on a sloped area. This section presents results already published by Notarpietro et al. (2011), results obtained applying a tomographic inversion to real observations taken on October 2010 in Italy, by a dense network of GNSS receivers.

3.1 State of the art

Several activities were carried out in the past in the field of neutral atmospheric tomography based on observations performed on GNSS signals. Starting from one of the first concept description given by Elosegui et al. (1999), the effectiveness of a 4-dimensional (4D) water vapour field tomographic reconstruction was assessed by Flores et al. (2000) on a 20x20x15 km atmospheric domain against ECMWF (European Centre for Medium-Range Weather Forecast) data. After that, several methods were applied to different kind of real or simulated GPS observables (obtained by more or less dense receiver networks), demonstrating the effectiveness of water vapour field reconstructions on different atmospheric volume sizes, with different resolutions, against radiosonde data, Numerical Weather Prediction models or other independent water vapour dataset. Some reference papers (the list is not exhaustive) are that of Hirahara (2000), Gradinarsky and Jarlemark (2004), Champollion (2005, 2009), Bi et al. (2006), Troller et al. (2006), Nilsson and Gradinarsky (2006).

In the framework of the European Space Agency project METAWAVE (Mitigation of Electromagnetic Transmission errors induced by Atmospheric Water Vapour Effects), we applied a new approach to the ZWDs estimated from the observations collected by a local network of GPS geodetic receivers deployed over a small area around the city of Como. Such new approach is based on an algorithm previously developed, on which we shown, from a simulative point of view only, the possibility to infer wet refractivity fields without using first guess atmospheric models and without adopting any a priori informations (Notarpietro et al., 2008). Such an algorithm has been applied to real measurements collected by a local network of GPS receivers. In what follows we will summarize the results we obtained.

3.2 Theoretical basis, retrieval technique, observables and validation approach

Basically, two different classes of algorithms can be applied to perform atmospheric tomography (and tomography in general). The first belongs to iterative reconstruction techniques (for example the Algebraic, the Multiplicative Algebraic or the Simultaneous Iterative Reconstruction Techniques, respectively called ART, MART and SIRT, see Herman 1980) which need a good first guess atmospheric model to converge at the "good" solution. The second belongs to the Least Square Inversion (or Generalized Inversion) techniques, which are "one-step" algorithms and do not need a first guess. Notarpietro et al. (2008), shown the possibility to infer Wet Refractivity fields without using first guess atmospheric models. The algorithm accomplishes the reconstruction in two consecutive steps. The first step allows the retrieval of a "raw" three dimensional wet refractivity distribution directly from Slant Wet Delays (SWD) observables $\Delta\Phi^{wet}$ (defined as equivalent optical length), which in turn depend on the wet refractivity (N_w) distribution along the ray path, in the way defined by the following equation:

$$\Delta\Phi^{wet} = 10^{-6} \int_{\text{ray-path}} N_w ds \quad (11)$$

Linearizing eq. 11 and considering the entire observation dataset, the following matrix equation turns out:

$$\mathbf{\Delta\Phi^{wet}} = 10^{-6}\mathbf{L}\cdot\mathbf{N_w} \qquad (12)$$

where **L** is the Data Kernel to be inverted to obtain the wet refractivity distribution, which is a matrix containing for each row, the lengths of each segment inside each voxel crossed by the generic rectilinear ray-path connecting the receiver and the satellite. This tomographic pre-processing step pertains to Least Square Inversion algorithms (Lawson & Hanson, 1974). It achieves the result through the constrained inversion (using Singular Value Decomposition) of the Tikonov-regularized Data Kernel matrix. Although the resolution obtainable with this pre-processing step is quite rough (the entire tropospheric volume has been divided into 2x2x20 voxels grid), this result is used as first guess for the algebraic technique used in the second phase of the proposed reconstruction algorithm. In particular we applied the SIRT technique to obtain the distribution of wet refractivity inside the tropospheric volume characterized by the final resolution (4x4x20 voxels grid).

With the aim of studying the potentialities of GNSS in the determination of local wet refractivity fields, needed for instance to correct InSAR derived landslide deformation maps, we used observations collected during a couple of weeks in 2008 by the MisT GPS network, defined by eight geodetic receivers that were deployed around the COMO Permanent Network station (which is placed in the North West part of Italy). This network was born for different purposes from the tomographic reconstruction of the wet refractivity field, and its design was not fully compliant with the requirements of this technique (details about each MisT station are reported in Table 1, while the MisT network topology is shown in Fig. 3). An attempt to improve the original design of the MisT network was done by performing a different daily campaign collecting data by two additional GPS portable receivers, named BISB and BOLE, placed at higher altitudes from the original network (respectively in the top of Monte Bisbino e Monte Boletto).

Station	Height	Receiver type
ANZA	280 m	Leica GRX1200
BRUN	738 m	Leica GX1200
CAST	286 m	Leica GRX1200
COMO	292 m	Topcon Odyssey
LAPR	349 m	Leica GX1200
PRCO	266 m	Leica GX1200
NAND	746 m	Leica GX1200
MGRA	353 m	Leica GX1200
DANI	614 m	Leica GX1200
BISB	1373 m	Topcon GB1000
BOLE	1199 m	Trimble 4700

Table 1. MisT GPS network description. Highligthed raws are those related the two "mountainous" receivers

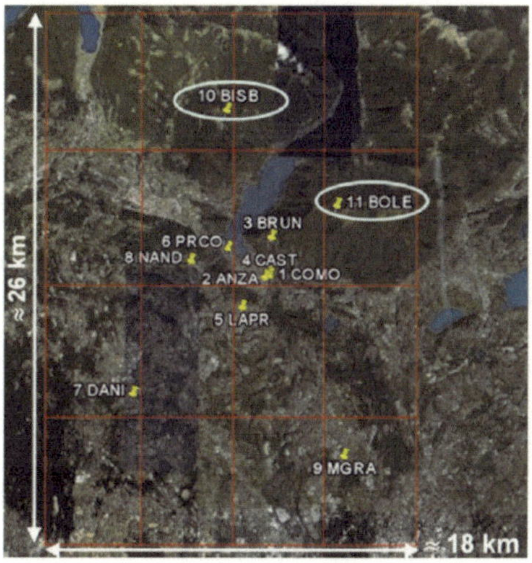

Fig. 3. Geographic distribution of the MisT network. The two "mountainous" GPS receivers are highlighted. The final volume discretization is also superimposed.

A daily multi-station adjustment of observations collected by the whole network was performed via the Bernese V0.5 software, to estimate jointly the station positions and the Hourly ZWDs parameters. These are basically averaged value of the tropospheric delay zenithal projection, affecting all the signals from the considered station to all the satellites in view, as they move along their orbits in 1 h time. Differences between the actual instantaneous slant delays and these averaged values projected back on the slant direction are to be found in the double difference adjustment residuals (this analysis is not described here). More precisely, carrier phase double differences were processed, all the single differences being formed with respect to the COMO reference station. The Bernese software models the tropospheric delay in each station-receiver phase measurement as the sum of a hydrostatic component and a wet one. The first can be modelled (and slanted toward the satellite position using the dry Niell's mapping function (Niell, 1996)) considering the Saastamoinen formulation (Davis et al., 1985) and interpolating surface pressure data (in time and space) obtained by 0.25°x0.25° ECMWF analysis. The second can be expressed as the product of an unknown parameter, the ZWD, by a known coefficient computed in our case from the wet Niell's mapping function. For each MisT station, input data were Hourly ZWDs, estimated during the week from October 12th to October 18th, 2008 and from November 13th to November 19th, 2008. Hourly ZWDs related to each MisT station, were then "geometrically" projected along the slant paths (using Niell's mapping functions) by upsampling at 1-min sample intervals the 15 min GPS satellites positions obtained from International GNSS Service (IGS) sp3 files and inverted using the developed tomographic procedure.

It has to be pointed out that the standard dataset adopted for tomographic reconstructions is built up by considering only 6 out of 9 MisT receivers. Firstly, COMO, ANZA and CAST are the three stations belonging to the so called MisT inner sub-network. We considered only ANZA among the three close stations of COMO, ANZA and CAST (deployed at distances less than 200 m from one-another), whose ZWDs are highly correlated (>95%). Moreover, ZWD data obtained processing NAND observations are used for self-consistency validation purposes ('leave-one-out' quality assessment) and are not included in the input dataset.

As we have previously stated, our tomographic approach is based on two consecutive reconstruction steps. The first one (data kernel generalized inversion) creates the first guess field for the second one (algebraic tomography), which doubles the horizontal resolution (from 2x2x20 to a 4x4x20 voxels grid, i.e. means 4.5x6.5x0.5 km^3). It has to be stressed that volume resolution is strictly related to the geometrical distribution of GNSS receivers and to the availability of observations. Higher resolutions would introduce an increasing number of voxels not crossed by any ray, thus worsening the final results. On the contrary, lower resolutions would imply a too coarse description of the field.

Considering the available observables we were able to obtain 168 or 144 Hourly wet refractivity maps (for the October or the November week respectively). Validation is carried out considering the difference between ZWD GNSS measurements taken over NAND receiver and corresponding ZWD estimates evaluated by vertically integrating the reconstructed wet refractivity maps. Considering the entire observing period, final statistics are thus based on 168 (144) ZWD differences (measured-estimated) distribution for the October (November) week and results are given in terms of their mean values and their rms values.

3.3 Results

In what follows, we will show results related to the so called baseline scenario and improvements obtained adding observations taken by mountainous receivers and from low elevation angles. Some hints about the impact of distance and height of the reconstruction error and about validation against independent data will be also given.

3.3.1 Baseline scenario results and effect of mountainous observation ingestion

The baseline scenario is that defined considering observations taken by the reduced MisT network formed by ANZA, BRUN, LAPR, PRCO, MGRA and DANI stations. For the October week, tomographic reconstructions were carried out considering ZWDs observed by the reduced MisT network observations taken during 12–18 October. The good agreement between measured and estimated ZWD time series evaluated above NAND during this period and for this scenario is shown in Fig. 4, while some statistics are given in column A of Table 2. Since data from the two mountainous receivers (BISB and BOLE in Fig. 3) were available only on 12th October, 2008, between 9.00 am and 7.00 pm, comparisons of measured and estimated ZWDs above NAND receiver were performed also considering observations taken by the reduced MisT network in this smaller period (column B of Table 2). A bias decrease of 0.4 mm is observed adding BOLE (1199 m a.s.l.) observations (see column C, Table 2) and of 1 mm adding both BOLE and BISB (1373 m a.s.l.) data (see column D, Table 2) in the input dataset. This demonstrate the necessity of measurements collected at higher altitudes which allows a best reconstruction of vertical refractivity gradients characterizing the first three atmospheric layers. The high rms error with respect the one characterizing the baseline result given in column B, Table 2, is probably due to the more noisy data acquired by the two portable mountainous receivers (this is also evidenced by the decrease in correlation observed between NAND ZWDs measurements and estimates).

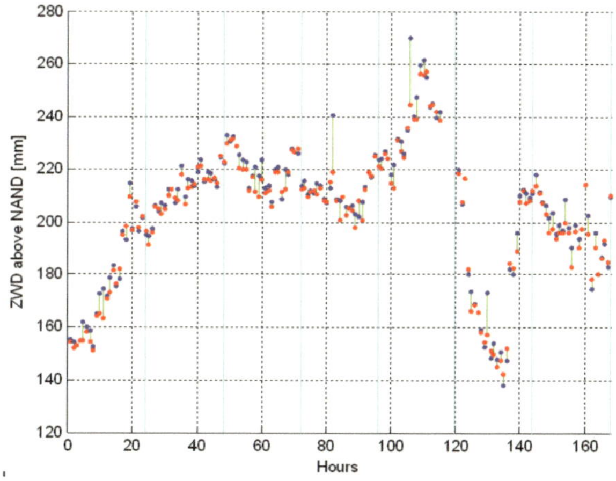

Fig. 4. Time series of ZWDs measured (blue dots) and estimated (red dots) after reconstruction above NAND station, for the baseline experiment.

	A.	*B.*	*C.*	*D.*
Mean [mm]	1.96	2.57	2.19	1.61
RMS [mm]	4.49	5.29	5.27	5.33
Corrcoeff	0.99	0.93	0.93	0.92

Table 2. Statistics of the ZWD difference (measured-estimated after reconstruction) over the NAND reference station.

3.3.2 Ingestion of low elevation observations

Considering the baseline scenario described in paragraph 3.3.1, it is clear that the improvement in the reconstruction of lower layers is strictly related to the availability of trajectories crossing (and discriminating) the lower tropospheric layers. In our tomographic reconstruction, only rays exiting from the top boundary of the analyzed 18x26 km^2x10 km volume were considered. In our case, the mean elevation angle was about 30°. Since the MisT network topography is fixed, to overcome this limit and therefore improving the retrieved field, we try to ingest also low elevation trajectories which enter from the lateral boundaries of the analyzed volume. Since SWDs associated to these rays contains both a contribution of the wet refractivity field inside the considered volume (namely, the inner volume) and outside the volume (the outer volume) up to 10 km height, we modelled and removed this last quantity from the SWDs associated to low elevation (< 30°) ray before entering the tomographic approach. The wet refractivity model considered in the outer volume was obtained considering three different approaches:

a. from a very coarse tomographic reconstruction performed on a bigger volume using the same GNSS experimental data (considering as input data those observed by the entire MisT network except those taken by the NAND receiver);
b. interpolating the CIRA-Q wet atmospheric climatologic model (Kirchengast et al., 1999) in the outer volume;
c. considering data taken by ECMWF analysis (91 pressure levels, 0.25°x0.25° grid resolution), collocated in time and space with the centre of each voxel belongs to the

outer volume (this was done by a bilinear space interpolation and a linear time interpolation of the meteorological data).

Results related to this analysis are summarized in Table 3. They confirm the importance of the availability of low elevation measurements issued from different altitudes to improve the estimation of vertical refractivity gradients in such a tomographic approach. It has to be noted that the availability of external independent information (atmospheric models or, better, meteorological data) for modelling the SWD component of low elevation observations in the outer volume seems to be necessary in this case. Because of the MisT network design (receivers not homogeneously distributed in the inner volume), the internal procedure based on the coarse tomographic reconstruction (case a)) is not very effective.

Statistics over 168 hours	*No low elevation observ.*	*a)*	*b)*	*c)*	*Case c) outlier rejection*
Mean [mm]	1.96	1.89	1.26	1.20	0.85
RMS [mm]	4.49	4.41	4.12	4.11	3.11

Table 3. Self-consistency results considering SWD derived by low elevation observations (taken during the October week) after the application of the outer volume wet refractivity modelling strategies a., b. and c.. Results are relative to the statistics of ZWD errors (measured-estimated after reconstruction) over the NAND reference station. Results related to the baseline scenario are reported in the first column as a reference. In the last column the evident outliers due to measurements (see blue dots in Fig. 4) were removed.

3.3.3 Impact of distance and height on reconstruction goodness

Results described previously are good, but are related to the baseline scenario. Considering this scenario, the validation has been performed above NAND receiver, which is in a good position since its baseline from the COMO master station is between the nearest and the farthest stations. In this further analysis we have considered all the measurements (ZWDs or ZTDs) available from the MisT network (8 receivers) during the entire week (excluding only the COMO receiver, see Fig. 3). Then we have excluded data (ZWDs or ZTDs) observed by one receiver per time, keeping such data as reference for the self-consistency validation purpose for that receiver. For each case we have run our tomographic reconstruction considering all the 168 Hourly ZWDs (or ZTDs) available per each station for the October week, mapping them into the slant directions and including also low elevation observations (following the procedure described in paragraph 3.3.2). The obtained 168 Wet Refractivity Maps (considering ZWDs as input to the tomography) or Total Refractivity Maps (considering ZTDs as input) have then been used to evaluate the ZWD and ZTD estimates above the reference receiver, which are compared with the ZWD and ZTD observations above that receiver. This analysis has been repeated for each receiver of the MisT network.

Root Mean Squares of ZWDs and ZTDs differences (measured-estimated after reconstruction) are then reported in function of the distance of the station from the COMO master station or in function of the height of the station. Such results are plotted in Fig. 5. The same analysis has been performed considering data taken by the MisT network extended to the two mountainous receivers during 12 October from 9:00 AM to 7:00 PM (only 10 Hourly averaged ZWDs or ZTDs observations are contemporaneously available to any receivers of the "extended" network). In this case results are shown in Fig. 6.

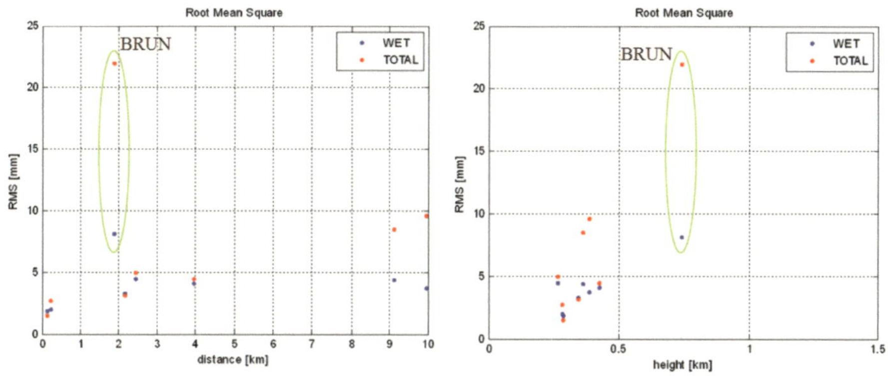

Fig. 5. rms of the differences between ZWDs (blue dots) or ZTDs (red dots) observed and estimated above each reference receiver, excluding data of that receiver from the input dataset before the reconstruction. All data observed by the MisT network during the entire week are taken into account. (Left) rms are plotted against the distance of the reference receiver from COMO master station. (Right) rms are plotted against the height of the reference receiver above WGS84. The degraded results obtained excluding BRUN receiver (which is the highest one) are highlighted.

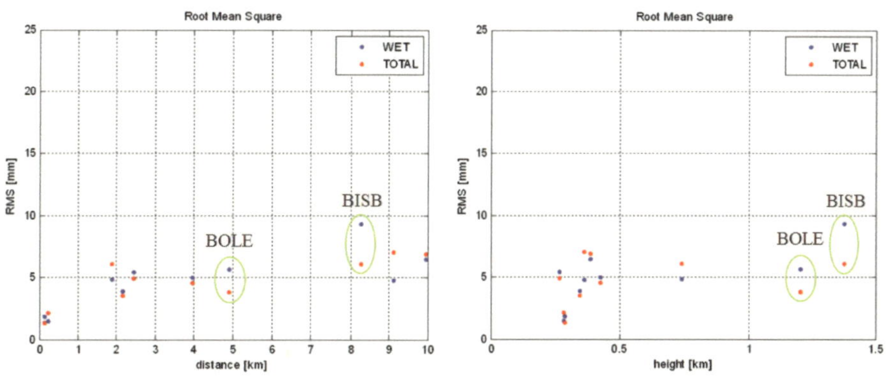

Fig. 6. Like Fig. 5, but considering all data observed by the MisT network and by the two mountainous receivers during the 10 hours of 12th October, 2008.

First of all this analysis confirms the impact of a good height displacement of receivers in the network. Even if MisT network topography has not been optimized for the geography of the analyzed area and for tomographic applications, if we consider the impact of height in the evaluation of propagation delays, we can say that the lack of receivers placed at higher altitudes will worsen final results. In particular, considering the original MisT network, where all the receivers are more or less placed in the same layer of the map (Fig. 5) we want to highlight that, if data observed at the highest receiver (namely BISB, which is placed in another vertical layer) are not given in input to the tomography, the rms of the difference between estimated and measured zenith delays (both Wet and Total) is generally doubled.

Things are better if we consider the MisT network plus the mountainous receivers. In this case the worsening is not so emphasized, since it is compensated by other receivers placed at similar altitudes (see Fig. 6). In both cases it seems that the results worsening follows a (more than) linear rule. It is absolutely not clear why the effects on the evaluation of Wet delays and Total delays are inverted, considering or not considering the mountainous receivers. It has to be noted that the network solution obtained for the mountainous receivers is not as accurate as that obtained for the other receivers, since the mountainous sensor positions have not been fixed. Moreover, results reflect 10 hours of observations instead of the entire week.

As far as the impact with distance is concerned, it is quite difficult to identify a clear relationship with results. Obviously if we exclude data observed by the nearest receivers (ANZA or CAST) to the reference one (COMO), results are better (rms is halved considering both the weekly data of the original MisT network and the 10 hours data of the MisT network plus mountainous receivers) than that we can obtain excluding one of the other (farther) receivers. But for all the other cases, it seems that final results are insensitive to distance. It is a surprising result since we expected a certain error correlation with distance. But the farthest receivers (MGRA and DANI) are placed in opposite positions with respect the map center and are the southest receivers (see Fig. 3). If we take into account low elevation observations (even if such observations are averaged, since they are obtained simply mapping hourly averaged Zenith observations into slant directions), rays related to the northern receivers (all the others) anyway interest the atmospheric volume above the southest receivers (and not viceversa, given the orbital positions of GPS satellites). And this could probably compensate the "distance" effect. Anyway, also in this case, further analysis and measurements are necessary to better understand if there is a clear relationship.

3.3.4 Validation against independent data

In order to assess the goodness of inferred wet refractivity fields in different points of the grid considering independent data, we also did a comparison of ZWDs obtained vertically integrating wet refractivity fields derived after tomographic reconstruction along each column of retrieved maps with those derived by ECMWF analysis co-located in the same points (and times), even if the ECMWF horizontal resolution (0.25°x0.25°) and time resolution (6 h) are too coarse with respect those characterizing our final maps.

Statistical comparisons were performed considering the 168 wet refractivity maps obtained using data observed by the reduced MisT network (plus NAND receiver) collected during the October week and considering the 144 maps obtained for the November one. Results are shown in Fig. 7, where the time series of both ZWDs estimated after tomographic reconstruction (blue lines) and evaluated using ECMWF data (red lines) are plotted for each column of our volume discretization. We classified the areas accordingly to the corresponding rms values (computed for each ZWD difference time series, after the average bias removal) using green, yellow and red colors. As expected, the northern part is where the agreement is worse. In that area we had no receiver and less satellites were in view in the north direction. On the other hands, in the southern area, agreement is better even if no receivers were present, thanks to the availability of a higher number of rays. The best area is obviously the central one.

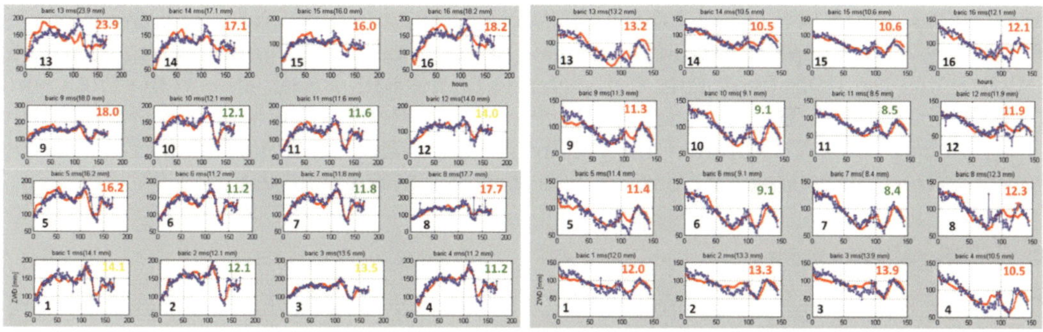

Fig. 7. Time series of ZWD obtained integrating ECMWF (red) collocated and estimated with tomography (blue) wet refractivity maps. Left: October week data; right: November week observations. The black numbers shown the column "number" inside the map.

Even if our main goal was to demonstrate the effectiveness in adopting tomographic reconstruction procedures for the evaluation of propagation delays inside water vapour fields, the real water vapour vertical variability and its time evolution is also well reproduced. Fig. 8(bottom) shows the time evolution of wet refractivity vertical profiles evaluated in the map centre (voxel 11 - see Fig. 7) during the overall October week, considering data taken by all the available MisT receivers. Unfortunately, no meaningful

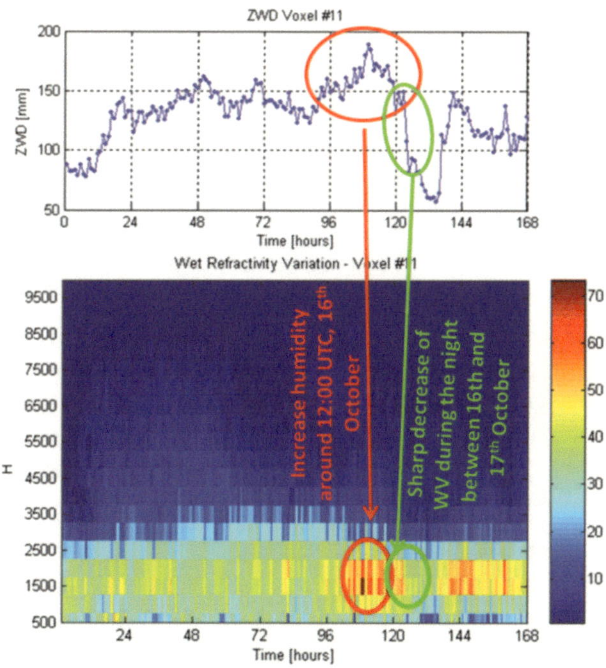

Fig. 8. Time evolution of wet refractivity distribution evaluated in the central column of the map (voxel 11) during the overall October week, considering data taken by all the available MisT receivers. Top: integrated wet refractivity along zenith (namely the ZWD time series). Bottom: vertical wet refractivity profile (measured in N-units) evolution (Heights are given in meters).

meteorological events happened during the observing period. Anyway, an increase of wet refractivity (water vapour concentration) can be evidenced between the 100th (4 am, 16th October) and the 120th (midnight, 16th October) hours, with a peak around the 110th hour (2 pm, 16th October). The increase is well reproduced in terms of integrated wet refractivity along zenith (see ZWD evolution in Fig. 8(top)). Moreover, meteorological data (not shown here) confirmed an increase of cloud covering during that time interval.

4. GNSS reflectometry

Other than for atmosphere monitoring, GNSS signals may be used to characterize the Earth surface. In this section this kind of remote sensing technique is described, considering two scenarios of observation: ocean and land.

The exploitation of GNSS signals reflected off the oceans allows to obtain altimetry measurements (sea surface heights), surface roughness from which wind intensity and direction is determined, sea-ice topography and its stratification. Additionally, land observations are used to determine the soil moisture content and to monitor the surface snow cover.

The most of performed experiments are based on code measurements, since signal phase coherence after reflections is not many times maintained, because smooth surfaces are rarely found in reality.

4.1 Description of observables, theoretical basis and retrieval technique

For remote sensing purposes, the reflected and direct GNSS signals coming from the same satellite are collected on bistatic radar geometry; at least two antennas are required: the first RHCP (Right Hand Circularly Polarized) and zenith looking in charge of receiving the direct signal, the second LHCP (Left Hand Circularly Polarized) and nadir looking used to track the reflections.

In order to be more precise, the overall system could be considered as a multistatic observing system, since up to 6/7 GNSS transmitters are contemporary visible by the receiver antenna.

Each reflection is geo-referenced knowing the geometry of acquisition, looking at the point where the GNSS signal is reflected under specular condition; for doing this, the observer coordinates are necessary. Therefore the direct signal is used not only as a reference but also for computing the position of the receiver.

Three acquisition scenarios are possible:

- Ground based: in this static configuration, the receiver is placed over mountains, towers and bridges and the collected measurements are used for testing the instrument functionalities and for monitoring small areas (i.e. coastal altimetry, local soil moisture content determination);
- On aircraft: the sensor is placed on aircrafts or rarely on balloons to demonstrate its performances and to monitor small regions with higher spatial resolution than space-based measurements. This dynamic configuration requires an evaluation of the Doppler shift due to the non-zero velocity of the aircraft; furthermore, this Doppler

shift improves the resolution on the surface by means of iso-Doppler lines computation.

- Space based: the sensor is placed on board a LEO satellite (400-800 km) with the aim of monitoring the entire Earth surface assuring a global coverage of the acquired reflections, which may be detected also very far from coastal zones (i.e. in the middle of the ocean); the Doppler shift experienced by the signal is the largest achievable among the three described scenarios.

The shape and extension of the footprint of the reflections depends on: the surface roughness, the sensor height above the Earth surface, the elevation of the reflected ray, the direction of the incidence plane respect to the receiver velocity.

The footprint must be considered lying on a plane tangent to the Earth surface in the specular reflection point. The distance of the specular reflection point from the receiver nadir increases when the elevation of the GNSS satellite decreases.

Inside the area interested by the reflection, the smallest resolution achievable from a geometrical point of view is determined by the cells generated by the intersections of the iso-delay and iso-Doppler lines.

Iso-delay lines are determined considering the points on the surface by which the reflected signal arrives at the receiver with the same delay. Generally speaking, these points are ellipses and are determined considering a single chip of the GNSS code as relative delay associated to each ellipse respect to the adjacent one (Martin-Neira, 1993).

Iso-Doppler lines are determined considering the hyperbolas on the surface where reflected signals come to the receiver with the same Doppler shift. The zero Doppler line is computed as the line passing through the receiver and orthogonal to its velocity direction (Martin-Neira, 1993).

Clearly, we cannot forget the antenna footprint, which acts as a filter in delay and Doppler on the surface looks. When the surface is smooth, the total power received is almost coming from the first Fresnel zone defined around the specular scattering point (Beckmann & Spizzichino, 1987). In this case, the computation of the cross-correlation between the reflected signal and the local GPS code replica gives a waveform simply delayed respect to the cross-correlation of the direct signal, but with the same triangle shape and a noise floor around.

When the surface is rough non-coherent reflections are expected and the use of the Fresnel zone to model the received power is ineffective. In this case, the glistening zone represents the source of scattered power (Beckmann & Spizzichino, 1987).

N scattering elements contained in the glistening zone are considered in determining the cross-correlation function

$$R_p(\Delta t_m) = \sum_{n=1}^{\infty} A_n \cdot e^{i(\phi_m - \phi_n)} \Lambda(\Delta t_m - \Delta t_n) \tag{13}$$

where Λ is the triangle cross-correlation function and the m index indicates the quantities referred to the modelled signal generated with the local GPS code replica. Through this formulation Rp becomes a summation of triangle functions weighted with the amplitude of the n_{th} element scattered field and delayed accordingly to the phase shift associated to each n_{th} scattering element. The final correlation function shape in this case is shown in Fig. 9.

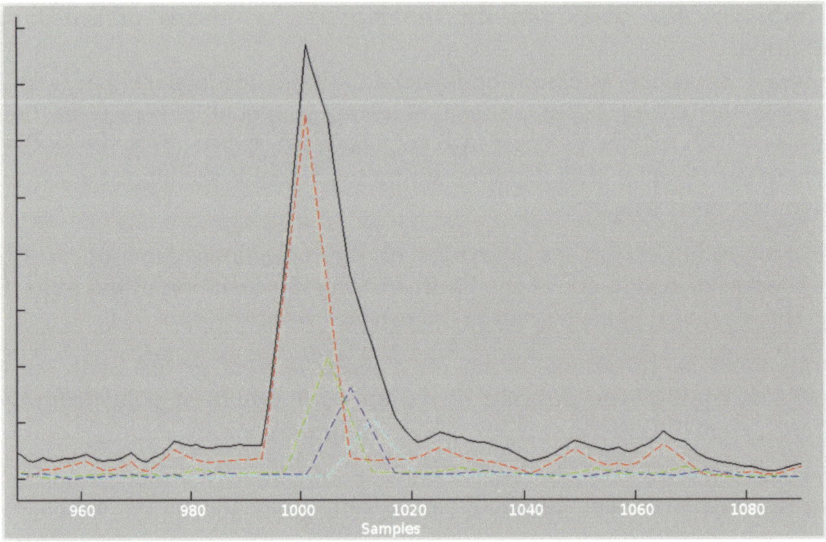

Fig. 9. Shape of the correlation function for non-coherent reflections (black); each triangle refers to the signals received by an isorange, 8 samples equal to 1 C/A chip

The basic observables are the delay of the reflected signal respect to the direct one, and the received power after reflection. Both observables are retrieved looking at the correlation function of the reflected signal and eventually comparing or normalizing it with the correspondent correlation of the direct signal.

The delay is used to determine the surface height, so is considered in case of GNSS signals reflected off water surfaces (Martin-Neira et al., 2001; Hajj & Zuffada, 2003). The height of the surface respect to the observer is retrieved in eq. 14 through the relative delay $\Delta\tau$, the speed of light c and the elevation angle of the reflection γ.

$$c\Delta\tau = 2h\sin\gamma \tag{14}$$

On the other hand, the reflected power is used to determine the surface reflectivity and the scattering cross section (Masters et al., 2004).

The surface reflectivity belongs to the coherent part of the scattered power that is measurable from the specular part of the received echo; it is used to determine the reflection coefficient that is related to the incident angle and dielectric constant. The dielectric constant is related to the soil composition and to its moisture content following empirical models or carefully calibrating the data (Masters et al., 2004).

The surface can be characterized looking at its roughness from the scattering cross section, since it contains the non-specular part of the reflected power. In this case we consider reflected power part calculated from the amplitude and the gradient of the correlation function on the right side of its maximum.

In order to retrieve surface winds over the sea, the shape of the non-specular echo is compared with a simulated one obtained using a sea surface model (Zavorotny & Voronovich, 2000; Elfouhaily et al., 2002).

4.2 State of the art

Winds retrieval and altimetry are the more consolidated applications, while soil moisture and ice monitoring are under-development.

Many instruments were developed up to now (Nogues-Correig et al.,2007) and the techniques of retrieval have been tested through many experimental activities. The early experiments deal basically with altimetry; measurements were collected either from a static position (Martin-Neira et al, 2001), from balloon (Cardellach et al., 2003) or from aircraft (Lowe et al, 2002). Other set of experiments were developed to retrieve the ocean surface state (Garrison et al., 2000), such as wind or sea roughness. Last but not least, the technique was demonstrated on board a small satellite, the UK-DMC (Gleason et al., 2005).

Nevertheless, nowadays no operative missions exist in this field.

From our point of view, during the SMAT-F1 project we developed a prototype based on a Software Defined Radio solution, using a navigation software receiver (Tsui, 2005). This is the NGene SW receiver, developed by NAVSAS group of Politecnico di Torino (Fantino et al., 2009). The instrument is highly reconfigurable, since collects raw I and Q IF samples of the incoming signals (direct and reflected). A sampling frequency of 8.1838 MHz is used, giving about 8 samples per C/A code chip.

Moreover, the small hardware architecture is made up of cheap COTS (Commercial Of The Shelf) components, with very low overall weight and power consumptions. These features make the system suitable to be easily placed on board aircrafts, also small U.A.V.s (Unmanned Aerial Vehicle) (Cucca et al., 2010).

4.3 Results

Using the described receiving system, we carried out two experiments. The first data collection has been made on a static position looking at the sea surface from a high cliff. The second was performed placing the receiver on an aircraft and acquiring GNSS signals reflected from rice fields.

4.3.1 Sea surface data collection

The first data collection was carried out on December 2010, from Sardinia Eastern coast at 157 m above the sea surface, near Cala Gonone. This region is characterized by high cliffs like those shown in Fig. 10.

Fig. 10. The Sardinia Eastern Coast near Cala Gonone (©Google Maps)

Considering the satellites in view, we pointed our antenna towards 170° of azimuth respect to geographical North. We successfully track reflected signals coming from GPS satellite 16 and GPS satellite 30. The geometry of acquisition was determined computing the iso-delay lines with ½ C/A code chip step and the specular reflection points, shown in Fig. 11 (in red for the 16th and in green for the 30th) together with the antenna footprint (depicted in light blue). All the points have been superimposed on ©Google static Maps and georeferenced in UTM. For both satellites the relative delay-doppler maps were computed over 1 s of non-coherent integration time and normalized from 0 to 1. Results are shown in Fig. 12. For satellite 30 (Fig. 12 (left)), the map is characterized by a very low noise, since the expected scattered signal is almost coherent and limited to one iso-range area, with no successive returns with delay greater than 1 chip (8 samples rising to the maximum, 8 samples going down to the noise floor).

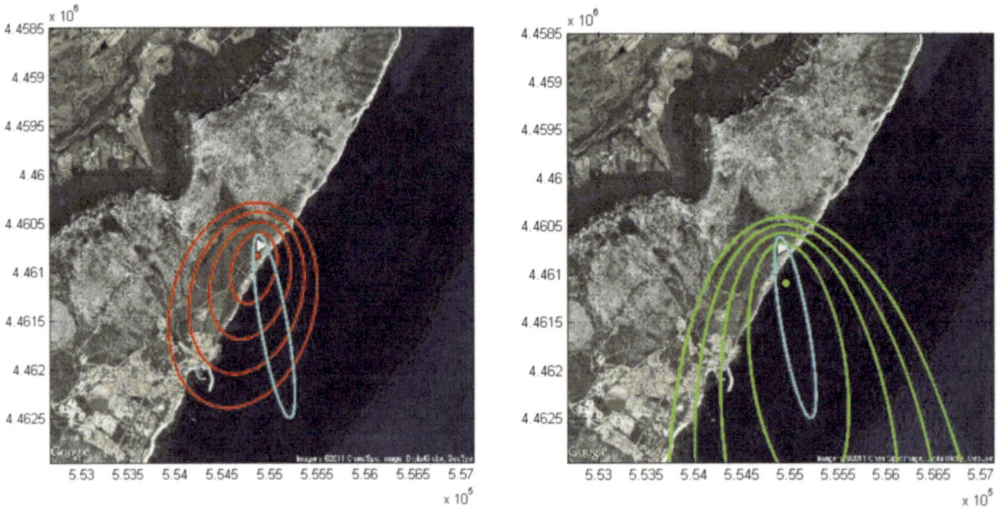

Fig. 11. Specular reflection point and iso-delay lines superimposed in UTM on ©Google static Maps (satellite 16 in red, satellite 30 in green)

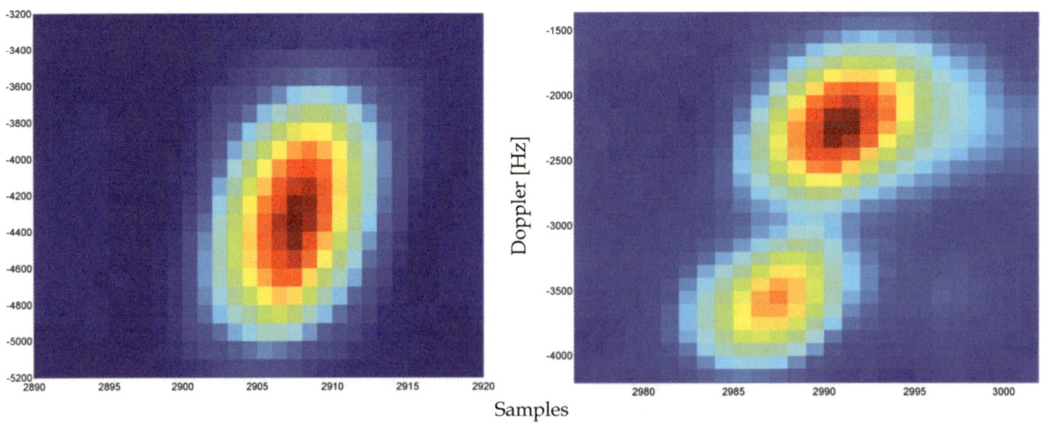

Fig. 12. Delay-Doppler maps for satellite 16 (right) and satellite 30 (left)

For satellite 16 (Fig. 12 (right)) two different echoes are visible. The first echo is coming from the coast (range closest to the observer and lower intensity due to scattering from the terrain); while the second is characterized by a greater delay, a different Doppler and successive returns lasting about 2 chips. In this case the correlation peak expires after 24 samples. This is a typical example of the capability to extract informations also from the un-coherent part of the signal.

Thus, our receiving system is able to track coherent and un-coherent reflections and to contemporary distinguish between echoes with different delays, Doppler shifts and intensity.

4.3.2 Rice fields data collection

During the second data collection of May 2011, an experiment performed flying over an area placed in the Piedmont region (north west part of Italy), the receiving system was placed on board a small aircraft in order to track reflections from rice fields. Since rice fields are flooded during this month, they are a perfect scenario to study reflection phenomena.

Like in the previous experiment, the geometry of reflections was analyzed and all the satellites with elevation lower than 33° were discarded, since below this elevation the specular reflections did not enter inside the -3 dB beam-width of the LHCP nadir looking antenna.

The signal to noise ratio detected from the reflected signal was normalized respect to the correspondent direct signal; moreover, we compute all the specular reflection points visible and to each point we associate the relative normalized signal to noise ratio.

On board the aircraft, a video camera was placed to see which fields were really flooded during the acquisition. The panoramic view extracted from the video was superimposed on ©Google Maps, together with the specular reflection points (Fig. 13). After the superposition, we have noticed a good agreement between the fields' state and the received power (see Fig. 13 and 14). The minimum received power correspondent to a low normalized signal to noise ratio is clearly associated to not flooded fields.

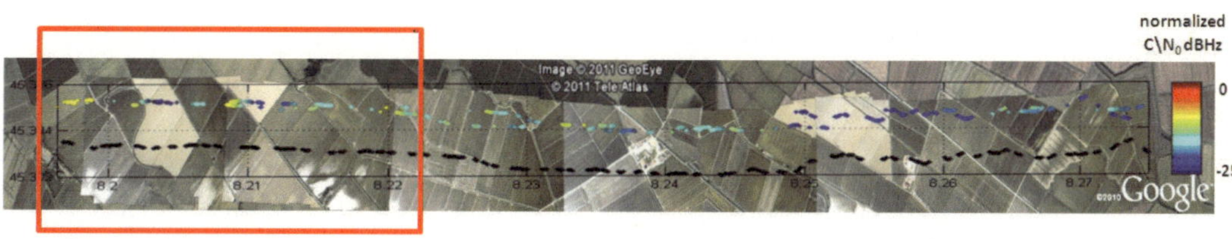

Fig. 13. Specular reflection points tracks for satellite 8 and 26 over Piedmont rice fields with relative normalized signal to noise ratio. See Fig. 13 for the red rectangle zoom.

Furthermore, we compare the signals of two different satellites with similar elevation but different azimuth; we notice a high correlation between the two specular reflection point tracks both from the qualitative (Fig. 13, Fig. 14) and the quantitative (Fig. 15) point of view. The quantitative comparison is performed considering the reflected power coming from the same longitude, considering a bean of 0.01°. Further investigations on this behavior are under development.

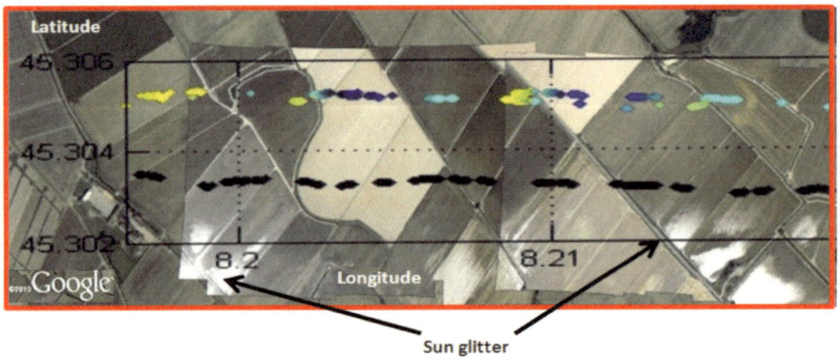

Fig. 14. Zoom of the specular reflection point tracks along the rice fields on ©Google Maps

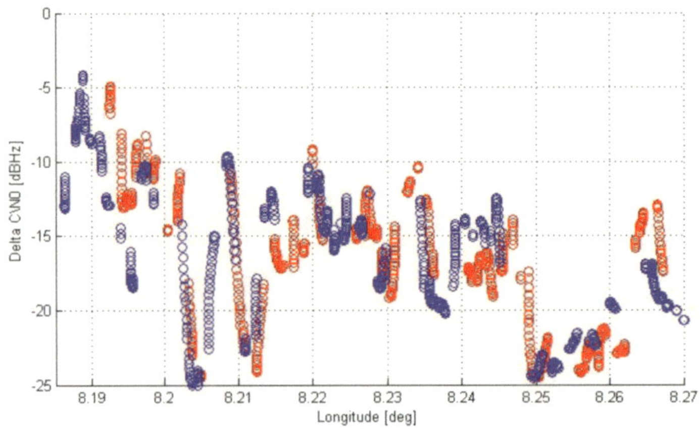

Fig. 15. Quantitative comparison of normalized signal to noise ratio for satellite 8 (red) and 26 (blue) considering reflection points with the same longitude

5. Conclusions and outlook

Scope of this chapter was to give an overview on some very powerful and quite recent Remote Sensing possibilities emerged exploiting GNSS observations, which complement the atmospheric and Earth's surface remote sensing traditionally performed by dedicated payloads and instrumentation.

GPS ground receivers can provide valuable and accurate information on integrated precipitable water vapor, considering that single receivers or fairly dense networks are available in many part of the world, providing a quite cheap and reliable source of information. As described previously, many investigations have been carried out in this respect to develop processing techniques, to validate the results through comparisons with independent sources and to exploit the final product. For instance, ZTD or IPWV data from a GPS ground based network can be assimilated into Numerical Weather Prediction models, or integrated with additional sources of IPWV to produce two-dimensional water vapour fields, leading to improved products.

As far as the tomographic approach for the retrieval of Neutral Atmospheric Refractivity maps is concerned, we demonstrated that it is possible (and with a good level of accuracy)

as long as some tricks are taken into account. In particular it has to be outlined that, in order to make neutral atmospheric tomography more effective, the choice of the GNSS network topology is a key aspect. A good horizontal receiver's distribution guarantees a good retrieval of horizontal gradients. A good vertical receiver's distribution guarantees also a good retrieval of vertical gradients. Even if our network topology was not optimal for tomographic purposes, the inclusion of measurements (even if not very accurate) performed by two receivers placed at higher heights and of the low elevation observations, demonstrate this aspect. Since a suitable vertical receiver distribution is difficult to implement, the availability of quasi-horizontal observations is necessary. Then, limb sounding Radio Occultation observations are necessary in order to guarantee good observations coming also from low elevation angles (this aspect has already been demonstrated by Foelsche and Kirchengast, 2001 and Notarpietro et al., 2008).

GNSS signals reflected off the Earth surface which represent an error source for navigation purposes, are instead useful for characterizing land and sea surfaces both from a monitoring and early-warning point of view. In particular the possibility of extracting information about the sea height and roughness, the soil moisture content, the snow and ice cover state have been successfully proven. Presently, no operative missions exist but many experimental activities have been carried out and the interest of national space agencies is constantly growing. From our point of view, we put some efforts in developing an instrument capable of collecting reflected GNSS signals, since we believe in the potentialities of this technique.

We definitely believe that the "expansion" of GNSS sources expected when also the European GALILEO, the Indian IRNSS and the Chinese BEIDOU navigation satellite systems will be deployed, together with the consequent availability of Radio Occultation observations, and the consequent availability of "vertical" and "horizontal" observations, will improve definitively all the techniques here presented.

6. Acknowledgements

For matherial contained in sections 2 and 3, authors are grateful to ESA for supporting the work in the framework of the METAWAVE project (Contract: ESTEC 21207/07/NL/HE). Work described in section 4 was supported by Regione Piemonte for funding the SMAT-F1 project, ISMB and Digisky for the development of the experimental campaign over rice fields and NAVSAS group for providing the NGene receiver.

7. References

Askne, J. & Nordius, H. (1987). Estimation of tropospheric delay for microwaves from surface weather data. *Radio Science*, Vol.22, No.3, pp. 379-386, ISSN 0048-6604

Anthes, R. A.; Ector, D.; Hunt, D. C.; Kuo, Y-H.; Rocken, C.; Schreiner, W. S.; Sokolovskiy, S. V.; Syndergaard, S.; Wee, T-K.; Zeng, Z. P.; Bernhardt, A.; Dymond, K. F.; Chen, Y.; Liu, H.; Manning, K.; Randel, W. J.; Trenberth, K. E.; Cucurull, L.; Healy, S. B.; Ho, S-P.; McCormick, C. T.; Meehan, K.; Thompson, D. C. & Yen N. L. (2008). The COSMIC/FORMOSAT-3 Mission: Early Results. *Bulletin of the American Meteorological Society*, Vol. 89, pp. 313-333, doi: 10.1175/BAMS-89-3-313

Basili, P.; Bonafoni, S.; Ferrara, R.; Ciotti, P.; Fionda, E. & Ambrosini, R. (2001). Atmospheric water vapour retrieval by means of both a GPS network and a microwave radiometer during an experimental campaign at Cagliari (Italy) in 1999. *IEEE Transaction on Geoscience and Remote Sensing*, Vol.39, No.11, pp. 2436-2443, ISSN 0196-2892

Basili, P.; Bonafoni, S.; Mattioli, V.; Ciotti, P. & Pierdicca, N. (2004). Mapping the atmospheric water vapor by integrating microwave radiometer and GPS measurements. *IEEE Transaction on Geoscience and Remote Sensing*, Vol.42, No.8, pp. 1657-1665, ISSN 0196-2892

Beckmann, P.; Spizzichino, A. (1987) *Scattering of Electromagnetic Waves from Rough Surfaces*, Artech House Publishers, ISBN 0-89006-238-2, New York, NY

Bevis, M.; Businger, S.; Herring, T.A.; Rocken, C.; Anthes, R.A. & Ware, R.H. (1992). GPS meteorology: remote sensing of atmospheric water vapour using the Global Positioning System. *Journal of Geophysical Research*, Vol.97, pp. 787-801, ISSN 0148-0227

Bevis, M.; Businger, S.; Chiswell, S.; Herring, T.A.; Anthes, R.A.; Rocken, C. & Ware, R.H. (1994). GPS meteorology: mapping zenith wet delays onto precipitable water. *Journal of Applied Meteorology*, Vol.33, pp. 379-386, ISSN 0894-8763

Bi, Y.; Mao, J. & Li, C. (2006). Preliminary results of 4-D water vapor tomography in the troposphere using GPS. *Adv. Atmos. Sci.*, Vol. 23, pp. 551–560

Brunner, F. K. & Welsch W. M. (1993). Effect of the troposphere on GPS measurements. *GPS World*, Vol.4, No.1, pp. 42-51, ISSN 1048-5104

Businger, S.; Chiswell, S.; Bevis, M.; Duan, J.; Anthes, R.; Rocken, C.; Ware, R.; Van Hove, T. & Solheim, F. (1996). The Promise of GPS in Atmospheric Monitoring. *Bulletin of the American Meteorological Society*, Vol. 77, pp. 5-18, ISSN 1520-0477

Cardellach, E.; Ruffini, G.; Pino, D.; Rius, A.; Komjathy, A.; Garrison, J.L. (2003). Mediterranean Ballon Experiment: ocean wind speed sensing from the stratosphere, using GPS reflections. *Remote Sensing of Environment*, Vol.88, No.3, (December 2003), pp. 351-362

Champollion, C.; Masson, F. & Bouin, M.-N. (2005). GPS water vapour tomography: preliminary results from the ESCOMPTE field experiment. *Atm. Res.*, Vol. 74, pp. 253–274

Champollion, C.; Flamant, C. & Bock, O. (2009). Mesoscale GPS tomography applied to the 12 June 2002 convective initiation event of IHOP_2002. *Q. J. R. Meteorol. Soc.*, Vol. 135, pp. 645–662, doi: 10.1002/qj.386

Coster, A.J.; Niell, A.E.; Burke, H.K. & Czerwinski, M. G. (1997). The Westford water vapour experiment: use of GPS to determine total precipitable water vapour, In: *Lincoln Laboratory Massachusetts Institute of Technology Technical report 1038*

Cucca, M.; Notarpietro, R.; Perona, G.; Fantino, M. (2010) Reflected GNSS signals received on unmanned aerial vehicle platform: a feasibility study. *Proceedings of IEEE Gold Remote Sensing Conference 2010*, Livorno, Italy, April 29-30, 2010

Davies, O.T. & Watson, P.A. (1998). Comparison of integrated precipitable water vapour obtained by GPS and radiosondes. *Electronics Letters*, Vol.34, pp. 645-646, ISSN 0013-5194

Davis, J. L.; Herring, L.T.A.; Shapiro, I.I.; Rogers, A.E. & Elgered, G. (1985). Geodesy by radio interferometry: Effects of atmospheric modelling errors on estimates of baseline length. *Radio Science*, Vol.20, pp. 1593-1607, ISSN 0048-6604

de Haan, S.; Holleman, I. & Holtslag A.A.M. (2009). Real-Time Water Vapor Maps from a GPS Surface Network: Construction, Validation, and Applications. *Journal of Applied Meteorology and Climatology*, Vol.48, No.7, pp. 1302-1316, ISSN 1558-8424

Duan, J.; Bevis, M.; Fang, P.; Bock, Y.; Chiswell, S.; Businger, S.; Rocken, C.; Solheim, F.; van Hove, T.; Ware, R.; McClusky, S.; Herring, T.A. & King, R.W. (1996). GPS meteorology: direct estimation of the absolute value of precipitable water. *Journal of Applied Meteorology*, Vol.35, pp.830-838, ISSN 0894-8763

Elgered, G.; Davis, J.L.; Herring, T.A. & Shapiro, I.I. (1991). Geodesy by radio interferometry: Water vapour radiometry for estimation of the wet delay. *Journal of Geophysical Research*, Vol.96, pp. 6541-6555, ISSN 0148-0227

Elfouhaily, T.; Thompson, D.R.; Linstrom, L. (2002). Delay-Doppler Analysis of bistatically reflected signals from the ocean surface: theory and application. *IEEE Transactions on Geoscience and Remote Sensing*, Vol.40, No. 3, (March 2002), pp. 560-573

Elosegui, P.; Davis, J.L. & Gradinarsky, L.P. (1999). Sensing atmospheric structure using small scale space geodetic networks. *Geophys. Res. Lett.*, Vol. 26, pp. 2445–2448, doi:10.1029/1999GL900585

Emardson, T.R.; Elgered, G. & Johansson, J. (1998). Three months of continuous monitoring of atmospheric water vapor with a network of Global Positioning System receivers. *Journal of Geophysical Research*, Vol.103, pp. 1807–1820, ISSN 0148-0227

Fantino, M.; Molino, A.; Mulassano, P.; Nicola, M.; Pini, M. (2009). N-Gene: A GPS and Galileo Fully Software Receiver., *Proceedings of SDR-Italy'09 Workshop – From Software Defined Radio to Cognitive Networks*, Pisa, Italy, July 2-3, 2009

Flores, A.; Ruffini, G. & Rius, A. (2000). 4D tropospheric tomography using GPS slant wet delays. *Ann. Geophys.* Vol. 18, pp. 223–234, ISSN: 14320576

Foelsche, U. & Kirchengast, G. (2001). Tropospheric water vapor imaging by combination of ground-based and spaceborne GNSS sounding data. *J. Geophy. Res.*, Vol. 106, pp. 27221–27231

Garrison, J.L.; Katzberg, S.J. (2000). The application of reflected GPS signals to ocean remote sensing. *Remote Sensing of Environment*, Vol.73, pp.175-187

Gleason, S.T.; Hodgart, S.; Yiping, S.; Gommenginger, C.; Mackin, S.; Adjrad, M.; Unwin, M. (2005). Detection and Processing of Bistatically Reflected GPS Signals from Low Earth Orbit for the Purpose of Ocean Remote Sensing. *IEEE Transactions on Geoscience and Remote Sensing,* Vol.43, No.6, (June 2005), pp. 1229-1241

Gutman, S.I. & Benjamin, S.G. (2001). The Role of Ground-Based GPS Meteorological Observations in Numerical Weather Prediction. *GPS Solutions*, Vol.4, No.4, pp. 16-24, ISSN 1080- 5370

Gradinarsky, L.P. & Jarlemark, P. (2004). Ground based GPS tomography of water vapor: analysis of simulated and real data. *J. Meteorol. Soc. Jpn.*, Vol. 82, pp. 551–560, ISSN:0026-1165

Hajj, G.A.; Kursinski, E.R.; Romans, L.J.; Bertiger, W.I. & Leroy S.S. (2001). A technical description of atmospheric sounding by GPS occultations. *J. of Atmosph. and Solar-Terr. Phys.*, Vol. 64.

Hajj, G.; Zuffada, C. (2003). Theoretical description of a bistatic system for ocean altimetry using the GPS signal. *Radio Science*, Vol.38, No.10, pp. 1-19

Herring, T.A. (1992). Modelling atmospheric delays in the analysis of space geodetic data, *Proceedings of the symposium on refraction of transatmospheric signals in geodesy*, Netherlands Geodetic Commission, J.C. De Munk and T.A. Spoelstra, (Ed),Publications on Geodesy, Delft, The Netherlands, No.36, pp. 157–164

Herman, G.T. (1980). *Image Reconstruction from Projections: The Fundamentals of Computerized Tomography*. Academic Press, 1980

Hirahara, K. (2000). Local GPS tropospheric tomography. *Earth Planets Space*, Vol. 52, pp. 935–939

Kirchengast, G.; Hafner, J. & Poetzi, W. (1999). *The CIRA86aQ_UoG model: an extension of the CIRA-86 monthly tables including humidity tables and a Fortran95 global moist air climatology model*. Tech. Rep. for ESA/ESTEC, vol. 8

Kursinski, R. (1994). Monitoring the Earth's Atmosphere with GPS. *GPS World*, Vol.5, No.3, pp. 50-54, ISSN 1048-5104

Kursinski, E. R.; G. A., Hajj; J. T., Schofield; R. P., Linfield & K. R. Hardy (1997). Observing Earth's atmosphere with radio occultation measurements using the Global Positioning System, *J. Geophys. Res.*, Vol 102 (D19), pp. 23,429–23,465, doi:10.1029/97JD01569

Ifadis, I.I. (1986). The atmospheric delay of radio waves: modeling the elevation dependence on a global scale, In: *Technical report 38L, Chalmers University of Technology, Goteborg ,Sweden*

Lindenbergh, R.; Keshin, M.; van der Marel, H. & Hanssen, R. (2008). High resolution spatio-temporal water vapour mapping using GPS and MERIS observations. International Journal of Remote Sensing, Vol.29, No.8, pp. 2393-2409, ISSN 0143-1161

Liou, Y. A.; Pavelyev, A. G.; Matyugov, S. S.; Yakovlev & Wickert J. (2010). *Radio Occultation Method for Remote Sensing oft he Atmosphere and Ionosphere*, InTech, ISBN 978-953-7619-60-2, Vukovar, Croatia.

Lawson, C. & Hanson, R. (1974). *Solving Least Squares Problems*. Prentice-Hall, Englewood Cliffs, New York

Lowe, S. T.; Zuffada, C.; Chao, Y.; Kroger, P.; Young, L.E.; LaBreque, J.L. (2002). 5-cm precision aircraft ocean altimetry using GPS reflections. *Geophysical Research Letters*, Vol.29, No.10, doi: 10.1029/2002GL014759

Luntama, J-P.; Kirchengast, G.; Borsche, M.; Foelsche, U.; Steiner, A.; Healy, S.; Von Engeln, A.; O'Clerigh, E. & Marquardt C. (2008). Prospects of the EPS GRAS Mission For Operational Atmospheric Applications. *Bulletin of the American Meteorological Society*. Vol. 89, pp. 1863-1875, doi: 10.1175/2008BAMS2399.1

Kleijer, F. (2004). *Troposphere Modeling and Filtering for Precise GPS Leveling,* NCG KNAW Netherlands Geodetic Commission Publication on Geodesy, ISBN 978-90-6132-284-9, Delft, The Netherlands

Macpherson, S.R.; Deblonde, G.; Aparicio, J.M. & Casati, B. (2008). Impact of NOAA Ground-Based GPS Observations on the Canadian Regional Analysis and Forecast System. *Monthly Weather Review*, Vol.136, pp. 2727–2746, ISSN 0027-0644

Martin-Neira, M. (1993). A passive reflectometry and interferometry system (PARIS): Application to ocean altimetry. *ESA Journal*, Vol.17, pp. 331-355

Martin-Neira, M.; Caparrini, M.; Font-Rossello, J.; Lannelongue, S.; Vallmitjana, C. S. (2001). The PARIS concept: an experimental demonstration of sea surface altimetry using GPS reflected signals. *IEEE Transactions on Geoscience and Remote Sensing,* Vol. 39, pp. 142-150, ISSN 0196-2892

Masters, D.; Axelrad, P.; Katzberg, S. (2004). Initial results of land-reflected GPS bistatic radar measurements in SMEX02. *Remote Sensing of Environment,* Vol.92, No.4, pp. 507-520, doi:10.1016/j.rse.2004.05.016

Melbourne, W.; Davis, E.; Duncan, C.; Hajj, G.; Hardy, K.; Kursinski, E.; Meehan, T. & Young L. (1994). *The application of spaceborne GPS to atmospheric limb sounding and global change monitoring,* Publication 94-18, Jet Propulsion Laboratory.

Morland, J. & Matzler, C. (2007). Spatial interpolation of GPS integrated water vapour measurements made in the Swiss Alps. *Meteorological Applications,* Vol.14, pp. 15-26, ISSN 1350-4827

Niell, A.E. (1996). Global mapping functions for the atmospheric delay at radio wavelengths. *Journal of Geophysical Research,* Vol.101, pp 3227-3246, ISSN 0148-0227

Nilsson, T. & Gradinarsky, L. (2006). Water vapor tomography using GPS phase observation: simulation results. *IEEE Trans. Geosci. Remote Sens.,* Vol. 44, pp. 2927–2941, doi: 10.1109/TGRS.2006.877755

Nogués-Correig, O.; Cardellach Galì, E.; Sanz Campderròs, J.; Rius, A. (2007). A GPS-Reflections Receiver that computes Doppler/Delay Maps in real time. *IEEE Transactions on Geoscience and Remote Sensing,* Vol.45, No.1, (January 2007), pp.156-174, ISSN 0196-2892

Notarpietro, R.; Cucca, M.; Gabella, M.; Venuti & Perona G. (2011). Tomographic reconstruction of wet and total refractivity fields from GNSS receiver networks. *Advances In Space Research,* pp. 16, 2011, ISSN: 0273-1177, doi: 10.1016/j.asr.2010.12.025

Notarpietro, R.; Gabella, M. & Perona, G. (2008). Tomographic reconstruction of neutral atmospheres using slant and horizontal wet delays achievable through the processing of signal observed from small GPS networks. *Ital. J. Remote Sens.,* Vol. 40, pp. 63–74, ISSN 1129-8596

Pierdicca, N.; Rocca, F.; Rommen, B.; Basili, P.; Bonafoni, S.; Cimini, D.; Ciotti, P.; Consalvi, F.; Ferretti, R.; Foster, W.; Marzano, F.S.; Mattioli, V.; Mazzoni, A.; Montopoli, M.; Notarpietro, R.; Padmanabhan, S.; Perissin, D.; Pichelli, E.; Reising, S.; Sahoo, S. & Venuti, G. (2009). Atmospheric water-vapor effects on spaceborne Interferometric SAR imaging: comparison with ground-based measurements and meteorological model simulations at different scales, *Proceedings of International Geosciences and Remote Sensing Symposium (IGARSS09),* pp. 320-323, ISBN 978-1-4244-3394-0, Cape Town, South Africa, July 12-17, 2009

Rocken, C.; Ware, R.H.; van Hove, T.; Solheim, F.; Alber, C. & Johnson, J. (1993). Sensing atmospheric water vapour with the Global Positioning System. *Geophysical Research Letters,* Vol.20, pp. 2631-2634, ISSN 0094-8276

Saastamoinen, J. (1972). Atmospheric Correction for the Troposphere and Stratosphere in Radio Ranging of Satellites, In: *The Use of Artificial Satellites for Geodesy,* S.W. Henriksen et al., (Ed), vol.15, 247-251, Geophysics Monograph Series, A.G.U., Washington, D.C.

Smith, T.L.; Benjamin, S.G; Schwartz, B.E. & Gutman, S.I. (2000). Using GPS-IPW in a 4-d data assimilation system. *Earth, Planets and Space*, Vol.52, pp. 921–926 ISSN 1343-8832

Thayer, G.D. (1974). An improved equation for the radio refractive index of air. *Radio Science,* Vol.9, pp. 803-807, ISSN 0048-6604

Treuhaft, R.N. & Lanyi, G.E. (1987). The effect of the dynamic wet troposphere on radio interferometric measurements. *Radio Science*, Vol.22, pp. 251-265, ISSN 0048-6604

Troller, M.; Geiger, A. & Brockmann, E. (2006). Tomographic determination of the spatial distribution of water vapour using GPS observations. *Adv. Space Res.*, Vol. 37, pp. 2211–2217

Tsui, J. (2005). *Fundamentals of Global Positioning System receivers: a software approach*, Wiley, ISBN 0-471-38154-3, New York, USA

Ware R.H.; Exner M.; Feng D.; Gorbunov M.; Hardy K.R.; Herman B.; Kuo Y.H.; Meehan T.K.; Melbourne W.G.; Rocken C.; Schreiner W.; Sokolovskiy S.V.;. Solheim, F.; Zou, X.; Anthes, R.; Businger, S.& Trenberth, K. (1996). GPS Sounding of the Atmosphere from Low Earth Orbit: Preliminary Results. *Bull. Am. Meteorol. Soc.*, Vol. 77, pp. 19-40

Ware, R.H.; Alber, C.; Rocken, C. & Solheim, F. (1997). Sensing integrated water vapour along GPS ray paths. *Geophysical Research Letters*, Vol.24, pp. 417-420, ISSN 0094-8276

Yuan L.; Anthes, R.A.; Ware, R.H.; Rocken, C.; Bonner, W.; Bevis M. & Businger, S. (1993). Sensing Climate Change Using the Global Positioning System. *Journal of Geophysical Research*, Vol.98, No.14, pp. 925-937, ISSN 0148-0227

Zavorotny, V.U. & Voronovich, A.G. (2000). Scattering of GPS Signals from the Ocean with Wind Remote Sensing Application. *IEEE Transactions on Geoscience and Remote Sensing,* Vol.38, pp. 951-964

9

Acceleration Visualization Marker Using Moiré Fringe for Remote Sensing

Takeshi Takakai
Hiroshima University
Japan

1. Introduction

Advances in metrology contribute to various fields, and many sensors are developed as a result, such as high-performance devices, compact and lightweight devices, and low-cost devices, depending on the specific application environment. The sensors used in the recent years to measure physical quantities commonly feature electrical elements, and when such sensors are attached to measurement points, it is necessary to supply electrical power at these points. This implies that external power must be supplied via wiring, or, that an internal electrical power source must be fitted into the sensor. Furthermore, wired or wireless methods are needed to transmit the measurement data from the sensor.

In some environments, this can be a limiting factor for the applicability of measurements. To solve this problem, measurement techniques that do not require electrical power to be supplied to measurement points have been developed, which utilize optical elements or ultrasound, for example. In an earlier paper, we proposed a mechanism that makes the use of moiré fringes to visualize a physical force without any need for an electrical power supply at the measurement points. We also demonstrated typical applications of this technique by fitting the mechanism to a robot gripper (Takaki, 2008) and an endoscopic surgical instrument (Takaki, 2010a).

In addition, a large number of research has been carried out on the maintenance and management of large structures such as industrial plants, buildings, or bridges, by measuring their physical behavior. In particular, there have been many studies on vibrational phenomena (Umemoto, 2010) (Yun, 2010) (Kim, 2010). However, providing wiring for all the sensors in such large structures is no easy matter. We therefore propose the use of markers that utilize moiré fringes to enable acceleration to be visualized and displayed, without the need for an electrical power supply at the measurement points. In this manner, it is possible to measure acceleration remotely, without any wiring, by capturing images of the markers with a camera.

There have been previous studies on measuring displacement by means of moiré fringes (Kobayashi, 1987) (Reid, 1984) (Basehore, 1981) (Meadows, 1970) (Takasaki, 1970). Although these techniques have the advantage of not requiring a direct supply of electrical power at the measurement points, they require the use of lasers or special light sources to enable the projection of stripe patterns in order to generate the moiré fringes. The technique proposed in the present paper is different in that it does not require any special light source, but instead utilizes ambient light.

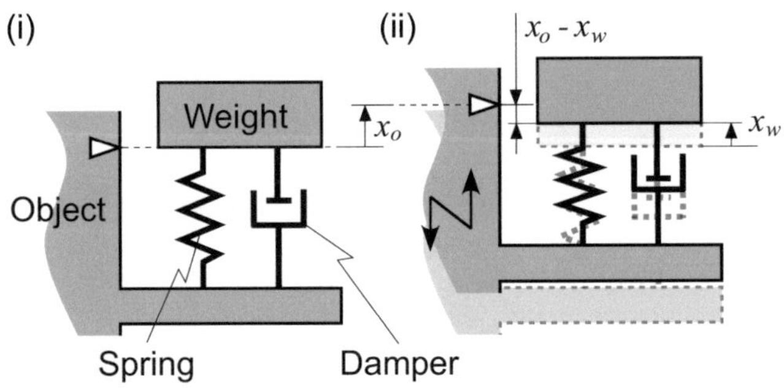

Fig. 1. Seismic system

In our proposed method, moiré fringes are generated by superimposing two glass plates, printed with parallel line gratings. Although a similar method was previously suggested for measuring displacement (Masanao, 1986), it has no provision for measuring acceleration or for the remote acquisition of data using a camera.

In the present study, we propose acceleration visualization markers (Takaki, 2010b) that enable the display of moiré fringes corresponding to the magnitude of the acceleration, and we demonstrate a method for acquiring acceleration data by means of the captured images of these markers . Chapter 2 of this report describes the principle by which the magnitude of the acceleration can be measured using moiré fringes. Chapter 3 describes the method for acquiring one-axis acceleration data using captured images of the markers, and Chapter 4 describes the method of upgrading the marker to an x- and y-axis acceleration visualization markers. Chapter 5 describes the developed markers, created using selected materials with careful attention paid to damping characteristics, and explains the mechanical characteristics of the markers. It is shown how acceleration data can be acquired using a high-speed video camera. Chapter 6 concludes this study.

2. Principle

2.1 Seismic system and acceleration

It is well known that acceleration can be measured using the seismic system (Holman, 2001), which consists of a spring, a damper, and a weight, as shown in Fig. 1 (i). Let us assume that a measurement object is under acceleration, as shown in Fig. 1 (ii), and that it moves by a displacement x_o. The displacement of the weight x_w is caused by the influence of the acceleration. We discuss the method of calculating the acceleration of the object $\ddot{x}_o$ from the relative displacement $x_o - x_w$. Let m, b, and k be the mass of the weight, viscosity of the damper, and spring constant of the spring, respectively, and these are constant. The sum of the forces acting on the weight is then

$$m\ddot{x}_w + b(\dot{x}_w - \dot{x}_o) + k(x_w - x_o) = 0, \tag{1}$$

where $\ddot{x}_w$, $\dot{x}_w$ and $\dot{x}_o$ are

$$\ddot{x}_w = \frac{d^2 x_w}{dt^2}, \quad \dot{x}_w = \frac{dx_w}{dt} \quad \text{and} \quad \dot{x}_o = \frac{dx_o}{dt}. \tag{2}$$

We consider the initial conditions as $x_w(0) = 0$, $\dot{x}_w(0) = 0$, and $x_o(0) = 0$. We then obtain the Laplace transform equation as follows:

$$ms^2 X_w(s) + bsX_w(s) + kX_w(s) = bsX_o(s) + kX_o(s). \tag{3}$$

Thus, $X_w(s)/X_o(s)$ is

$$\frac{X_w(s)}{X_o(s)} = \frac{bs + k}{ms^2 + bs + k}. \tag{4}$$

The transfer function of the seismic system with input $x_o - x_w$ and output $\ddot{x}_o$ is written as

$$\begin{aligned} G(s) = \mathcal{L}\left(\frac{x_w - x_o}{\ddot{x}_o}\right) &= \frac{X_w(s) - X_o(s)}{s^2 X_o(s)} \\ &= \frac{1}{s^2}\left(\frac{X_w(s)}{X_o(s)} - 1\right) \\ &= \frac{-1}{s^2 + 2\zeta\omega_n s + \omega_n^2}, \end{aligned} \tag{5}$$

where ζ is the dimensionless damping ratio and ω_n is the natural angular frequency of the system. ζ and ω_n are given by

$$\zeta = \frac{b}{2\sqrt{mk}} \quad \text{and} \quad \omega_n = \sqrt{\frac{k}{m}} \tag{6}$$

The transfer function of a system $G(s)$ can be described in the frequency domain as

$$G(j\omega) = \frac{-(1/\omega_n)^2}{1 - (\omega/\omega_n)^2 + 2\zeta(\omega/\omega_n)j} \tag{7}$$

The magnitude $|G(j\omega)|$ $(= |(x_w - x_o)/\ddot{x}_o|)$ and the phase angle ϕ are respectively written as

$$|G(j\omega)| = \frac{(1/\omega_n)^2}{\sqrt{\left(1 - (\omega/\omega_n)^2\right)^2 + \left(2\zeta\,(\omega/\omega_n)^2\right)^2}}$$

and

$$\phi = -\tan^{-1}\frac{2\zeta(\omega/\omega_n)}{1 - (\omega/\omega_n)^2} - \pi \tag{8}$$

When $\omega \ll \omega_n$, as shown in Fig. 2, $|G(j\omega)|$ and ϕ are approximately given by

$$|G(j\omega)| = \left|\frac{x_w - x_o}{\ddot{x}_o}\right| \simeq \frac{1}{\omega_n^2} \quad \text{and} \quad \phi \simeq -\pi \tag{9}$$

Therefore, the relationship between the relative displacement $x_w - x_o$ and the acceleration $\ddot{x}_o$ can be written as

$$\ddot{x}_o \simeq \omega_n^2(x_o - x_w). \tag{10}$$

According to this equation, the natural angular frequency ω_n can be obtained from constant values of the mass of the weight m and spring constant k, as we can see by Eq. 6; therefore, ω_n^2 is a constant. If the relative displacement $x_o - x_w$ is magnified sufficiently, the acceleration $\ddot{x}_o$ also becomes perceivable. However, in general, the relative displacement $x_o - x_w$ is too small to observe. Therefore, it is necessary to use a technology that can magnify the relative displacement $x_o - x_w$. We have focused on the use of a moiré fringe to magnify the relative displacement in this case.

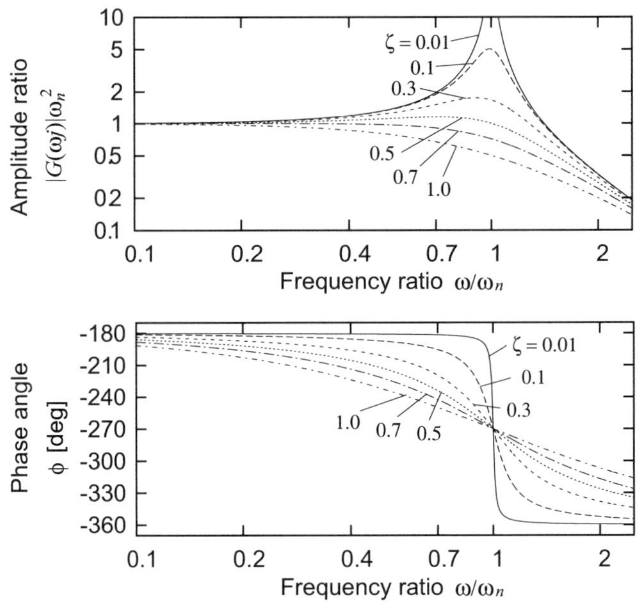

Fig. 2. Frequency response of the seismic system

2.2 Moiré fringe

Let us understand the concept of a moiré fringe (Kobayashi, 1987). As shown in Fig. 3 (i), line gratings 1 and 2 have the same pitch p_g and line grating 2 is inclined at a small angle φ and superimposed on line grating 1; a fringe known as the moiré fringe appears at a large pitch p_m ($> p_g$). The pitch p_m is larger than the pitch p_g of line gratings 1 and 2. The relationship between the pitches is given by

$$p_m = \frac{1}{2 \sin \frac{\varphi}{2}} \, p_g. \tag{11}$$

As shown in Fig. 3 (ii), when line grating 1 is moved in the direction (x) at pitch p_g, the moiré fringe moves in the direction (X) at pitch p_m. Therefore, the displacement can be displayed visually at a magnification of $1/2\sin(\varphi/2)$. This magnification is defined as M. When the relative displacement of the line gratings is $x_o - x_w$, the displacement of the moiré fringe can be described by the following equation:

$$x_m = M(x_o - x_w). \tag{12}$$

2.3 Structure of the acceleration visualization marker

To obtain a constant magnification M using moiré fringes, as described in Section 2.2, even if a relative displacement $x_o - x_w$ occurs, the angle φ must be maintained as a constant. To satisfy this requirement, two elastic plates of the same shape are used, as shown in Fig. 4 (i). This structure permits a relative displacement $x_o - x_w$ without any change in the angle φ, as shown in Fig. 4 (ii). Moreover, the elasticity and the damping capacity of the elastic plates function as the spring and the damper of the seismic system, respectively. If a weight is installed in this structure, it becomes a seismic system, and the acceleration $\ddot{x}_o$ can be calculated from the relative displacement $x_o - x_w$.

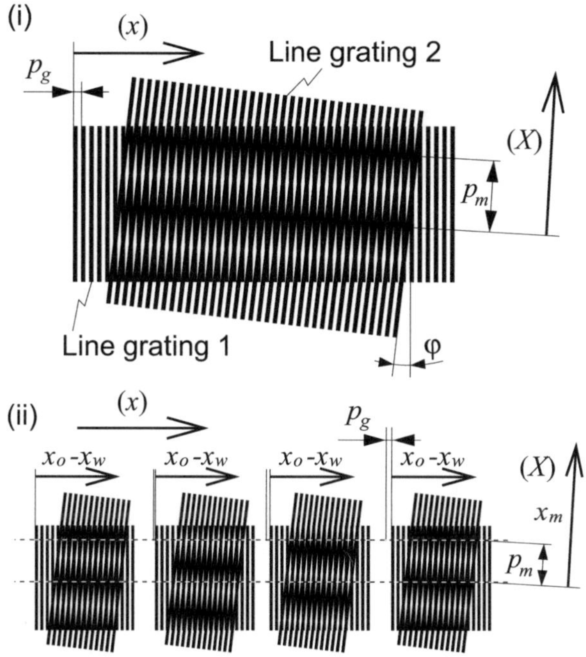

Fig. 3. A moiré fringe

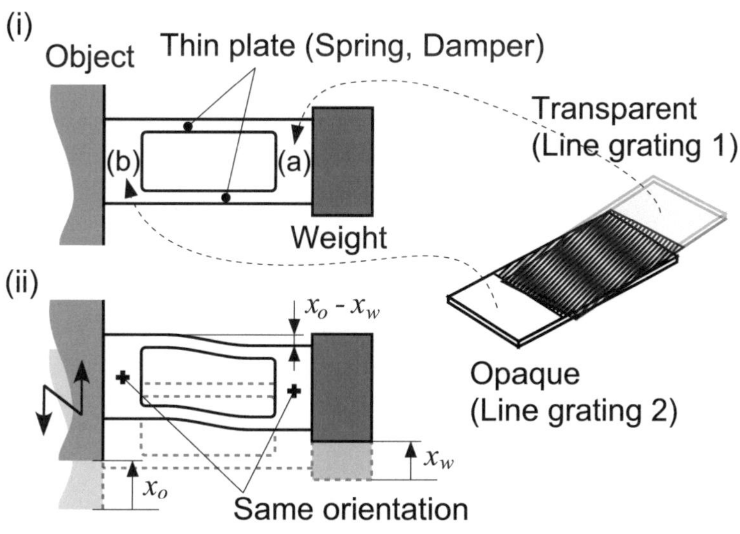

Fig. 4. Structure of the acceleration visualization marker

Line gratings 1 and 2 are respectively printed on transparent and opaque glass plates and fixed at locations (a) and (b) as shown in Fig. 4. The relative displacement $x_o - x_w$ produced by the acceleration $\ddot{x}_o$ is displayed by the moiré fringe at magnification M. Therefore, the magnitude of the acceleration $\ddot{x}_o$ can be confirmed visually. These acceleration visualization elements combine to form the acceleration visualization marker.

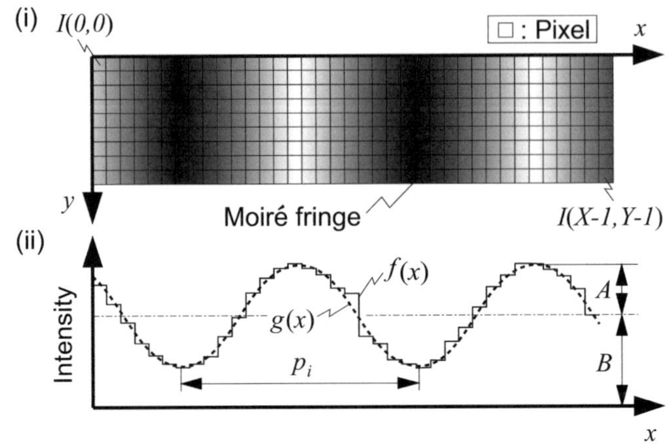

Fig. 5. Intensity of a moiré fringe and a fitted curve

3. Method of extracting acceleration value by image processing

3.1 Fitted sine curve

The image of a moiré fringe is trimmed from an original image taken by a camera, and the x- and y-axes are defined as shown in Fig. 5 (i). The size of the trimmed image is (X, Y), and the brightness value of the pixel at (x, y) is defined as $I(x, y)$. $f(x)$ is the average of the brightness value along the y-axis. $f(x)$ can be written as

$$f(x) = \frac{\sum_{k=0}^{Y-1} I(x,k)}{Y} \tag{13}$$

$g(x)$ is a fitted sine curve of $f(x)$. $g(x)$ can be written as follows:

$$g(x) = A\sin(\frac{2\pi}{p_i}x + \theta) + B \tag{14}$$

Figure 5 (ii) shows the difference between $f(x)$ and $g(x)$ in a example case. p_i, A, B, and θ are the pitch, amplitude of the brightness value, offset of the brightness value, and phase of the moiré fringe in the trimmed image, respectively. p_i can be obtained from an autocorrelation analysis of $f(x)$, and A, B, and θ can be obtained using the least square method.

3.2 Phase of fitted sine curve and displacement of moiré fringe

Figure 6 (i) shows the image of a moiré fringe when no acceleration is applied to the acceleration visualization marker. The brightness value of this moiré fringe is fitted to $g(x)$, and the phase in this state is assumed to be θ_0, as indicated by (a) in Fig. 6 (iii). When acceleration is applied to the marker, a relative displacement of $x_o - x_w$ occurs, and the moiré fringe moves by x_m, which can be calculated from Eq. (12). In the image, when a unit length corresponds to l pixels, the moiré fringe moves by lx_m, as shown in Fig. 6 (ii). When the phase θ shifts by $\Delta\theta$, as indicated by (b) in Fig. 6 (iii), the relationship between these two terms is given by

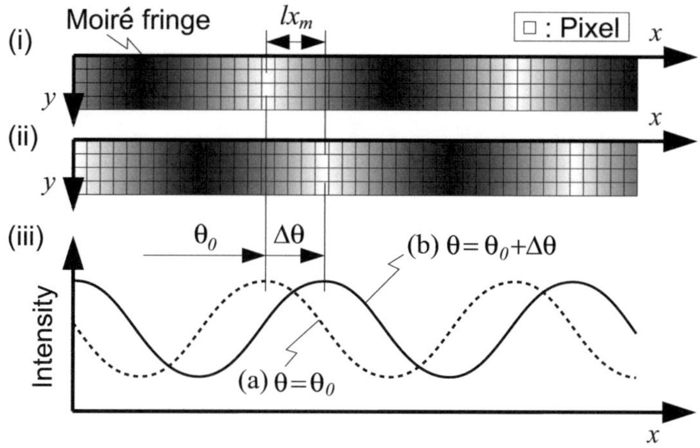

Fig. 6. Phase of a moiré fringe

$$lx_m = \frac{p_i}{2\pi}\Delta\theta \tag{15}$$

From Eqs. (10), (12), and (15), the acceleration $\ddot{x}_o$ is given by

$$\ddot{x}_o \simeq \frac{\omega_n^2 p_i}{2\pi l M}\Delta\theta \tag{16}$$

Here, phase $\Delta\theta$ can take the value $\Delta\theta + 2\pi n$ (n is an integer) because a sine curve is a periodic function. Therefore, it is necessary to obtain the value of n. Let $\Delta\theta_n$ and $\Delta\theta_{n-1}$ be the phase $\Delta\theta$ calculated from current image data and one frame of previous image data, respectively. When the frame rate of the video camera is high, the difference between $\Delta\theta_n$ and $\Delta\theta_{n-1}$ takes a small value, and it can be assumed that

$$|\Delta\theta_n - \Delta\theta_{n-1}| < \pi. \tag{17}$$

When $\Delta\theta_{n-1}$ is known, the value of n can be known because the range of $\Delta\theta_n$ is limited.

4. x- and y-axis acceleration visualization marker

The previous chapter described the one-axis maker. This chapter describes a method of upgrading the marker to an x- and y-axis acceleration marker. Fig. 7(i) shows a moiré fringe having the same configuration as that shown in Fig. 3. When line gratings 1 and 2 in Fig. 7(i) are rotated by 90°, the moiré fringe is also rotated by 90°, as shown in Fig. 7(ii). This moiré fringe moves in the direction (Y) when line grating 1′ is moved in the direction (y). Fig. 7(iii) shows a square grating which can be obtained by combining the line gratings shown in Fig. 7 (i) and Fig. 7 (ii), and a square-shaped moiré fringe is observed. When square grating 1 is moved in the directions (x) and (y), the square-shaped moiré fringe moves in the directions (X) and (Y). Therefore, even a slight displacement along the x- and y-axes can be magnified and displayed visually.

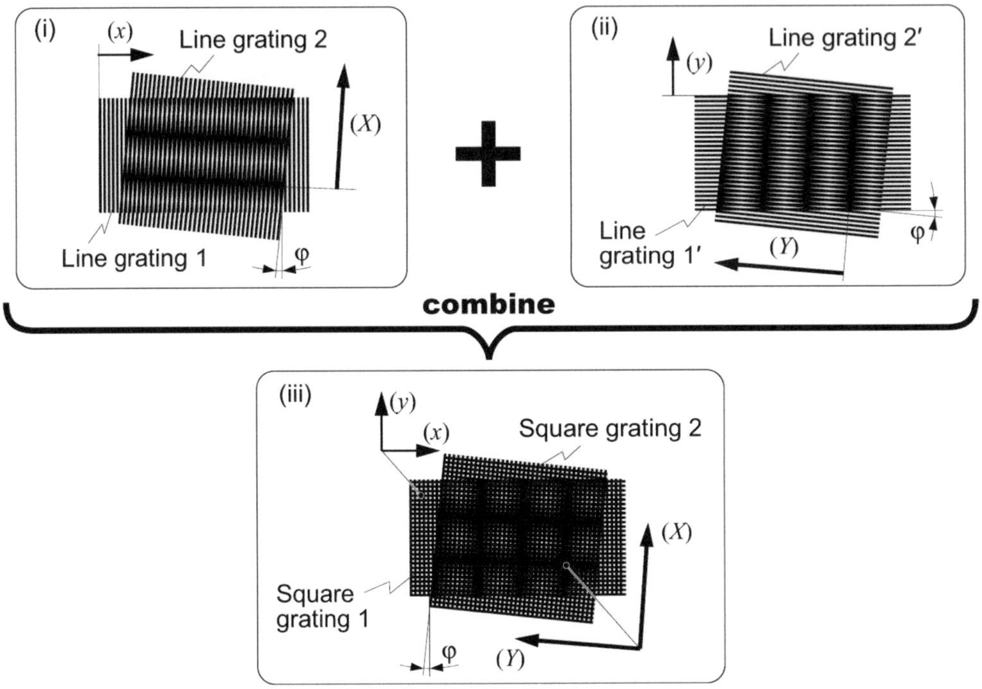

Fig. 7. Combination of two moiré fringes

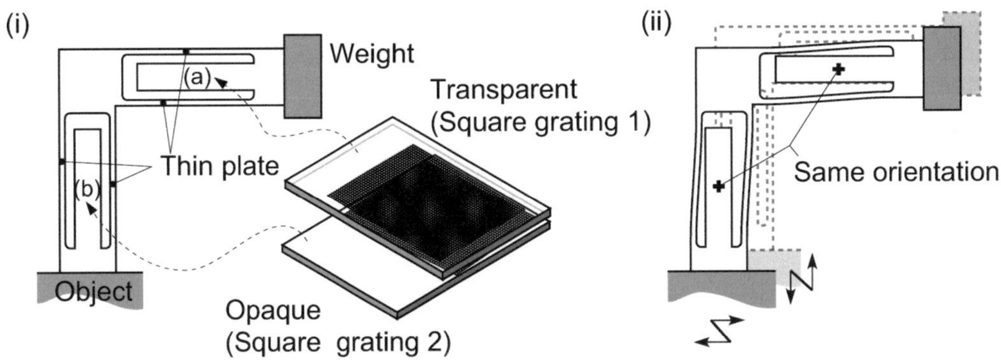

Fig. 8. Structure of the x- and y-axis acceleration visualization marker

For the same reasons as those described in Section 2.3, the angle φ must be maintained as a constant. To satisfy this requirement, two horizontal elastic plates and two vertical elastic plates are used, as shown in Fig. 8(i). This structure permits x- and y-axis displacement without any change in the angle φ, as shown in Fig. 8(ii). Line gratings 1 and 2 are respectively printed on opaque and transparent glass plates and fixed at locations (a) and (b) shown in Fig. 8. The x- and y-axis relative displacements produced by x- and y-axis acceleration components are displayed by the moiré fringe at magnification M. If a weight is installed in this structure, it becomes a seismic system, and the x- and y-axis acceleration values can be calculated using same algorithm as that described in Section 3.

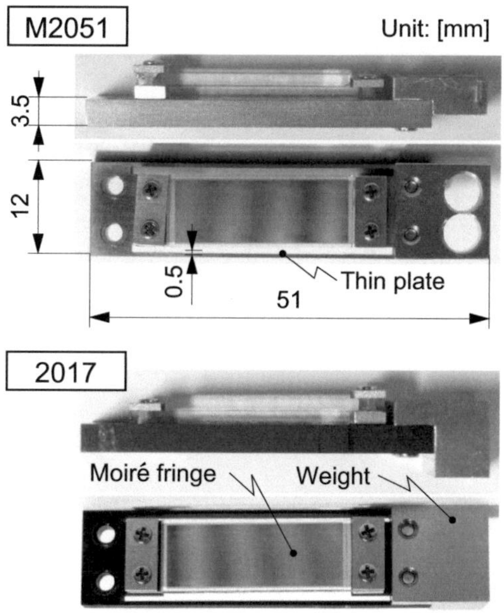

Fig. 9. Developed acceleration visualization marker

5. Experiment

5.1 Developed 1-axis acceleration visualization marker

The damping characteristic of the elastic plates influences the performance of the acceleration visualization marker. We selected two materials for the elastic plates: M2052 and 2017. M2052 includes manganese (73%), copper (20%), nickel (5%), and iron (2%), and it has a high damping capacity (Kawahara, 1993a) (Kawahara, 1993b). 2017 is an aluminum base alloy and its damping capacity is low. Figure 9 shows the developed 1-axis acceleration visualization marker, and, as shown, the shape of the elastic plates is the same.

The pitch of the line grating p_g is 0.02 mm, and its line thickness is 0.01 mm. The pitch of the moiré fringe p_m of the developed marker by using M2052 is 6.1 mm, and the relative displacement $x_o - x_w$ can be displayed visually at a magnification M of 303. The total mass is 11.7 g. The values of p_m, M, and the total mass in the case where 2017 is used are 5.5 mm, 277, and 13.0 g, respectively.

For a comparison of the accuracies of the acceleration values obtained using the marker and calculated using the algorithm described in Section 2.1, the same natural angular frequency ω_n needs to be maintained. To adjust the natural angular frequency ω_n, we machined the weight and adjusted its mass. Therefore, the shape of the weight became different. Details related to the natural angular frequency ω_n are described in Section 5.5.

5.2 Natural angle frequency and damping capacity of the 1-axis marker

To examine the mechanical characteristics of the developed 1-axis acceleration visualization marker, the marker was freely vibrated and the displacement of the weight was measured with a laser displacement sensor (KEYENCE, LK-G30). The experimental result is shown in Fig. 10. The natural angular frequency ω_n of the markers made M2052 is 409 rad/s (=65.1

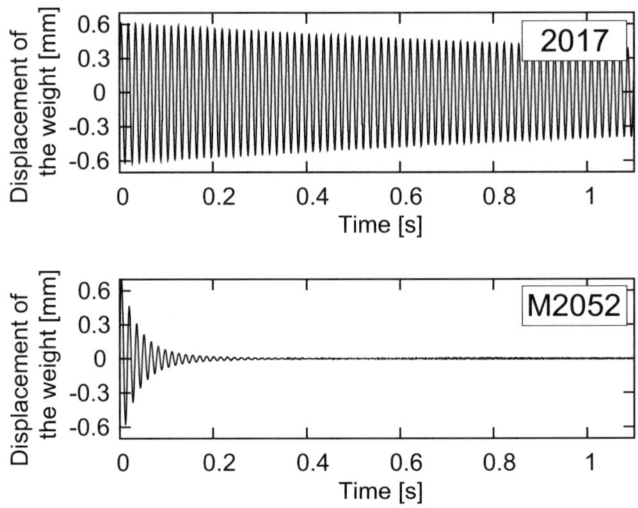

Fig. 10. Damped free vibration

Hz), and the damping ratio ζ is 0.0473. Values of ω_n and ζ obtained when marker are made using 2017 are 406 rad/s (=64.6 Hz) and 0.0011, respectively. The vibration of the composed marker of M2052 attenuates faster than that composed of 2017.

5.3 Acceleration measurement using the 1-axis marker

Using image processing, the proposed marker was verified to be able to provide an accurate value of acceleration. Figure 11 shows the experimental setup. The developed 1-axis markers made using M2052 and 2017 are attached to a vibration exciter. For comparison, a conventional 3-axis acceleration sensor (Freescale Semicondutor, MMA7260Q) is also attached to the vibration exciter. A high-speed camera (Photron, FASTCAM-1024PCI) takes images of the marker from a distance of 470 mm at 2000 fps. A distance of 1 mm corresponds to 6.0 pixels in the taken image, and the size of the image is 1024$\times$512 pixels. The amplitude of the vibration exciter is measured by the laser displacement sensor. A LED is used to achieve the synchronization of the high-speed camera, the laser displacement sensor, and the acceleration sensor.

Figure 12 shows the image of a moiré fringe trimmed from the image obtained from the high-speed camera. Its size is 125$\times$100 pixels. Figure 13 shows the average of the brightness value along the y-axis, $f(x)$, and the fitted sine curve $g(x)$. Figure 14 shows the acceleration values obtained from the markers made using M2052 and 2017 and from the acceleration sensor when the vibration exciter vibrates at 13 Hz. The amplitude of the vibration exciter is 0.58 mm.

The acceleration value obtained from the marker made using M2052 is close to that obtained from the acceleration sensor. However, the corresponding value obtained from the marker made using 2017 has an additional acceleration component at 65 Hz. The root mean square errors for the proposed method using M2052 or 2017 as materials for the marker and the 3-axis acceleration sensor are 0.24 m/s^2 and 1.4 m/s^2, respectively. Higher accuracy can be obtained from the marker made using M2052 than from that made using 2017.

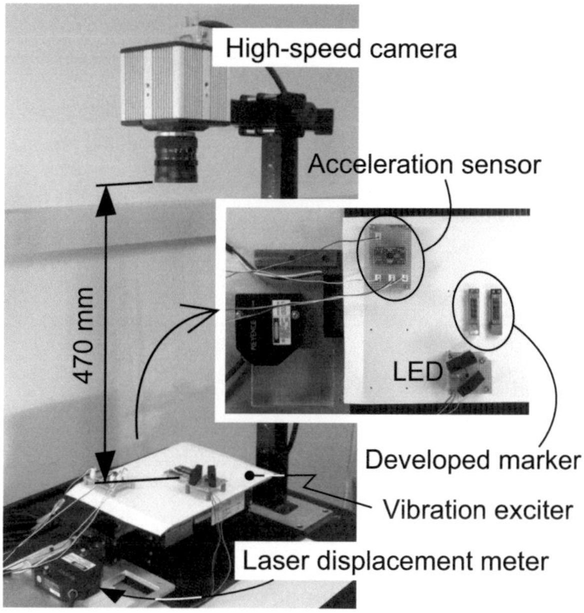

Fig. 11. Experimental setup for developed markers

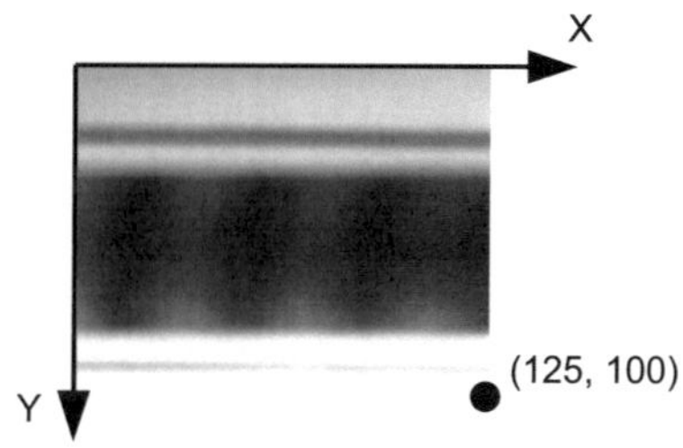

Fig. 12. Image of a moiré fringe

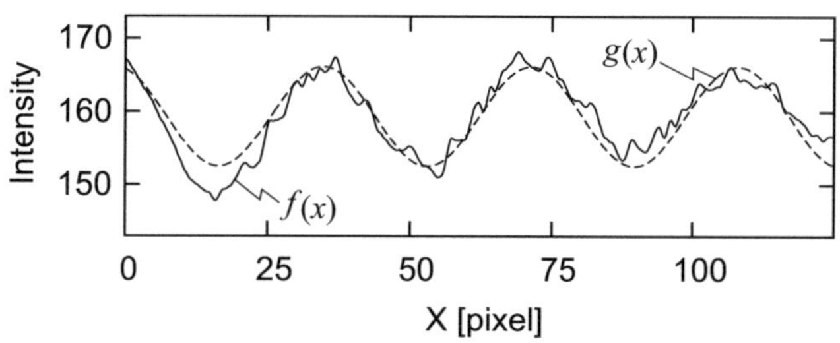

Fig. 13. Intensity of an image and a fitted curve

Figure 15 shows the power spectra of the acceleration values obtained from the acceleration sensor and the proposed markers made using M2052 and 2017. All the power spectrums have a peak at 13 Hz. This peak corresponds to the frequency of the vibration exciter. The natural angular frequencies ω_n of the two developed markers are both approximately 400 rad/s (=65 Hz). The power spectrum of the marker made using M2052 is close to that of the acceleration sensor at 65 Hz. However, the power spectrum of the marker made using 2017 has a strong sharp peak at 65 Hz. This is because it vibrates sympathetically with the slight vibration of 65

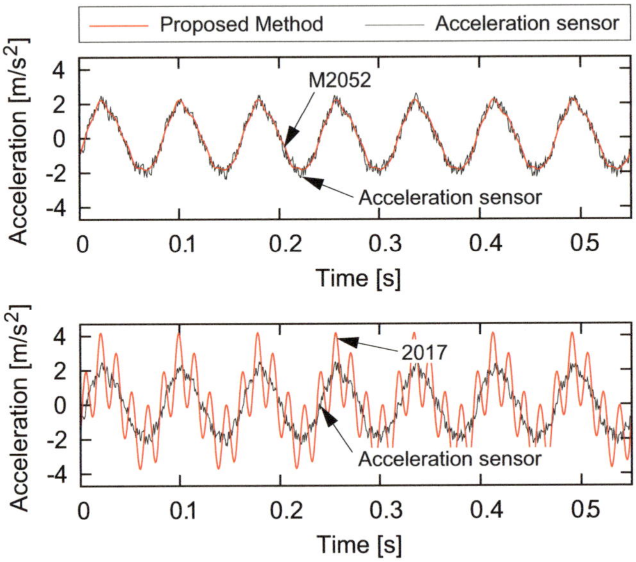

Fig. 14. Experimental result showing acceleration values

Hz included in the vibration exciter and does not attenuate because its damping ratio is small, as described in Section 5.5. Therefore, the high-damping material M2052 is more suitable for the acceleration visualization marker than the low-damping material 2017.

5.4 Developed x- and y-axis acceleration visualization marker

Figure 16 shows the developed x- and y-axis acceleration visualization marker. The material used for the elastic plates is M2052. The pitch of the line grating p_g is 0.03 mm, and its line thickness is 0.01 mm. The pitch of the moiré fringe p_m is 8.9 mm, and the relative displacement $x_o - x_w$ can be displayed visually at a magnification of 298. The total mass is 33 g.

5.5 Natural angle frequency and damping capacity of the x- and y-axis acceleration visualization marker

To obtain the natural angular frequency ω_n and the damping ratio ζ of the developed x- and y-axis acceleration visualization marker, the marker was freely vibrated and the displacement of the weight was measured with the laser displacement sensor. The experimental results are shown in Fig. 17. The natural angular frequencies ω_n for the x- and y-axes are 300 rad/s (=47.7 Hz) and 323 rad/s (=51.4 Hz), respectively, and the damping ratios ζ are 0.114 and 0.093, respectively.

5.6 Acceleration measurement using the x- and y-acceleration visualization marker

The developed x- and y-axis acceleration visualization marker could provide an accurate value of x- and y-axis acceleration, as confirmed by using image processing. Figure 18 shows the experimental setup. The developed x- and y-axis marker is attached to a vibration exciter. For comparison, the 3-axis acceleration sensor is also attached to the vibration exciter. The camera takes images of the marker from a distance of 600 mm at 2000 fps. A distance of 1 mm corresponds to 3.6 pixels in the taken image, and the size of the image is 512×512 pixels. The

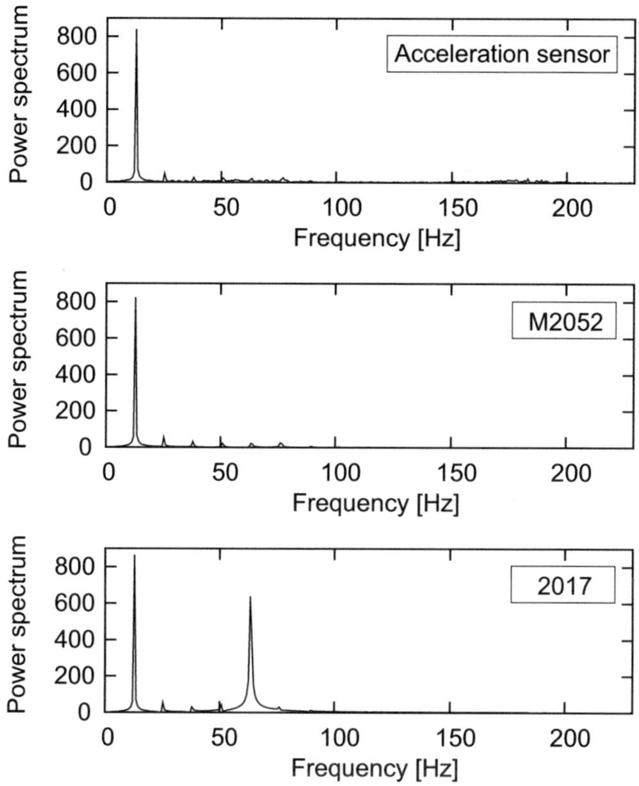

Fig. 15. Power spectra of acceleration values obtained using different devices

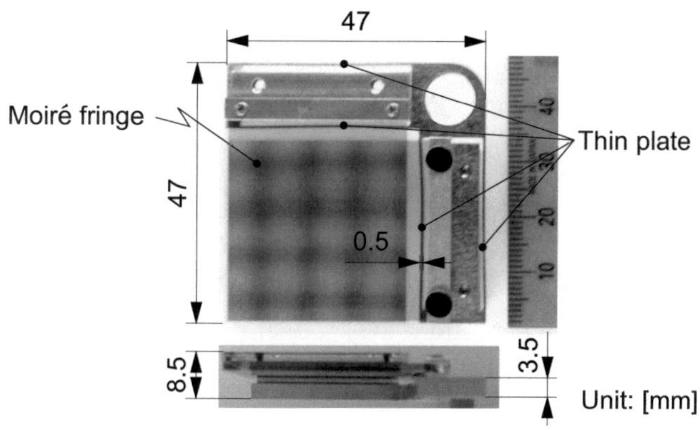

Fig. 16. Developed x- and y-axis acceleration visualization marker

amplitude of the vibration exciter is measured using the laser displacement sensor. A LED is used to obtain the synchronization of the high-speed camera, the laser displacement sensor, and the acceleration sensor.

Figure 19 shows the acceleration values obtained from the x- and y-axis markers and from the acceleration sensor when the amplitude of the vibration exciter is less than 1 mm. The acceleration value obtained from the markers is close to that obtained from the acceleration

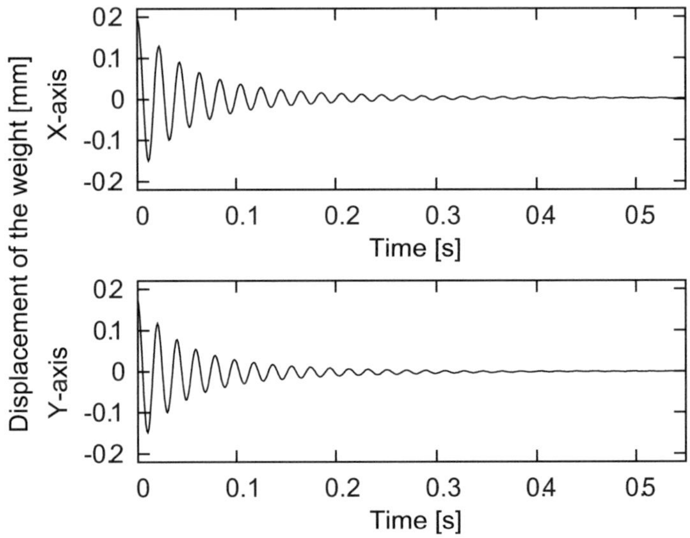

Fig. 17. Mechanical characteristic of the x- and y-axis acceleration visualization marker

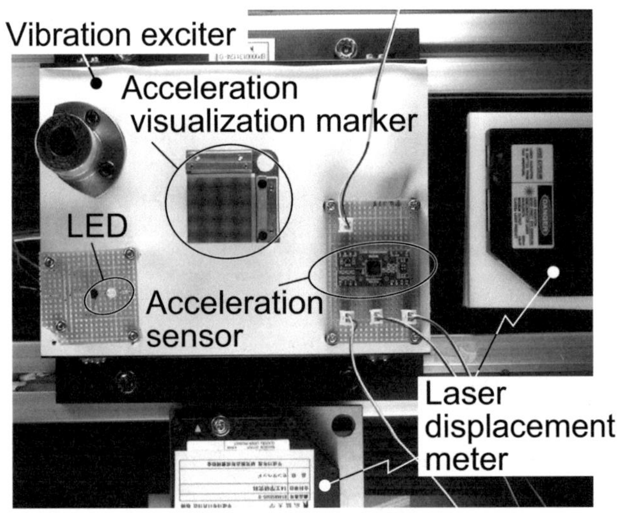

Fig. 18. Experimental setup of the x- and y-axis acceleration visualization marker

sensor. The root mean square errors for the x- and y-axis acceleration values given by the proposed method and those given by the 3-axis acceleration sensor are 0.22 m/s^2 and 0.23 m/s^2, respectively.

Figure 20 shows the power spectra of the signals shown in Fig. 19. The natural angular frequencies ω_n of the x-axis, 323 rad/s (=51.4 Hz), and y-axis, 300 rad/s (=47.7 Hz), are not observed in the power spectra for x- and y-axis acceleration values obtained using the proposed marker, and the obtained values are close to the acceleration value given by the acceleration sensor.

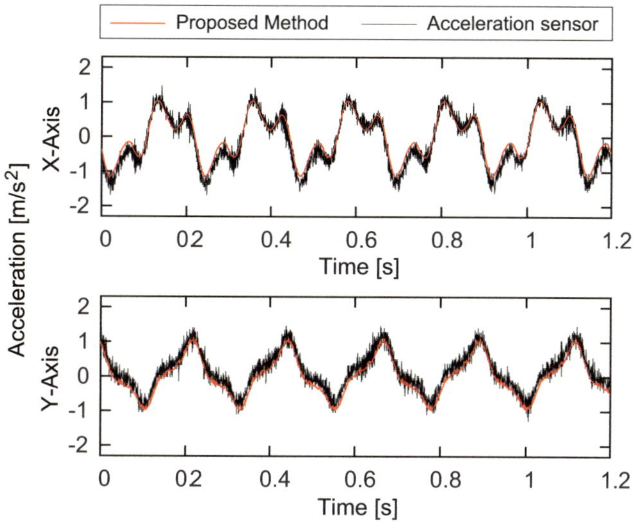

Fig. 19. Experimental result for the x- and y-axis acceleration visualization marker

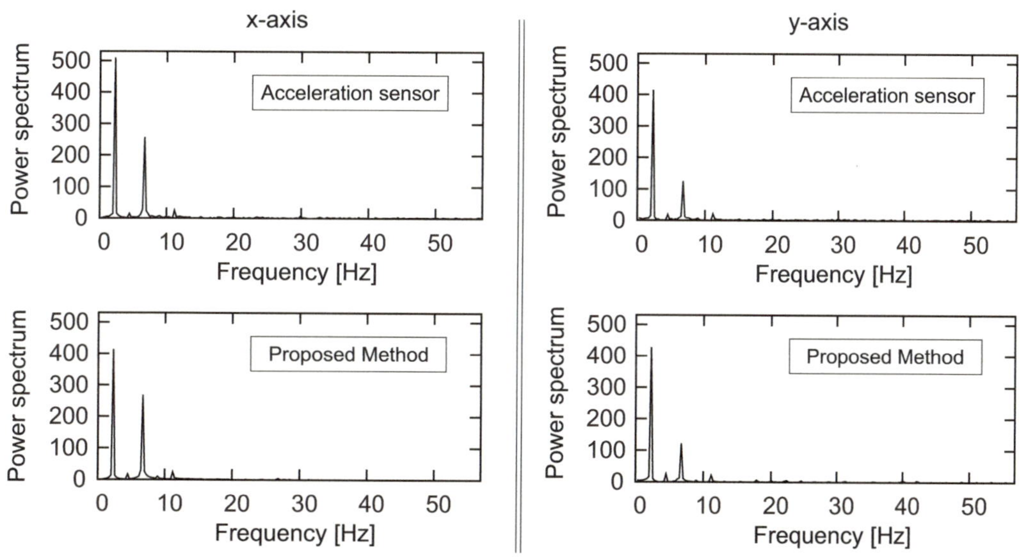

Fig. 20. Power spectra for the x- and y-axis acceleration visualization marker

6. Conclusion

This paper presents an acceleration visualization marker that uses a moiré fringe. It can enable the visualization of acceleration without the use of electrical elements such as amplifiers and strain gauges and can provide an accurate value of acceleration using image processing. Our future work will involve the measurement of the acceleration value from a remote place located more than 100 m away by using a telephoto lens.

7. References

Takaki, T.; Omasa, Y. & Ishii, I. (2008). Force Visualization Mechanism using Moiré Fringe for Robot Grippers, *Proceedings of the 28th Annual Conference of the Robotics Society of Japan*, 1K3-07, 2008. (in Japanese)

Takaki, T.; Omasa, Y.; Ishii, I.; Kawahara, T.; & Okajima, M. (2010). Force Visualization Mechanism Using a Moiré Fringe Applied to Endoscopic Surgical Instruments, *Proceedings of the 2010 IEEE International Conference on Robotics and Automation*, pp. 3648-3653, 2010.

Umemoto, S.; Fujii, M.; Miyamoto, N.; Okamoto, T.; Hara, T.; Ito, H. & Fujino, Y. (2010). Deflection measurement for bridges with frequency-shifted feedback laser, *Proceedings of the Bridge Maintenance, Safety, Management and Life-cycle Optimization*, pp. 2570-2574, 2010.

Yun, C. -B.; Soho, H.; Jung, H. J.; Spencer, B. F. & Nagayama, T. (2010). Wireless sensing technologies for bridge monitoring and assessment, *Proceedings of the Bridge Maintenance, Safety, Management and Life-cycle Optimization*, pp. 113-132, 2010.

Kim, C. W.; Kawatani, M.; Ozaki, R.; Makihata, N.; & Kano, M. (2010). Low-cost wireless sensor node for vibration monitoring of infrastructures, *Proceedings of the Bridge Maintenance, Safety, Management and Life-cycle Optimization*, pp. 780-786, 2010.

Kobayashi, A. S. (1987). *Handbook on Experimental mechanics*, Prentech-hall, ISBN 0-13-377706-5.

Reid, G. T. (1984). Moiré firnges in metrology, *Optics and Lasers in Engineering*, Vol. 5, pp. 63-93, 1984.

Basehore, M. L.; & Post, D. (1981). Moiré method for in-plane and out-of-plane displacement measurements *Applied Optics*, Vol. 21, No. 9, pp. 321-328, 1981.

Meadows, D. M.; Johnson, W. O.; & Allen, J. B. (1970). Generation of Surface Contours by Moiré Patterns, *Applied Optics*, Vol. 9, No. 4, pp. 942-947, 1970.

Takasaki, H. (1970). Moiré Topography, *Applied Optics*, Vol. 9, No. 6, pp. 1467-1472, 1970.

Takaki, T.; Omasa, Y.; & Ishii, I. (2010). Acceleration Visualization Marker using Moiré Fringe for Remote Sensing, *Transactions of the Japan Society of Mechanical Engineers, Series C*, Vol. 76, No. 770, pp. 2592-2597, 2010. (in Japanese)

Masanao, M. (1986). *Sensing Techniques of Mechanical Quantities*, Corona Publishing Co., Ltd., ISBN 4-627-61161-7. (in Japanese)

Holman, J. P. (2001). *Experimental Method for Engineers*, Thomes Casson, ISBN 0-07-366055-8.

Kawahara, K.; Sakuma, N.; & Nishizaki, Y. (1993). Effect of Third Elements on Damping Capacity of Mn-20Cu Alloy, *Journal of the Japan Institute of metals*, Vol. 57, No. 9, pp. 1089-1096, 1993. (in Japanese)

Kawahara, K.; Sakuma, N.; & Nishizaki, Y. (1993). Effect of Fourth Elements on Damping Capacity of Mn-20Cu-5Ni Alloy, *Journal of the Japan Institute of metals*, Vol. 57, No. 9, pp. 1097-1100, 1993. (in Japanese)

10

Looking at Remote Sensing the Timing of an Organisation's Point of View and the Anticipation of Today's Problems

Y. A. Polkanov
Private
Belarus

1. Introduction

Any remote measurement involves recording a signal from some sort of continuous medium of atmosphere in general. This is a signal which possesses a certain temporary structure, and in turn, this temporary structure bears some information on the spatial inhomogeneities of the continuous medium's structure and the arrangement of its specific properties (e.g. optical, microphysical, etc.). The nature of these structures depends upon the thermodynamic processes in the environment and the sustainability of these processes. Thermodynamics is the inevitable factor for their participation and it demands an account of the processes having obviously extended character. The classical approach assumes some property of the environment at a certain point in time and at a certain point in the medium. In accordance with this, today's remote measurements use the digitisation of a received signal with certain stable time step of digitisation. All efforts have been consolidated so as to receive the medium-sized digital signal samples which have been reduced to an acceptable size. Such an approach has at its core a logical contradiction - information of the properties of an extended environment trying to get at the point where it actually is not. There is something that is subject to consideration absolutely from other positions and the use of other tools. Measurements should be conducted in a certain 'visible' volume which provides the effect of the 'presence' of the medium and which has a specific thermodynamic 'meaning'; that is, that it has some of the 'thermodynamic memory'. These volumes should be comparable (in length) to the length of all zones' (lines') measurements. However, this generates a new contradiction which arises when the discretisation signal is read out. How should one get a spatial resolution close to the size of the inhomogeneity with the signal time's discretization using intervals commensurate with the length of the track measurements? This contradiction can be resolved only indirectly, using a principle that can be called a kind of 'principle of relativity'. Here, we use a pair of discrete samples which have a common border and a second boundary which is different to the desired step of discretization. This approach provides for the possibility of studying the environment and its irregularities while maintaining the required signal/noise ratio.

The internal logic of this approach abstracts the properties of the medium at the point and then moves on to the study of the environment as a self-organising system. The 'test body' of such research is the structure of the inhomogeneities of the medium. The nature of this structure is directly dependent upon the thermodynamic stability of the environment. Changes within the structure of the inhomogeneities are more mobile and are preceded by changes in the thermodynamic state of the environment as a whole. We take this as an axiom. As such, the structure of the inhomogeneities is central to the prediction of processes within the environment. This becomes especially important during the development process of a catastrophic scenario. Their nonlinear nature makes standard methods for the analysis of irregularities ineffective because of the number of initial assumptions, which often only apply to the environment in the classical sense. Therefore, I propose a structural-statistical method for analysing the structure of inhomogeneities.

2. Methodological approach

2.1 Measurements

The results of the actual measurements of laser systems for the remote-sensing of the atmosphere are used to verify the proposed approach. Currently, the laser systems for remote-sensing use high-power pulsed lasers, and the backscattering signal is written with a certain sampling step corresponding to the required spatial resolution. Moreover, the growth of the length of the track-sensing leads to a disproportionate growth of the power source and the dynamic range of the incoming signal. It also causes the multiple scattering effects which can be difficult to take into account.

The new approach is based upon the use of a low-power radiation source (for example, a source of white light) within the specified parameters of the gating. The dark pulse of the continuous light source has a duration equal to usual laser pulse lidar (about 10^{-8} c). The time interval between the dark pulses is close to the time of the radiation propagation in an area where we can neglect the multiple scattering. The digitisation of the remote-sensing signal can be performed with standard digital systems (a constant gate) as well as with systems based on the proposed approach (an increasing gate).

I propose to restore the average characteristics of the medium to long sections of a length close to the length of the track measurements. This will significantly increase the accuracy of the reconstruction of the properties of the real heterogeneous medium. Signal processing assumes the creation of the registration system with an increasing time-step gate of the incoming signal (the one-dimensional case).

The comparative calculation of the required radiation power was held for a given signal/noise ratio for different average atmosphere extinction coefficients $\sigma = 10^{-2},..., 1$ km^{-1} (the old and new systems).

For high transparency ($\sigma = 10^{-2}$ km^{-1}), the maximum length of the zone of measurement is chosen equal to the length of the layer of a dense atmosphere, significantly affecting the scattering signal (L_{max} = 30 km). To muddy the atmosphere, this distance is set by the condition that the optical depth does not exceed $\tau = 2\sigma L = 10$. This allows us to consider the scattering of the signal with an accuracy of 0.05% and to neglect the signal over large distances. Single scattering occurs with the condition $\tau = 2\sigma L < 3$ (Kovalev, V. A., 1973; Ablavskij, L. M. and Kruglov, P. A., 1974). The calculations were made on the assumption of single scattering. The data obtained is used only so as to illustrate the detected trends ($\tau > 3$).

All estimates are carried out based on an expression derived from the lidar equation for systems I and II (Polkanov, Y. A. and Ashkinadze, D. A., 1988):

$$n_i = AW_1 (e^{-2\sigma li} (1 - e^{-2\sigma lc}))/L^2 \quad (1)$$

$$n_{i1,i2} = AW_2 \sum_{i=l_T/l_s}^{L_{1,2}/l_s} \left(e^{-2\sigma li}\left(1 - e^{-2\sigma lc}\right)\right) / (l_i + l_s / 2)^2 \quad (2)$$

Where n_i, n_{i1},i_2 - obtained discrete values from the scattering signal (number of photon counts); A – a coefficient which brings together the supporting equipment characteristics; $W_{1,2}$ – the power of the laser radiation; L – the distance from the centre section of the route, by which the signal is recorded; l_i - the distance from the system to this site; l_s – the length of the section; l_T - the length of the shadow zone of the lidar where the signal is not recorded (600 m). We assume for the system that II $L_2 > L_1$, $(n_{i2} - n_{i1}) = n_i$. An advanced assessment of the relative measurement error of the signal (δ_i, δ_{ix} for System I and System II) was conducted on the basis the expressions (Polkanov, Y. A. et al., 1985; Polkanov, Y. A. et al., 2004):

$$\delta_i = t_\beta((n_i - n_n)^{1/2})/ n_i \quad (3)$$

$$\delta_{ix} = t_\beta((n_{ix} - 2B_x n_n)^{1/2})/ n_{ix} \quad (4)$$

Where t_β – the coefficient equal to the probability of the matching error computed to its actual value (if $t_\beta = 2$, the probability is equal to 0.95). The necessity of this evaluation is due to the appearance depending δ_{ix} (t) for system II (signal/noise = const). This is due to the progressive rise in the value of the time intervals recording the scattering signal (with the digitisation step – t_s). The level of background illumination takes into account the introduction of the coefficient B = f (t) in (4). The measurement error for individuals counts the signal and background-level measurement errors, becoming comparable for large intervals of T_S. They are significantly higher than the level of internal noise (in. ns.) receiving system (in this case, $n_{in.ns} \sim 0.1$, t_s = 0,4 ms). Moreover, the summed value of the signal increases to a certain point in time, reaching a maximum level of accumulated signal (Kovalev, V. A., 1973; Ablavskij, L. M. and Kruglov, P. A., 1974). However, the level of background illumination increases linearly with time. The calculations used the results of the actual measurement system I (n_i, n_b, σ). The coefficient A in (1) is also evaluated and used in subsequent calculations for the system II (2).

2.2 Processing

The following processing scheme was assumed: the initial signal (as a time function) → the generalised structure of a signal → an elementary cell of the signal structure. The multiplication of such cells allows the complete restoration of the characteristic structures in the supervised space.

The indicator of the time stability of the signal structure was the dispersion of the components of the elementary cell of a signal structure. If the dispersion exceeds an interval between elements of the revealed cell then the structure is unstable. The correlation of the generalised frequency structure of a horizontal signal and the generalised parameter which fixes the thermodynamic stability of the environment is a characteristic sign of the self-organising of the environment.

The basis for the reception of new results is a series of works on the laser sounding of the atmosphere in stable nightime conditions. This has allowed the development of certain methods for the structural-statistical processing of an initial remote signal. The aim is to reveal the signs of the steady organisation of the frequency structure of environmental inhomogeneities.

The generalised regular structure comes from the summary of the sequence of the discrete readouts. They were received by the scanning of the investigated volume of the environment in a horizontal plane to a set of directions and with the set angular permission (Polkanov, Y. A. et al., 1989).

During the following stage, the signal is represented in the form of a regular structure of local maxima and minima. There was a separate analysis of the 'plus' and 'minus' structures (Polkanov, Y. A. and Kudinov. V. N., 1989).

These components behave as whole object and are registered as a uniform regular structure (type harmonious) only in the case of a steadily vertical stratified environment. When the infringement of the stability of the stratification of environmental communication between the 'plus' and 'minus' structures decreases, they become increasingly independent of one another other. The degree of such dependence can be characterised by a certain numerical parameter (Polkanov, Y. A. et al., 1991; Polkanov, Y. A. et al., 2009).

The thermodynamic stability of the environment and its stratification can be characterised numerically by a special generalised parameter on the basis of Richardson's number. With the infringement of the thermodynamic stability of the environment, this parameter adopts wavy characteristics on a vertical plane. The length of such a 'wave' with the falling of the environmental stability was decreased.

The integrated regular structure of vertical thermodynamic distribution is an indicator of such stratification of the environment.

It is possible to speak about the communication of the optical structure horizontal stability with the vertical stability of the thermodynamic structure of the environment and its stratification as being an indicator of such stability (Polkanov, Y. A. et al. 1989).

Besides this, the infringement of the stability of the environment leads to the infringement of the stability of the revealed structure and the occurrence of obvious anomalies within the structure (Polkanov, Y. A. et al., 1991; Polkanov, Y. A. et al., 2008) whose behaviour can provide information on the direction of the reorganisation (self-organisation) of the environment.

3. Update of the concept of signal/noise ratio

It transpired that the signal/background noise ratio (S/N) is ambiguous due to the accuracy of the measurement of the scattering signal by the use of the extended strobe. Indeed, when t_i = const and S/N = const for the system I, this automatically means the constancy accuracy of the scattering signal (∂ = const) from strobe to strobe. For example, we set ∂ = 10% for $\sigma \sim 0.1$ km^{-1} for the basic equipment (system I) with S/N = 10. For the systems of type II, the signal is accumulated over time intervals the value of which is not constant, but rather varies in such a way that satisfies the condition: t_i (n) = (t_i (n-1) + t_i)) is t_i (n)> t_i (n-1). The essential point here is the rise of the level of the recorded background illumination with the increasing duration of the strobe. The scattering signal increases from the strobe to strobe – in general – to the so-called maximum accumulated signal (Kovalev, V. A., 1973).

The background illumination level is significantly higher than the corresponding internal noise receiver (in.ns = 0.1 for t_i = 0.4 ms). It exceeds the signal of system I, with a point, but is comparable with the level of the scattering of the signal of system II ($\tau < 3$). In this case, the accuracy of the scattering signal and the background are similar, and they can be used as useful signals on an equal basis.

In fact, we have a mixture of two signals - the scattering signal and the background signal. Their value increases from strobe to strobe and the first of them (S) rises to a certain level (W_{max}) whilst the second of them (b) increases linearly with time and indefinitely.

In these circumstances, the accuracy of the scattering signal increases when S/N = const (1 because a strobe the length of the time of registration is increasing.

Thus, there is a new dependence - ∂ (t) which was previously unavailable for system I . Table 1 lists the measurement error depending upon the distance l_s (n) corresponding to the interval gating t_s (n), if S/N = 10 = const, for σ = 0.1 km^{-1}, n_b = 50, t_i = 0.4 s.

L (km)	1	2	3	4	5	10
δ,%	3,7	1,7	1,3	1,2	1,1	0,8

Table 1. The measurement error decreases with increasing interval gating.

Model calculations showed that the measurement accuracy of the scattering signal for system II is several times higher than the measurement accuracy for system I. This means that for the same radiation power of remote systems, greater measurement accuracy is achieved for systems of type II through special time organisation and its recording of the digitised signal ($\partial_{II} \neq \text{const} \leq \partial_I = \text{const}$).

We can talk about the actual incompleteness of the concept of the signal/background ratio for the registration systems of type II when the strobe length (a single reference signal) depends upon the position of the laser pulse on a remote line sensing. Moreover, it is possible that the signal/background ratio is less than unity but that the measurement accuracy remains high. This is possible when the signal/internal noise ratio (S/in.ns) and the background/internal noise ratio (b/in.ns) is much higher than 1. An example of such situations is provided by Table 2.

L(km)	1	2	3	4	5	10
S/N	1,23	0,81	0,65	0,55	0,49	0,33

Table 2. Signal/background ratio, depending upon the length of the strobe (km) and where the measurement error δ = 10% (const).

The obtained simulation results suggest that the measurement accuracy was higher than expected, if only to carry out the calculation of the signal/background ratio for systems of type II.

4. Data analysis and simulation results

Model calculations based on the data obtained by the laser probing of the atmosphere by means of system I with an output power equal to 0.67 mW (Ashkinadze, D. A., Belobrovik, V. P., Spiridovich, A. L., Kugeiko, M. M. and Polknov, Y. A., 1980; Ashkinadze, D. A and Polkanov, Y. A., 1980; Polkanov, Y. A. et al., 1985; Polkanov, Y. A and Ashkinadze D. A., 1988; Polkanov, Y. A. et al., 1991).

The results of real lidar measurements are used to model the time organization of the proposed emission and detection. Lidar has the following characteristics:

- Radiation source:

Radiation energy E = 0.01 J;

Pulse duration T_o = 15 ns;

Pulse repetition frequency f = 50 Hz.

- Receiving system:

Diameter of the receiving mirror D = 0.1 m;

Operation of a photomultiplier tube (PMT) - an account of the photons;

Quantum efficiency of PMT η = 0,1;

- Recording equipment:

Time interval signal detection in single channel t_i = 0,4 mkc;

Number of cycles of signal m - 3000;

Total measurement time t = 60 s (Polkanov, Y. A. et al., 1985).

The measurement conditions corresponded to the registration of a Poisson flow of the signal photons (Polkanov, Y. A., 1983). The number of the cycles of the accumulation provided a measurement error of no worse than 50%.

Fig. 1. The lidar scheme, with a separated transmitter and receiver.

The simulation results are presented as a set of tables.

4.1 The simulation results of the proposed temporal organisation of the detected signal

We shall call this remote sensing system (lidar) as a base 'System I' (the old system), and a system with increasing intervals of registration (strob) 'System II' (the new system).

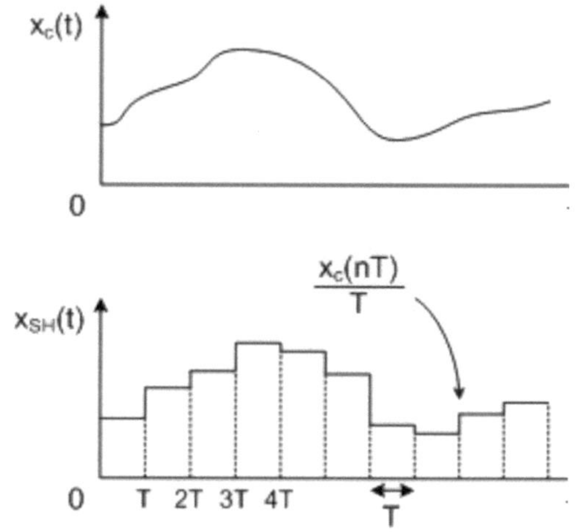

Fig. 2. Discrete-time signal $x_c(t)$ processing for System I.

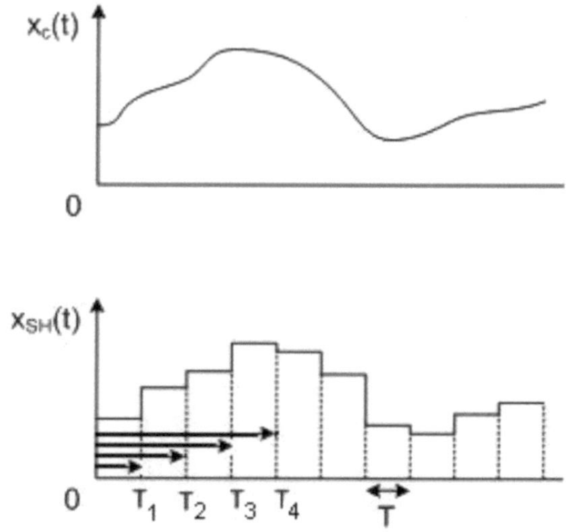

Fig. 3. Discrete-time signal $x_c(t)$ processing for System II ($T_1 = T$, $T_2 = 2T$, $T_3 = 3T$, $T_4 = 4T$).

The calculation of the signal/background ratio and the corresponding measurement errors of the scattering signal is carried to the appropriate conditions of 'twilight' (∂_1) and 'cloudy day' (∂_2) when the level of background illumination increases by two orders of magnitude. The following table shows the dynamic range (DR) and signal/background ratio for the a wide range length of the path sounding (L) for each value of the extinction coefficient (σ) from the real range. The level of illumination is selected for the corresponding conditions with a high transparency of the atmosphere ($\sim 10^{-2}$ km^{-1}).

Fig. 4. Organisation model of the discrete-time signal processing.

σ (km^{-1})	L(km)	1	2	3	4	5	10	15	30	DR
	s/n	19,00	9,40	6,50	5,00	4,00	3,00	1,30	0,60	29
0.01	∂_1 (%)	2,60	2,10	2,01	2,03	2,06	2,36	2,62	3,29	29
	∂_2 (%)	8,34	8,87	9,94	11,10	12,15	17,21	20,69	28,75	29
	s/n	164,30	82,30	54,80	40,00	31,80	15,10	9,90	4,80	34
0.1	∂_1 (%)	0,84	0,66	0,63	0,63	0,63	0,63	0,65	0,66	34
	∂_2 (%)	1,24	1,25	1,36	1,51	1,65	2,29	2,79	2,97	34
	s/n	363,60	166,70	106,00	75,50	59,40	27,70	18,10	15,90	23
0.3	∂_1 (%)	0,56	0,46	0,44	0,44	0,44	0,44	0,45	0,46	23
	∂_2 (%)	0,70	0,66	0,63	0,87	0,95	1,26	1,52	1,61	23
	s/n	419,80	156,80	94,80	66,70	52,10				8
1.0	∂_1 (%)	0,52	0,46	0,46	0,47	0,47				8
	∂_2 (%)	0,63	0,77	0,87	0,98	1,06				8
σ(km^{-1})	L(km)	0.1	0.2	0.3	0.4	0.5				8
	s/n	420000	156000	95000	66000	52000				8
10.0	∂_1 (%)	0,05	0,05	0,05	0,05	0,05				8
	∂_2 (%)	0,05	0,05	0,05	0,05	0,05				8

Table 3. The measurement error (∂) of the signal/noise ratio (S/N) and the dynamic range (DR), depending upon the length of the path sounding (L) and the extinction coefficient of the medium (σ).

The measurement error of these conditions is calculated by formula (1) and does not exceed a few percent. For most cases, we can assume that it will be less then common instrument errors of the detecting apparatus. The error increases with the daytime measurements (at the same transparency), but by no more than an order of magnitude; its increase is insignificant for the extinction coefficient range σ = 1-10 km^{-1}. The dynamic range of the signal/background ratio is small and varies with changing conditions in the atmosphere; it is much smaller than in the case of system I (DR = 8 - 34).

The data obtained suggests the following conclusions:

1. The use of 'growing' of the proposed type of strobe allows for the measurement of the single-scattering signal with a high precision. Measurements become possible in the daytime.
2. Small dynamic range of the signal/background ratio will simplify the recording equipment without compromising the accuracy of the measurement by eliminating any redundant requirements for its performance.
3. The use of this approach shifts the problem of increasing the measurement accuracy from the area associated with the environment to the area associated only with the instrumental capabilities of the remote systems (i.e. they are more controlled).
4. An additional advantage of the developed approach is the small dynamic range of change of the error signal scattering, depending upon the distance to the considered section of the remote sensor.

The accuracy varies slightly from a strobe to strobe on most of the track soundings. This provides significant advantages for the correctness of the subsequent interpretation of the data.

Fig. 5. Normalised error of measurement, depending upon the length of the path sounding.

The analysis of the dynamic range of the scattering signal also shows the advantages of system II. The dynamic range of the signal does not exceed the value 10^2, compared with value 10^6 for systems of type I. Table 5 shows the value (DR) of the scattering signal and the background level for system II. In our case, the range for backlight can reach 10^3, remaining several orders of magnitude lower than for the scattering signal of system I.

σ(km^{-1})	**L(km)**	**1**	**30**	**DR**
0.01	signal	6462	15600	72
0.01	background (noise)	850	25000	900
σ (km^{-1})	**L(km)**	**1**	**2**	**DR**
0.1	signal	57580	113730	59
0.1	background (noise)	850	25000	900
σ(km^{-11})	**L(km)**	**1**	**17**	**DR**
0.3	signal	127465	217625	29
0.3	background (noise)	850	14150	283
σ (km^{-1})	**L(km)**	**1**	**2**	**DR**
1	signal	147150	190310	6,5
1	background (noise)	850	4150	24,4
σ (km^{-1})	**L(km)**	**0,1**	**0,5**	**DR**
10	signal	14739000	19004096	6,4
10	background (noise)	85	415	24,4

Table 4. The number of the signal count (n_s) and the background (n_b) for the photon counting mode, and the dynamic range (DR), depending upon the extinction coefficient of the medium (σ).

σ(km^{-1})	**L(km)**	**1**	**2**	**3**	**4**	**5**	**10**	**15**	**30**
	Δl_s(m)	10	43	97	18	29	125	320	169
0.01	Δt_s(ns)	67	287	647	120	193	233	2130	1130
	ΔP_{nbg}	8	36	81	15	24	104	26	141
	Δl_s(m)	2	6	16	36	70	40	500	9700
0.1	Δt_s(ns)	13	40	107	240	467	266	3300	64700
	ΔP_{bg}	2	5	13	30	58	733	417	8083
	Δl_s(m)	0,6	4,4	18,0	60,0	190,0			
0.3	Δt_s(ns)	4	29	120	400	1270			
	ΔP_{nng}	065	367	15,0	50,0	158,0			
	Δl_s(m)	0,7	21,7	730,0	50,0		**N ex.n. = 0,3**		
1.0	Δt_s(ns)	4,7	145,0	4870,0	333,0				
	ΔP_{nbg}	066	18,0	608,0	42		**ΔP_s > ΔPex.n.**		
σ(km^{-1})	**L(km)**	**0.1**	**0.2**	**0.3**	**0.4**		**ΔP_s = 100, 10, 1**		
	Δl_s(m)	-	0,3	3,9	6,0				
10.0	Δt_s(ns)	-	2	26	40,0				
	ΔP_{nbg}	-	0,2	2,6	5,0				

Table 5. The required increase of the signal sampling interval (space, time) which provided the desired signal increase and its corresponding background increase.

To estimate the limiting possibilities of system II, we calculated the allowable spatial resolution of the remote sensing under various conditions in the atmosphere. The calculation was performed as follows:

1. A constant increment of the scattering signal ΔP_s is posed.
2. The increment area sounding (ΔL) which provided a signal increment (ΔP_s) for certain values of the extinction coefficient (σ) and the length of track is then identified.
3. The increment ΔL thereby obtained is taken as the minimum spatial discretisation step track at a distance L.
4. The necessary step time sampling rate is determined for the recording equipment (Δt_s) on the basis of the obtained values, ΔL.
5. The increment background illumination (the number of the background count ΔP_b) is determined on the basis of the intervals' increment, Δt_s

To estimate the limiting possibilities of system II, we calculated the valid value of the spatial resolutions under various conditions in the atmosphere. The calculation was performed as follows: the value of ΔP_s given as the number of samples (100, 10, 1), with the transition from one value to another. The value ΔP_s does not exceed step ΔL spatial discretisation achieved the basic apparatus in version of the system I. The calculation results for $W_0 = 0.67$ mW are shown by Table 6. The increments ΔP_s certainly took higher increments due to the internal noise receiver. This was the case for $\sigma = 10^{-2}$ km^{-1} to L = 30 km, for $\sigma = 0.1$ km^{-1} to L = 15km, for $\sigma = 0.3$ km^{-1} to L = 5 km, and for $\sigma = 1$ km^{-1} to L = 4 km.

We have exceeded ΔP_s over ΔP_b in all cases (to dusk) when $\Delta Pc = 100, 10$. This is much less than was the case for system I. The simulation results suggest that there is a real opportunity to provide the increment of the scattering signal on the increment of the recorded background illumination ($\Delta P_s > \Delta P_b$) for a wide range of conditions by the adjustment of the values Δt. At the same time, the allowed (minimum) time increments Δt_s (increments for the individual remote-sensing signal samples) do not exceed - in this case - hundreds of nanoseconds (in the zone of single scattering).

4.2 The simulation results of the proposed organisation of the sounding signal radiation

The proposed approach can be applied not only to the organisation of the temporary registration of the incoming signal (in the case of passive systems), but also to the temporary organisation of the radiation of the sounding signal (in the case of active systems).

We consider three types of organisation of the radiation source:

- Pulsed light source (laser) with a pulse substantially shorter than the sounding track (type I).
- Pulsed light source (laser) with a pulse substantially equal the sounding track (type II).
- Long pulsed light source with a repetition-rate that ensures the duration of the interval between the pulses is equal or near to the pulse length of type I (type III), dark pulse laser (Mingming, Feng, Kevin L. Silverman, Richard P. Mirin and Steven T. Cundiff, 2010).

Again, asAs before, the basic system is taken to be a real system of type I (V. E. Zuev, M. V. Kabanov, 1977) with a constant duration of strobe (t_s = 0.4 ms) and the characteristics

	1	2	3
Duration of the radiation impulse	15 ns	18 µs	18 µs
Length of the radiation impulse	4.5 m	5.4 km	5.4 km
Radiation Energy	0,01 J	18 µsJ	18 µJ
Radiation impulse power	667 kW	1 W	1 W
Registration strobe duration	0.4 µs	0.4 µs	0.4 µs
Registration strobe length	60 m	60 m	60 m
Strobe numbers on a line	90	90	90
Line length	5.4 km	5.4 km	5.4 km
Measurement total time	60 s	60 s	60 s
Number of the accumulation cycles	3000	514000	3300000
Frequency of the impulses	50 Hz	8.57 kHz	55.5 kHz
Spatial interval between impulses	» 5 km	30 km	« 5 km
Total radiation energy	30 J	9.25 J	60 J
Average radiation power	0.5 W	0.15 W	1.0 W
Background readout number in a strobe	50	8570	55500
Signal readout number in a strobe (min)	83	4258	182600
Measurement error	33%	7%	0.6%
Background readout number/parcel	0.017	0.017	0.017
Number of signal readout number/parcel (min)	0.028	0.008	0.055

Table 6. The calculated characteristics of the equivalent remote-sensing systems of type I, II and III.

described above. In addition, we used data obtained by probing the system in advanced atmospherics with the extinction coefficient $\sigma = 0.1$ km^{-1}.

The comparative evaluation of the above types of systems was carried out under the assumption used that in the future there would be a a source of continuous light source radiation with a radiated power ~ 1W, since this energy is easily attainable at the present level of the laser system development. We select a maximum . The length of the route maximises the accumulated signal for the conditions of a single scattering ($\tau = 2\sigma l \leq 3$). For system III, an assumption is introduced – the interval between pulses (60 m) does not affect the accumulated signal for distances greater than the path length of the maximum accumulation ($L_{max} \sim \tau = 3$) (Polkanov, Y. A. et al., 2007; Polkanov, Y. A. et al., 2008).

The temporal organisation of the remote-sensing signal reception, the level of background illumination, and the total measurement time is expected the same for all the simulated systems. This data is shown by Table 7, which summarises all of the necessary characteristics for comparison.

The analysis of this data allows us can conclude that to provide the necessary signal levels due to the growth of the pulse repetition rate, the frequency of system II should be raised to 8.57 kHz, and that of for system III to 55.5 kHz. These limitations are needed so as to exclude the presence on the track sensing of the two light pulses (for systems I and II) or the dark pulses (for system III).

The computed frequencies provided a growing number of background counts (compared with system I) for system II (171 times) and system III (- 1100 times). Accordingly, the number of signal photon counts was increased in 51 and 2200 times. This allows us to reduce the measurement error from 33% to 7% and 0.6% respectively for systems II and III. The evaluation shows that the use of systemsusing a system of type II andor type III - even with a radiation source with a capacity of 1 Watt - can significantly improve the measurement accuracy of the scattering signal, relative to the system I (radiation power ~ 1 Watt).

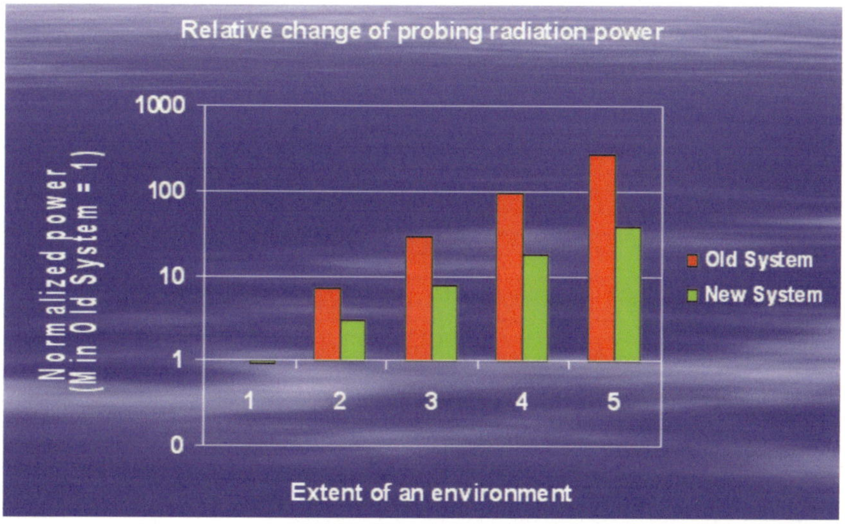

Fig. 6. Normalised power systems I (old) and II (new) as a function of environmental conditions (σ=0,01-1 km^{-1}) at the same signal/noise ratio.

This is achieved through the formation of the continuous emission of long pulses (dark pulses) with a high repetition rate. It allows for a fixed measurement time (60 s) registering a much larger number of photons. Thus, we can reduce the required power of the radiation source. It is interesting estimating the maximum possible repetition rate laser pulses for system II. The pulse length varies from one pulse to the next the length of the registration strobe (single reference signal) changes from one strobe to another gate. To eliminate the effect of the scattering signal from the previous pulse, the interval between pulses (l_Δ) was chosen according to the condition: σ (l_Δ + MDV) = 7.5. In this case, the contribution from the previous pulses in the signal did not exceed 10% of the maximum accumulated signal. The values of the maximum possible repetition-rate of system II is represented in the table below.

Here, the following notation was used: $l\tau_0$ – the pulse duration τ_0; l_Δ - the interval between pulses in meters; f – the frequency of pulses; M – the number of the accumulation cycles; E_o

$l\tau_0$ (km)	1	2	3	4	5	10	15	30
τ_0 (μs)	3,3	6,7	10,0	13,3	16,7	33.0	50,0	100,0
σ (km^{-1})				0,01 - 0,1				
lΔ (km)				30				
f (kHz)	9,7	9,4	9,1	8,8	8,6	7,5	6,7	5,0
M	581 000	562 000	546 000	528 000	516 000	450 000	400 000	300 000
E_0 (μJ)	51,6	53,4	54,9	56,8	58,1	66,7	74,6	100,0
W_0(W)	15,6	8,0	5,5	4,3	3,5	2,0	1,5	1,0
σ (km^{-1})				0,3				
lΔ (km)				11,7				
f (kHz)	23,6	21,9	20,4	19,1	17,9	1`3,9	11,2	10,4
M	1 420 000	1 310 000	1 220 000	1 150 000	1 080 000	830 000	670 000	620 000
E_0 (μJ)	21,0	23,0	24,0	26,0	28,0	36,0	45,0	48,0
W_0(W)	6,4	3,4	2,5	2,0	1,7	1,1	0,9	0,8
σ (km^{-1})				1.0				
lΔ (km)				3,5				
f (kHz)	66,7	54,5	46,1	40,0	35,3			
M	4 000 000	3 300 000	2 800 000	2 400 000	2 100 000			
E_0 (μJ)	7,5	9,2	10,8	12,5	14,2			
W0(W)	2,3	1,4	1,1	0,9	0,8	fE_0 =const		
σ (km^{-1})				10.0				
lΔ (km)				0,35		W_0 = 0,5 W		
f (kHz)	666,7	545,4	461,5	400,0	352,3			
M	40 000 000	33 000 000	28 000 000	24 000 000	21 000 000	$E^{(2)}_{\Sigma} = E^{(1)}_{\Sigma}$		
E_0 (μJ)	0,7	0,9	1,1	1,2	1,4			
W_0(W)	2,3	1,4	1,1	0,9	0,8			

Table 7. The maximum possible pulse repetition frequency (f) of the radiation remote-sensing systems (type II) depending upon the environment (σ).

- the energy of the radiation. This provides an accuracy that is not worse than the accuracy of the measurement system II for the same values of the extinction coefficient (σ). The number of emitted photons is equal in all the simulated cases. This corresponds to the radiation energy of system I for a full-time measurement (60 sec), which corresponds to the average power W_0 = 0.5 W. This allows us to visually compare systems with different types of organisation of the radiation source. Likewise, we assessed the limiting frequencies of the pulses of radiation systems for system III. The data obtained is summarised in the following table:

σ (km^{-1})	$l\tau_0$	τ_0 (μs)	lΔ (m)	f (kHz)	M	E_0 (μJ)	W_0 (W)
0,01	30,0	100	60	10	600 000	50,0	0,5
0,1	30,0	100	60	10	600 000	50,0	0,5
0,3	16,7	55,7	60	18	1 100 000	28,0	0,5
1,0	5,0	16,7	60	60	3 600 000	8,4	0,5
10,0	0,5	1,7	60	600	36 000 000	0,9	0,5

Table 8. The maximum possible pulse repetition frequency of the radiation remote-sensing systems depending upon the environment (type III).

The necessary energy radiation does not exceed ten microjoules at the limiting frequencies. This suggests the use of low-power lasers as radiation sources in systems II and III, with optical shutters which open with a given frequency (f). The maximum frequency is obtained at ~ 1 MHz, but it has a range of 10-100 kHz in most cases. This is achieved by conventional optical shutters.

In the above conditions, the maximum pulse power of system II does not exceed 18W. For system III, the power is equal to 0.5W which is sufficient to achieve a measurement error not worse than tenths of a percent, excluding the errors caused by the instrument.

4.3 The simulation results of the signal structure stability of remote-sensing

We investigated the behaviour of three sample models in relation to the signal from a self-organising environment. The behaviour of the three sample models was analyzed. It has a 16-17 readout and a digitisation step - 30 minutes, with total duration of measurements from 12 to 17 days.

The averaging of the intervals between local maxima and minima gives the generalised intervals of the structure of the inhomogeneities (M+, M-).

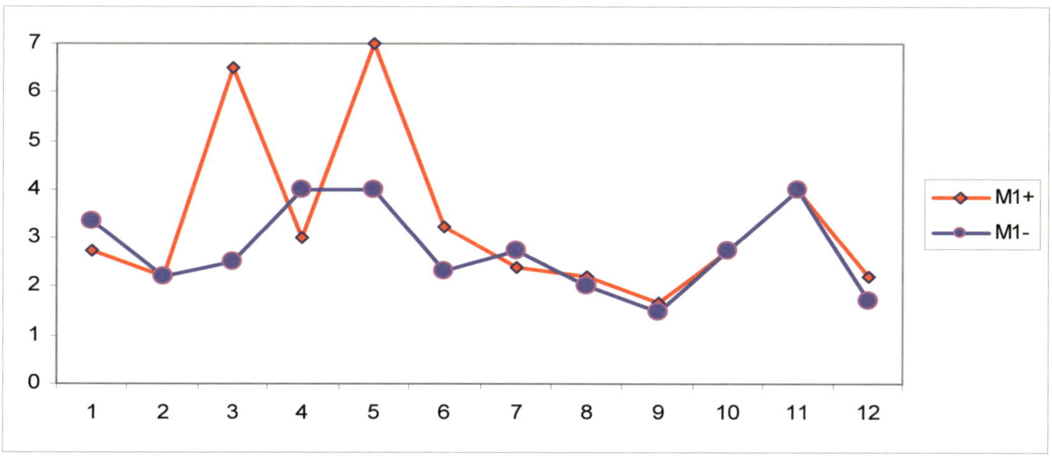

Fig. 7. Results of the interval definition between the elements of the generalised structure of different types, 'plus' and 'minus' (M+, M-) for some areas (1-12).

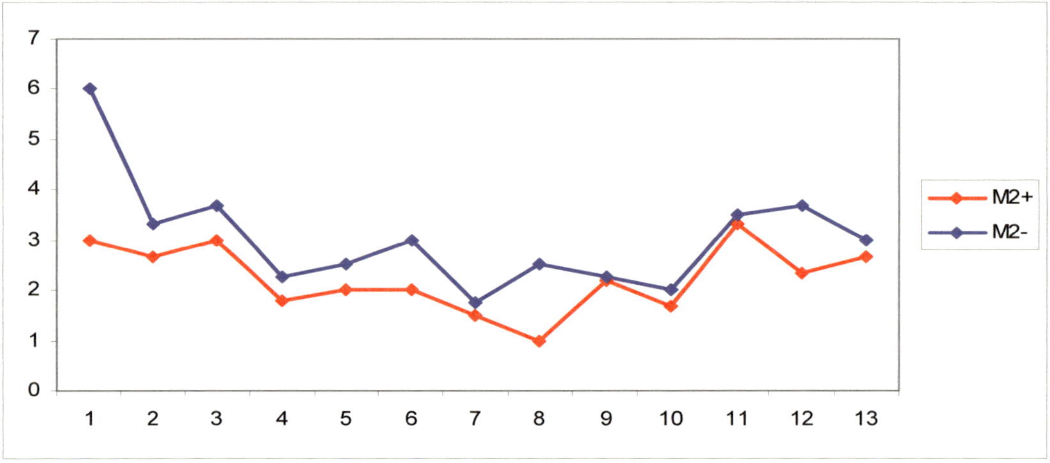

Fig. 8. Results of the interval definition between the elements of the generalised structure of different types, 'plus' and 'minus' (M+, M-) for some areas (1-13).

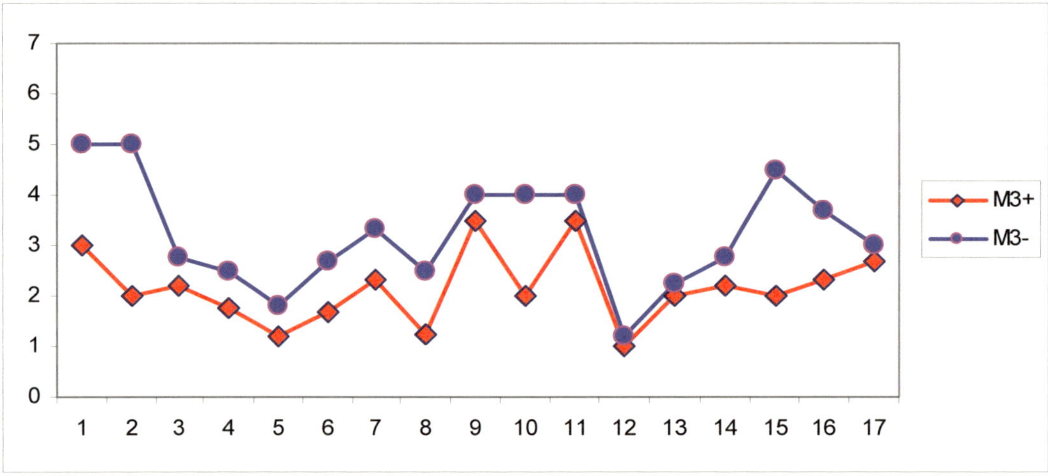

Fig. 9. Results of the interval definition between the elements of the generalised structure of different types, 'plus' and 'minus' (M+, M-) for some areas (1-17).

The interval size changes between the elements of the generalised structure as 'plus' or 'minus' has a complex character:

For the first sample, the peak growth of the interval sizes for the 'plus' structure (several times) in the third and fifth day is observed. It takes place against a wavy course of the 'minus' structure signal. The character of the change of the 'plus' and 'minus' structures actually coincides with each other in the range of the 9-12 day.

For the second sample, the waviness, falling down character of the dependence, with some subsequent general lifting and the constant prevalence (leadership) of the 'minus' signal structure, is characterised.

For the third sample, we see the integral character and the mutual position of the structures, which repeats the second sample at more of the pulse character of the 'minus' signal structure.

This is probably an estimation of the revealed structure of the corresponding dispersion (D+, D-).

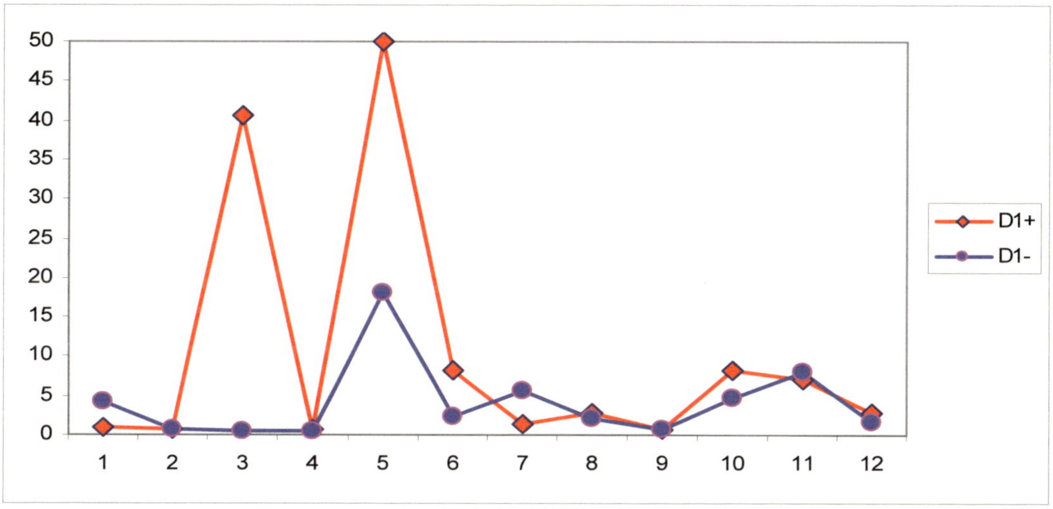

Fig. 10. The results of the dispersion definition between the elements of the generalised structure of different types, 'plus' and 'minus' (D+, D-) for some areas (1-12).

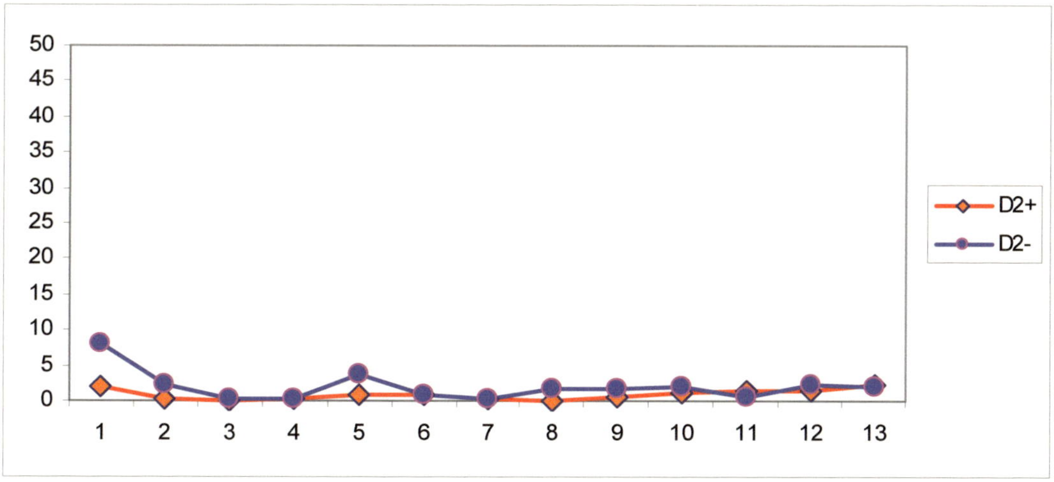

Fig. 11. Results of the dispersion definition between the elements of the generalised structure of different types, 'plus' and 'minus' (D+, D-) for some areas (1-13).

Change of the dispersion of an interval between the elements of the signal structure 'plus' and 'minus' has the following character:

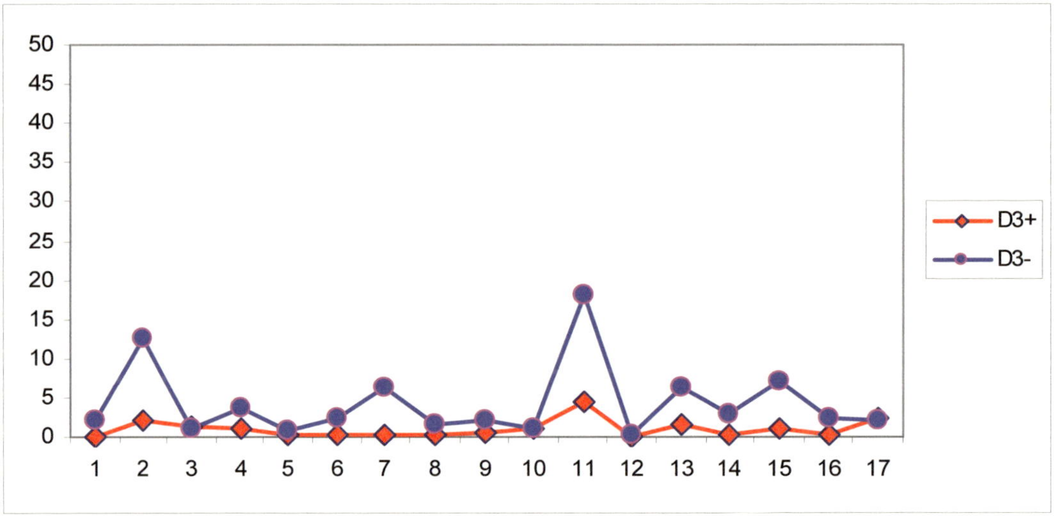

Fig. 12. Results of the dispersion definition between the elements of the generalised structure of the different types, 'plus' and 'minus' (D+, D-) for some areas (1-17).

For the first sample, the behaviour was similar to the behaviour of the intervals, i.e. the peak growth of the dispersion (instability) of the plus' structure intervals (several times) for the third and fifth day, and against a wavy course the 'minus' structure is characterised.

For the second sample, as well as for intervals, a poorly wavy character of dependence, with a constant prevalence (leadership) 'minus' structure is characterised.

For the third sample, the general character and mutual position of the structures has more pulse character, with emissions in the behaviour structure 'minus' for the second and eleventh day.

For a fuller analysis, additional characteristics have been used:

W - The regularity index; the average probability of the sample regular 'plus', 'minus' structure filling.

S - The connectivity index, the generalised difference of the probabilities of the sample regular 'plus', 'minus' structure filling.

The regularity index (W) of the frequency of the regular structure of the inhomogeneities shows that the probability of filling of the regular sample of the inhomogeneities at a certain interval (I = 1-13) is close to 0.5 only in the case where there is a sufficiently steady structure.

The characteristic tendency - the general course of curve W (I) is a little below the line 0,5. Essentially, the different behaviour of the regularity index (W) for the 'plus' and 'minus' structures is observed.

The W (I) of the 'minus' structure has a wavy character and actually does not reach the values 0,5. The W (I) of the 'plus' of the structure has a peaking characteristic and can reach values essentially more than 0,5 i.e. to fill all the sample.

It is probably necessary to draw a conclusion - the general excess of the level of probability 0,5 'plus' to for "plus" structures with big emissions W (I) 'plus' against the smooth behaviour of W (I) 'minus' can be a criterion for the displacement of the general course of an analysed signal towards its lifting.

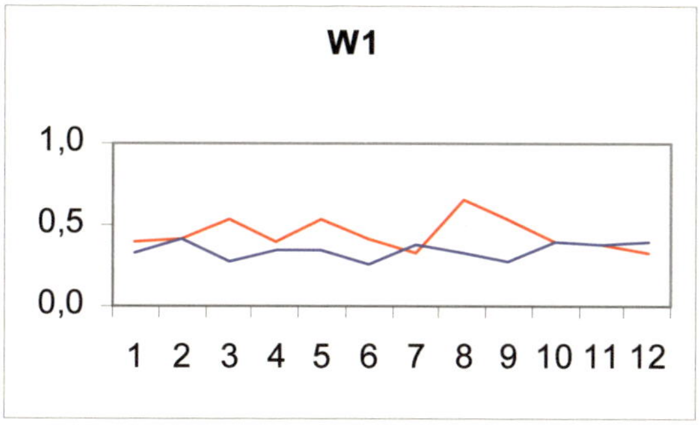

Fig. 13. The results of the calculation of the regularity index for the three situations presented above for the regular structure (W1).

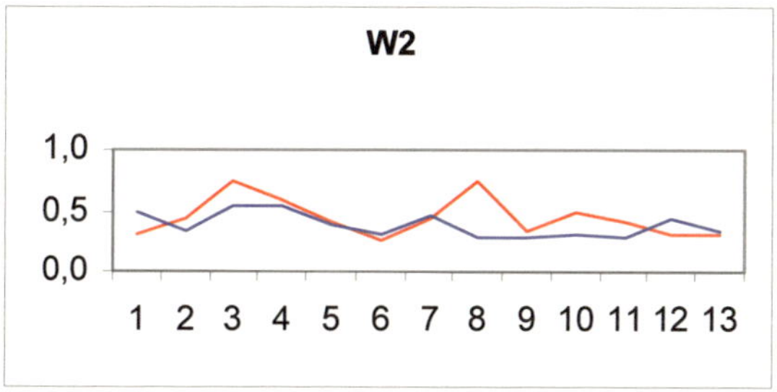

Fig. 14. The results of the calculation of the regularity index for the three situations presented above for the regular structure (W2).

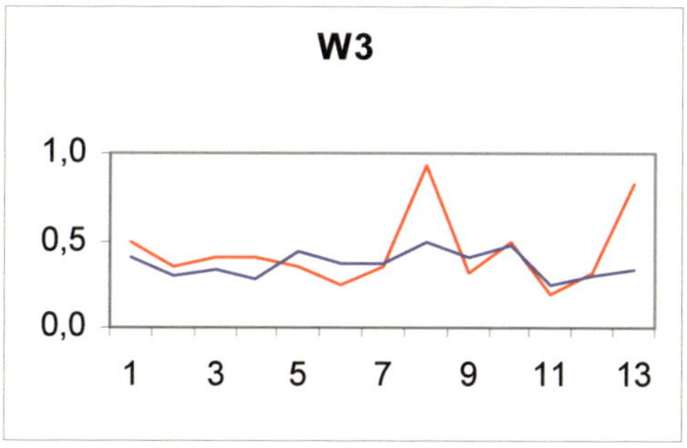

Fig. 15. The results of the calculation of the regularity index for the three situations presented above for the regular structure (W3).

The connectivity index (S) of the 'plus' and 'minus' structures is equal to zero and corresponds to a case of the behaviour synchronisation of the 'plus' and 'minus' structures, as a uniform structure of a harmonious type.

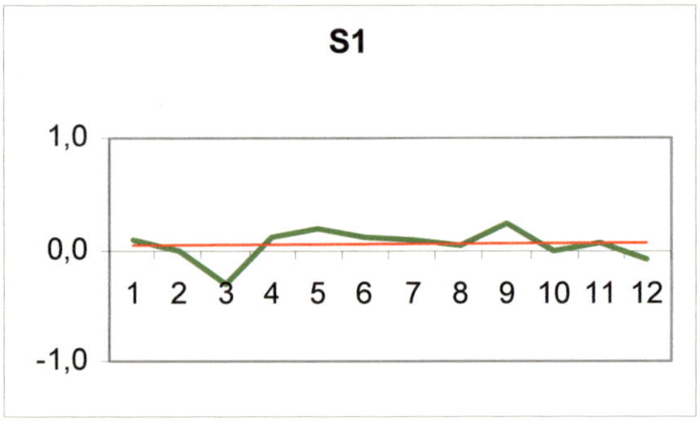

Fig. 16. The results of the calculation of the regularity index for the three situations presented above for the regular structure (W1).

The displacement of the connectivity index (S) in the 'plus' and 'minus' zone specifies on increase in the influence of 'plus' and 'minus' structures with an increase in the independence of their behaviour, rather than each other.

The displacement of the connectivity index (S) in a 'plus' zone can be interpreted as the presence of the leader-structure of the 'plus' type.

The displacement of an index of connectivity (S) in the 'minus' zone can be interpreted as the presence of the leader- structure of the 'minus' type.

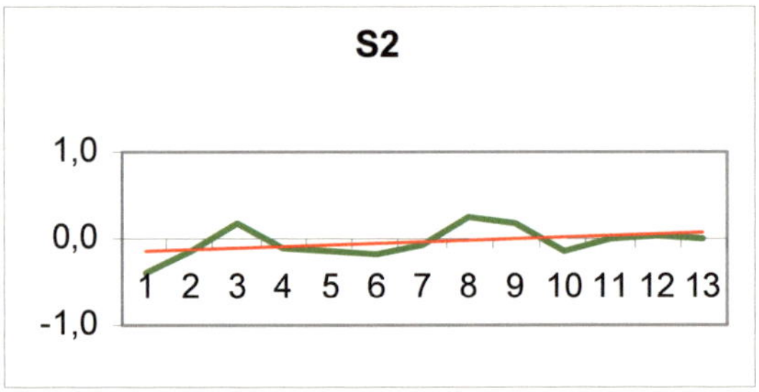

Fig. 17. The results of the calculation of the regularity index for the three situations presented above for the regular structure (W2).

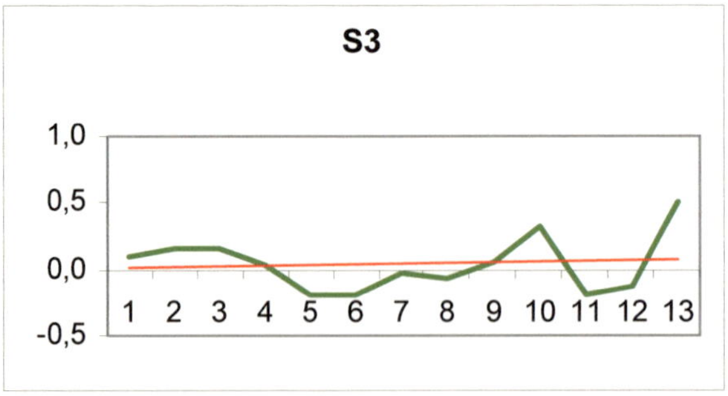

Fig. 18. The results of the calculation of the connectivity index for the three situations presented above for the regular structure (W3).

On the basis of the assumptions made, it is possible to draw the following conclusions:

For first sample, the connectivity index S1 (I) specifies the stable leader-structure of the 'plus' type.

For the second sample, the connectivity index S1 (I) specifies the transition of leadership from the structure of the 'minus' type to the structure of the 'plus' type.

For the third sample, the of connectivity index S3 (I) specifies the steady growth of the leadership of the structure of the 'plus' type.

The conclusion about the displacement of the general course of an analysed signal towards its general growth proves to be true in the presence of the 'plus' of leader-structures in all three samples of a signal, as the above results show.

5. Conclusion

5.1 Measurements

These estimates allow us to make some significant findings:

1. Increasing the accuracy of the measurement of the scattering signal can be achieved through the use of the described methods of the signal processing, long laser impulses or dark laser impulses.
2. This approach allows more accurate measurement of the scattering signal, by at least an order of magnitude, as well as measurements during the daytime up to distances comparable with the meteorological visibility range (MDV), including the area of multiple scattering.
3. The proposed system of remote measurement organisation allows us to solve existing contradictions and provides a specified signal/noise ratio under a wide range of conditions and at different times for remote tracks, and it allows the more accurate linking of the principles of recording equipment with the methodology of interpreting the data.
4. The application of the proposed approach to the principles of the construction of lidar systems allows us to use low-power light sources and, in a large measure, to get rid of hardware errors caused by shock loads on the receiving system.
5. This organisation of remote sensing systems allows us to pass from the problem of signal detection with high accuracy to the problem of minimising the distortion of the received signal, which is caused by instrumental factors.
6. The system with the probe interval (dark pulse) between impulses (type III) is the greatest prospect and it retains all of the advantages of systems of type II but with more performance.

The results obtained allow for a new approach to the problem of reconstructing the characteristics of the environment based upon remote sensing. It is not correctly solved for a real, heterogeneous environment, largely due to the exclusion of the consideration of thermodynamic processes (Polkanov, Y. A. et al., 1991).

5.2 Processing

1. "Leadership 'plus' structure" and "Leadership 'minus' structure" specifies, accordingly, the general lifting or falling of the signal.

2. "Leadership interception" specifies the tendency of change for the signal, from lifting to falling or the reverse.
3. Changing character of the structure means that the stability of the leadership of corresponding structure is subject to regular fluctuations ('plus' → 'minus' or 'minus' → 'plus').
4. The pointed character of the structure speaks about the obvious local tendency to change of the leadership of the corresponding structure.
5. The regularity of the revealed structure has an alternating character.
6. The higher the regularity, the higher the relative stability of the corresponding structure.
7. The higher the bond of the 'plus' and 'minus' structures, the higher their overall stability, even if intermittency has obvious characteristic.
8. The regular structure is the reason for the event both of a time interval corresponding to the taken measurements and that taken out of it, i.e. in the short-term and in the long-term plan.
9. The abnormal structure was the reason for the event, and not only in a time interval corresponding to the taken measurements, i.e. only in the short-term plan, as the regular structure smoothes out a special filtration.

For today's problems with remote sensing, it is necessary to apply the described methods of conflict resolution logic to the problem of reconstructing the characteristics of the environment. Then, the problem of measuring and processing the measurement results will be examined as a single complex. The extension of this logic leads to new tasks - the assessment of the impact of thermodynamic processes on the structure of inhomogeneities in the medium and their self-organisation and the development of criteria for the stability of such structures as indicators of self-protection and self-organisation (Polkanov, Y. A. et al., 200).

6. References

Ablavskij L.M. and Kruglov P.A. Main Geophysical Observatory Labour(in Russian), v.340, p.25, 1974

Ashkinadze D.A. and Belobrovik V.P. and Spiridovich A. L. and Kugeiko M.M. and Polknov Y.A. *Lidar determination of the dynamics of the fields of aerosol pollution in cities.* In the book.: VI Proc. Sympos. the laser and acoustic sensing. Proc. Part 1. Tomsk, 1980. 152-155.

Kovalev V.A. Main Geophysical Observatory Labour (in Russian), v.312, p.128, 1973

Mingming Feng, Kevin L. Silverman, Richard P. Mirin, and Steven T. Cundiff. *Dark pulse quantum dot diode laser.* Optics Express, Vol. 18, Issue 13, pp. 13385-13395, 2010 doi:10.1364/OE.18.013385 http://www.opticsinfobase.org/abstract.cfm?URI=oe-18-13-13385

Polkanov Y.A. *The opportunities of a mode of the photon account at registration of the dispersion signal of laser radiation in an atmosphere.* Radiotechnika i Electronuka (Radio engineering and electronics), 1983, v. XXVIII, № 10, p. 2080-2082. http://adsabs.harvard.edu/abs/1989Prib...32....6P

Polkanov Y.A. *On regular structure of scattering inhomogeneities for optical radiation in the atmosphere.* Izvestia Akademii Nayk SSSR. Seriya Rhizika Atmospheryi I Okeana

(News of the Academy of sciences of the USSR. Physics of an atmosphere and ocean), 1985 v. 21, № 7, p. 720-725.

Polkanov Y.A. and Ashkinadze D.A. *The opportunities of increase of the signal accuracy of optical radiation dispersion in an atmosphere*. Radiotechnika i Electronuka (Radio engineering and Electronics). 1988, v. 12, p. 2599-2603.
http://adsabs.harvard.edu/abs/1988RaEl...33.2599P

Polkanov Y.A. and Kudinov V. N. *A possibility for the analysis of the periodic structure of a complex signal*. Priborostroenie. 1989. V.32. № 4. P.6-11.
http://adsabs.harvard.edu/abs/1989Prib...32....6P

Polkanov Y.A. *One opportunity of the periodic structure analysis of a complex signal*. Izvestia Vysshich Uchebnych Zavedenij, Priborostroenie (News of high schools. Instrument making). 1989, v. 32, № 4, p. 6-11.

Polkanov Y.A. The *dynamic structure of a scattering signal and its connection with meteorological situation*. Izvestia Akademii Nayk SSSR. Seriya Optika Atmospheryi I Okeana (News of the Academy of sciences of the USSR. Physics of an atmosphere and ocean), 1989 v. 25, № 6, p. 599-603.

Polkanov Y.A. *Matchig between a change in the structure of atmospheric optical inhomogeneities and a set of meteorological parameters*. Meteorologiya I Gidrologiya (Meteorology and hydrology) 1991, № 3, p. 39-48.
http://cat.inist.fr/?aModele=afficheN&cpsidt=5237241

Polkanov Y.A. *A possibility of detecting anomalous inhomogeneties of the atmosphere (the Method of a nonlinear filtration of the return dispersion signal)* Izvestia Akademii Nayk SSSR. Seriya Optika Atmospheryi I Okeana (News of the Academy of sciences of the USSR. Physics of an atmosphere and ocean), 1992 v. 5, № 7, p. 720-725.
http://pdf.aiaa.org/jaPreview/AIAAJ/1994/PVJAPRE48291.pdf (Publication Date: 04/1994)

Polkanov Y.A. *Accuracy of the representation of the real scattering signal in terms of the lidar equation*. Journal of Applied Spectroscopy. Springer New York. 0021-9037 (Print) 1573-8647 (Online). Volume 37, Number 3 / September, 1982. p. 1091-1095. SpringerLink Date Tuesday, December 07, 2004.
http://www.springerlink.com/content/g532t57564377366/

Polkanov Y.A. *Sounding of the environment by means of the un-impulse of the low-power continuous source*. Proceedings of SPIE -- Volume 6750 Lidar Technologies, Techniques, and Measurements for Atmospheric Remote Sensing III, Upendra N. Singh, Gelsomina Pappalardo, Editors, 67501H (Oct. 3, 2007) (published online Oct. 3, 2007)
http://spiedl.aip.org/getabs/servlet/GetabsServlet?prog=normal&id=PSISDG006 75000000167501H000001&idtype=cvips&gifs=yes

Polkanov Y.A. *Medium sensing by low-power strobe pulse*. Proceedings of SPIE -- Volume 6936 Fourteenth International Symposium on Atmospheric and Ocean Optics/Atmospheric Physics, Gennadii G. Matvienko, Victor A. Banakh, Editors, 693613 (Feb. 14, 2008) (published online Feb. 14, 2008)
http://spiedl.aip.org/getabs/servlet/GetabsServlet?prog=normal&id=PSISDG006 936000001693613000001&idtype=cvips&gifs=yes

Polkanov Y.A. *Lidar measurements for the short-term forecast of meteorological stability* (Proceedings Paper) http://spie.org/x648.html?product_id=783368 (Publication Date: 02/2008)

Polkanov Y.A. *The structurally-statistical remote analysis of the self-organizing processes*.[7479-25] p. 31 Lidar Technologies, Techniques, and Measurements for Atmospheric Remote Sensing V Conference 7479 - Proceedings of SPIE Volume 7479Dates: Monday-Tuesday 31 August - 1 September 2009 http://www.google.com/search?client=opera&rls=ru&q=spie+2009+Polkanov+1.+The+structurally-statistical+remote+analysis+of+the+self-organizing+processes+in+an+complex+systems,+Yury+Polkanov&sourceid=opera&ie=utf-8&oe=utf-8

V. E. Zuev and M. V. Kabanov, *Transfer of Optical Signals in the Earth's Atmosphere* (in Russian), Sovetskoe Radio, Mosow 1977